Mobile Communication and Society

The Information Revolution & Global Politics
William J. Drake & Ernest J. Wilson III, editors
www.mitpress.mit.edu/IRGP-series

The Information Revolution and Developing Countries
Ernest Wilson, 2004

Human Rights in the Global Information Society
edited by Rikke Frank Jørgensen, 2006

Mobile Communication and Society: A Global Perspective
Manuel Castells, Mireia Fernández-Ardèvol, Jack Linchuan Qiu, and Araba Sey, 2007

Mobile Communication and Society

A Global Perspective

Manuel Castells, Mireia Fernández-Ardèvol, Jack Linchuan Qiu, and Araba Sey

A Project of the Annenberg Research Network on International Communication

The MIT Press
Cambridge, Massachusetts
London, England

MIT Press books may be purchased at special quantity discounts for business or sales promotional use. For information, please email special_sales@mitpress.mit.edu or write to Special Sales Department, The MIT Press, 55 Hayward Street, Cambridge, MA 02142.

This book was set in Stone Sans and Stone Serif by SNP Best-set Typesetter Ltd., Hong Kong. Printed and bound in the United States of America.

Library of Congress Cataloging-in-Publication Data
Mobile communication and society : a global perspective : a project of the Annenberg Research Network on international communication / Manuel Castells . . . [et al.].
    p. cm. – (The information revolution & global politics)
Includes bibliographical references and index.
ISBN 0-262-03355-0 (alk. paper)
1. Communication–Social aspects. 2. Mobile communication systems–Social aspects. 3. Information technology–Social aspects. I. Castells, Manuel. II. Series.
HM1206.M62   2006
303.48'33–dc22

2006044970

10 9 8 7 6 5 4 3 2 1

Contents

List of Figures

List of Tables

Acknowledgments

This book is the result of a joint research effort conducted over two years on three continents, with the support and collaboration of people and institutions in different countries. The authors wish to acknowledge the decisive support received from the Annenberg Foundation and the University of Southern California's Annenberg School for Communication in Los Angeles for the research that led to this book. We also acknowledge the support provided by the Internet Interdisciplinary Institute of the Open University of Catalonia in Barcelona. The School of Journalism and Communication, Chinese University of Hong Kong, also helped the work of our team. We are particularly grateful to Melody Lutz and Deb Lawler in Los Angeles, and Anna Sanchez-Juarez in Barcelona, for their help in the organization of the research and the preparation of the book.

We have benefited from the collegial environment provided by the Annenberg Research Network on International Communication. This was the intellectual foundation for much of our discussion on these matters. We thank in particular Jonathan Aronson, François Bar, Hernan Galperin, Jonathan Taplin, Jeff Cole, and Namkee Park. The final version of this book has been improved by the comments received at the time of our presentation of this research at the international conference on "Wireless Communication in a Global Perspective: Policies and Prospects," held at the Annenberg School of Communication, Los Angeles, in October 2004. We thank all participants at this outstanding conference for their most valuable insights.

Additional help in securing information came from the Canadian IDRC; the Telefonica Foundation in Spain, Argentina, and Chile; the Government of Chile; Agência Nacional de Telecomunicações (ANATEL) in Brazil; and Apoyo-Opinion and Mercado Bolivia, in Bolivia. A number of colleagues have commented on various drafts of this book and have corrected some of our errors, as well as suggesting ideas and providing information. Among them are Ernie Wilson, François Bar, Hernan Galperin, Jonathan Aronson, Jonathan Taplin, Jeff Cole, Larry Gross, Yuezhi Zhao, Carolyn Cartier, Mizuko Ito, James Katz, Owuraku Sakyi-Dawson, Kwasi Ansu-Kyeremeh, Renato Felix, Eduard Fernandez, Joseba Gomez, Xandra Troyano, Nuria Villagrasa, Luis Garay, Antonio Aranibar, Marcelo Branco, Madanmohan Rao, James Gomez, Paul Lee, Chin-Chuan Lee, Angel Lin, Chen Yun, Fu Li, and Peter White. In addition, a number of researchers around the world have graciously given us access to the preliminary results of their ongoing research projects. Our belated thanks go to Kalpana David, Jonathan Donner, Mary Gergen, Nalini Kojamraju, Rich Ling, Sebastian Ureta, Scott Campbell, Leopoldina Fortunati, Shi-Dong Kim, and Kyong-Won Yoon.

Keiko Mori helped our data collection in the Japanese language, as did Namkee Park and Seungyoon Lee in Korean, and Meritxell Roca in Russian. Manuel Castells wishes to express his deepest gratitude to Alexandra Konovalova and to Clara Millan Castells for their invaluable help in providing their own perspective as teenage users of mobile phones.

To all of them, and to our supporting institutions, goes our deepest gratitude for their generous help.

Los Angeles, Barcelona, Hong Kong, February 2006

Mobile Communication and Society

Opening: Our Networks, Our Lives

Wireless communication networks are diffusing around the world faster than any other communication technology to date. Because communication is at the heart of human activity in all spheres of life, the advent of this technology, allowing multimodal communication from anywhere to anywhere where there is appropriate infrastructure, raises a wide range of fundamental questions. How is family life affected by the ability of its members, including children, to pursue fairly independent activities and yet be constantly in touch? Is the office on-the-run coming into existence when people can reach their working environment and their professional partners from anywhere and at any time? Is the classroom transformed by the ability of students to communicate simultaneously face to face, with their laptops, and with their cell phones? Is the technological ability to perform multitasking anywhere further compressing time in our hurried existence? Does mobile communication favor the development of a new youth culture that makes peer-to-peer networks the backbone of an alternative way of life, with its own language, based on texting and multimodal communication, and its own set of values? How distinctive is this youth culture vis à vis the culture of society at large? And is it an age-specific state of mind or the harbinger of new patterns of behavior?

Are mobile phones expressions of identity, gadgets of fashion, tools of life, or all of these things? Mobile communication is said to enhance the autonomy of individuals, enabling them to set up their own connections, bypassing the mass media and the channels of communication controlled by institutions and organizations. How real is this autonomy? Are social constraints reintroduced in the new communication patterns prevailing under the conditions of wireless technology? And how is this autonomy reflected in the sociopolitical realm? Are "flash" political mobilizations, often reported in the media as effected by people's mobile communication power, truly spontaneous? And how much do they actually modify power relations in our society? Are time and space being transcended in social practice because of the ability to do everything from everywhere thanks to this capacity for ubiquitous, perpetual contact?

What are the new inequalities introduced by differential access to the infrastructure of wireless communication in a world based on connectivity? Can developing countries leapfrog the deployment of fixed telecommunication systems, jumping directly into a global communication system based on satellite and wireless telecommunication systems of different sorts? How much is this new connectivity contributing to development in an interdependent world in which knowledge, information, and communication are the key sources of wealth and power? How does the interplay between new communication technologies and people's lives vary according to cultures and institutions in different regions of the world, and among different social groups? Do we find emerging patterns of behavior and of social organization that seem to be common to various social contexts, as they are associated with the new forms of multimodal, wireless communication?

The answers to these questions affect our lives. They also condition public policies, business strategies, and people's decisions about deploying and using a new, powerful technological system of communication. Yet, because of the speed of technological change, and an eagerness to win competitive advantage in the new system of communication, decisions are made all over the world without much understanding of the social, economic, and political implications of wireless communication technologies. The assumptions underlying these decisions are often unwarranted.

Indeed, we know from the history of technology, including the history of the Internet, that people and organizations end up using technology for purposes very different from those initially sought or conceived by the designers of the technology. Furthermore, the more a technology is interactive, the more it is likely that the users become the producers of the technology in their actual practice. Thus, society needs to address responsibly the questions raised by these new technologies. And research can contribute to providing some answers to these questions. To look for these answers, we need knowledge based on observation and analysis. Rather than projecting dreams or fears of the kind of society that will result in the future from the widespread use of wireless communication, we must root ourselves in the observation of the present, using the traditional, standard tools of scholarly research in order to analyze and understand the social implications of wireless communication technology. People, insti-

tutions, and business have suffered enough from the prophecies of futur-
ologists and visionaries who promise and project whatever comes into their
minds on the basis of anecdotal observation and ill-understood developments. We take exception to such approaches.

Instead, the purpose of this book is to use social research to answer the
questions surrounding the transformation of human communication by
the rise and diffusion of wireless digital communication technologies. Our
answers, however tentative, will rely on the stock of available knowledge
on this topic in different areas of the world. While gathering our data, we
would have preferred to consider exclusively information and analyses produced within the rigorous standards of academic research. In fact, this constitutes a good proportion of the material examined here. However,
because of the novelty of the phenomenon and the slow motion of academic research in uncovering new fields of inquiry, the amount of comparative knowledge on this subject is too limited to grasp empirically the
emerging trends that are transforming communicative practices. Thus, we
have extended our data sources to reliable media reports and to statistics
and elaboration coming from government institutions and consulting
firms.

We have made every possible effort to assess the validity of the sources
employed and state clearly the limits of our knowledge in each particular
instance. Overall, we believe that the analysis presented in this book is a
reasonably accurate representation of what is happening in the world at
large, as it results from the hundreds of studies that we have consulted
from multiple sources, with the help of several institutions and organizations and of many experts in the field. Following a standard methodolog-
ical rule in social science, when a pattern of conduct (for example, the
substantial enhancement of individual and collective autonomy by wireless communication capability) repeats itself in several studies in several
contexts, we consider it plausible that the observation properly reflects the
new realm of social practice. We have applied this principle to all key topics
that characterize the field of inquiry on wireless communication. As a
result of our work, we have identified the emergence of a pattern of communication linked to the new technological conditions: a pattern whose
contours will appear gradually, chapter by chapter, throughout this book,
until it takes full shape in a conclusion that will bring together the analytical threads found in each of the themes under study.

The perspective of our analysis is global because so is the object of our
inquiry: wireless communication. Yet there are limits to the information
available in some contexts, as well as limits derived from our partial

knowledge of languages, in spite of the multicultural background of the authors of this book and the help we have received from assistants and colleagues in other languages, such as Japanese, Korean, and Russian. The first chapter provides a global overview of the diffusion of wireless communication. Yet, in the subsequent, analytical chapters, the different regions of the world have an uneven presence, depending on the stock of knowledge accumulated in each of them, and made available to us. Thus, we know a good deal about Norway because of the quality of Norwegian research in this field, while we know little about Nigeria because of the scant reliable evidence on this important country. Furthermore, in spite of having consulted hundreds of studies on all regions of the world, we have certainly not exhausted current information on the matter. However, *our aim is primarily analytical*, not encyclopedic – among other things, because the diffusion of wireless communication is proceeding so fast that purely descriptive data may soon become obsolete. *What we intend to do in this book is to construct an empirically grounded argument on the social logic embedded in wireless communication, and on the shaping of this logic by users and uses in various cultural and institutional contexts – an argument whose analytical value should stand by itself.*

Our emphasis on a cross-cultural approach comes essentially from our determination to avoid ethnocentrism in building this argument. The limitation of our work is that we cannot analyze what has not been studied, and we adamantly refuse to speculate without a minimum level of reliable evidence. We are dependent on the work of other researchers. So, as some important questions have barely been touched upon by research, this is reflected in our analysis. We believe, nonetheless, that by assessing empirically and analytically the emergence of wireless communication patterns at an early stage in the development of the new communication system we can help to build a cumulative body of knowledge that will evolve with the technology itself. Furthermore, these early studies, and our assessment of them, may be socially useful, as people, civil society, business, public services, and policy-makers adapt their communication technology strategies to the demands of society.

Let us provide you, the reader, with a road map for this book. We start (chapter 1) with a statistical overview of the diffusion of wireless communication in the past decade in different areas and countries of the world, while trying to account for differences in the rate of diffusion. Then (chapter 2) we present data on the patterns of social differentiation in the diffusion of the technology, and explain the causes and consequences of such differentiation: who has access to wireless communication, who has not, who has less, and why. Next (chapter 3), we study the social uses and social effects of wireless communication in different domains of human activity, grouped under the heading of "everyday life": from family life to the transformation of work, and to the emergence of social problems in the communication networks, problems such as security, surveillance, spam, scams, and digital virus epidemics.

We then enter into specific consideration of some major themes that have clearly appeared to be essential in the course of our research. The first one (chapter 4) is the deep connection between wireless communication and the rise of a youth culture (what we call a mobile youth culture) in most of the areas under our observation. The second theme (chapter 5) refers to changes in the practice of time and space resulting from wireless communication. The third (chapter 6) is the process of transformation of language by texting and multimodality. The fourth (chapter 7) is the growing importance of wireless communication in the processes of sociopolitical mobilization, particularly outside formal politics, a topic that we have considered by focusing on case studies of protest movements in a variety of contexts. The fifth theme, which we treat in some depth (chapter 8), is the relationship between communication and development in the framework of the new technological paradigm, focusing on developing countries. Finally, we summarize and elaborate the main trends resulting from our observations in a concluding chapter that provides tentative answers to the questions raised in this introduction.

Each chapter blends into the analysis data and research findings from a wide variety of areas and cultures across the world. We have tried to identify the specificity of each context in which the practices have been observed. But we have also found common trends and similarities across cultures that vindicate the singularity of wireless communication as a sociotechnical process. This tension between the local and the global, the cultural and the technological, is present throughout the analyses presented in our book. Therefore, for each topic we move freely from Europe to China, from the United States to Japan, from Africa to Latin America, when we find recurrent themes that manifest themselves in a variety of languages and cultures. The division into chapters follows a thematic logic, but it results in a great deal of asymmetry in the lengths of different chapters. This is simply because of the difference in the amount of research material found on each theme. We have kept the specificity of each theme, regardless of the length of each chapter, to emphasize its importance, and to call to the attention of researchers important topics on which knowledge is still insufficient at this point.

As this book relies on hundreds of references and tables, we have made an effort to make it readable by placing most of our information in appendixes which can be found on the website of this book at MIT Press (http://mitpress.mit.edu/Castells_Mobile). We have also indicated in the text the specific sources that support our analysis. Thus, interested readers may consult the full set of data, methods, and references beyond the print boundaries of this book by reaching to its virtual extension on the Web. Welcome to the hypertext.

Finally, throughout this book we refer to the concept of the "network society," and use the expression "the mobile network society," to emphasize the diffusion of the networking logic in all domains of social life by means of wireless communication technology. For an understanding of the network society in general, and for discussion of network theory, we refer to previous work by researchers in this field (Castells 2000a, b, 2004; Monge and Contractor 2003). The mobile network society is simply the enhancement of the social structure conceptualized as the network society by new, wireless communication technologies. For the specific nature of this enhancement, we refer to the analyses presented in this book.

Altogether, we hope that this research effort, within the limits of our current knowledge, will contribute to setting a tone for the future analysis and assessment of a fundamental trend that is redefining the relationship between communication, technology, and society around the world by unwiring the networks of our lives.

1 The Diffusion of Wireless Communication in the World

Wireless communication has diffused faster than any other communication technology in history. But it has done so differentially. This chapter presents an assessment of the growth of wireless communication globally, regionally, and in specific countries. We begin with the most pervasive technology – the mobile phone – and later include other forms of wireless technology, insofar as comparative data are available. After presenting the data, we will try to identify the factors that account for the uneven pattern of diffusion, according to various studies and expert opinion.[1]

Global Diffusion of Mobile Telephony

Within the span of about ten years, mobile telephony has moved from being the technology of a privileged few to an essentially mainstream technology. Mobile telephony really began to take off worldwide in the mid-1990s (figure 1.1), when the ratio of mobile to mainline telephones went up from about 1:34 (1991) to about 1:8 (1995).[2] By 2000, there was one mobile phone for less than two mainlines; and by 2003, mobile-phone subscriptions had overtaken mainline subscriptions for the first time. Both mobile (1,748 million lines by 2004) and fixed (1,198 million lines by 2004) telephone subscriptions have continued to rise, though mobiles at a faster rate, effectively doubling the number of lines available worldwide. In most countries, mobile phones have not yet become substitutes for wired phones, but rather act as a complement to the traditional system of telephony. In some (primarily developing) countries, however, mobile phones are serving as a technological substitute for fixed lines, and, to an increasing extent, certain groups of people in developed countries are also substituting mobile phones for economic reasons.[3]

This rapid growth in mobile telephony has not occurred uniformly around the world, however. By 2004, out of 182 countries, only 31

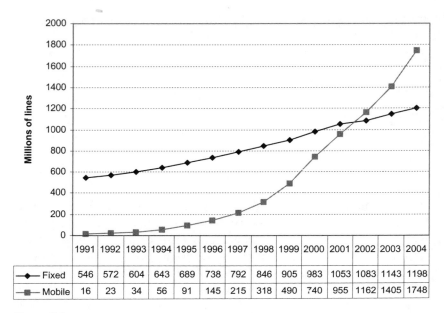

Figure 1.1
Global diffusion of fixed and mobile telephone lines. *Source*: ITU statistics (www.itu.int).

countries had penetration rates greater than 80 percent, and more than half had rates below 20 percent (figure 1.2). A majority (32 percent) had penetration rates below 10 percent. Thus, globally, the number of mobile phones per 100 inhabitants was only 27.75 in 2004, though this was still higher than the rate for fixed lines (19.04).

Regional Diffusion of Mobile Telephony

Mobile telephony has diffused at very different rates in the various regions of the world. However, to get an accurate picture of growth trends, a distinction must be made between subscriber growth (i.e., the number of subscribers) and penetration rates (i.e., the number of mobile-phone subscribers per 100 inhabitants). Subscriber growth has been rapid in almost all regions and countries; penetration rates, on the other hand, vary widely. In 2003, in pure subscription terms, at regional level, the Asian Pacific region had the greatest number of mobile-phone subscribers (730 million); at market level, the unified European Union market had the greatest number (321 million); and at country level, China and the United

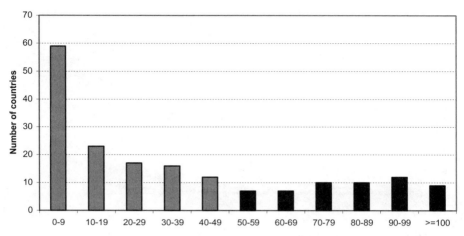

Figure 1.2
Mobile-phone subscribers per 100 inhabitants: number of countries in each range, 2003. *Source*: ITU statistics (www.itu.int). See online statistical annex (appendix 1) for chart data.

States, with their large populations, had the highest numbers of subscribers (334 million and 181 million subscribers respectively). Of total subscribers, 41 percent were in the Asian Pacific, 32 percent in Europe, 21 percent in the Americas, 4 percent in Africa, and 1 percent in Oceania. In terms of mobile-phone penetration, however, by 2004, Europe led (71.5 percent), followed by North America (66 percent) and Oceania (62.74 percent; figure 1.3). These three regions each had more than one mobile telephone subscriber per two persons. They were followed distantly by the rest of the American continent (30.2 percent), Asia (18.94 percent), and Africa (9 percent), all of which had less than half the penetration rates of the top three regions.

Looking at the trends in figure 1.3, North America had the highest penetration figures in 1992, surpassing its closest competitor (Oceania) by almost double. By 2001, however, Europe had taken over the leading position, following a growth spurt between 1997 and 2000. As we will show in subsequent sections, this shift can be attributed to a variety of factors in each region which slowed down or speeded up the general adoption of mobile phones.

Most regions (except parts of Europe and North America) display a variety of penetration levels, suggesting that mobile-phone diffusion has not necessarily developed according to geographic or regional boundaries

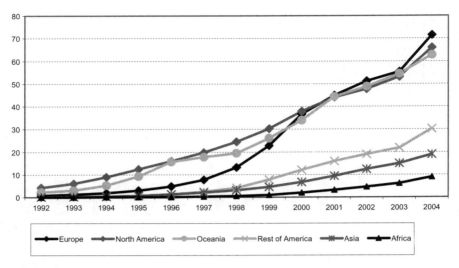

Figure 1.3
Mobile-phone subscribers per 100 inhabitants: world regions, 1992–2004. *Notes*:
North America = Canada + United States. Some figures refer to earlier years. *Source*:
ITU statistics (www.itu.int). See online statistical annex (appendix 2) for chart data.

(TNS 2002b). In any one region, one can find countries with very low
mobile diffusion and others with relatively high diffusion rates.[4] For
example, in Southeast Asia, Singapore had 89 percent penetration of
mobile phones in 2004, while India was at 4 percent. In Latin America,
the penetration rate in Chile was 62 percent, Brazil 36 percent, and Cuba
0.7 percent. Mobile-phone penetration in Bahrain (Middle East) was 88
percent, Tunisia 38 percent, and Syria 13 percent. A similar situation per-
tained in Africa, where penetration rates ranged from less than 1 percent
(for example, Burundi) to 60 percent (Seychelles).

While the above data show variety in regional diffusion of mobile
phones, Kauffman and Techatassanasoontorn (2005) have empirically
identified a "regional contagion" effect, which results in subscriber growth
in one country being strongly correlated with changes in subscriber growth
in its geographic region. According to this hypothesis, regional contagion
may be high, medium, or low, possibly depending on the amount of cross-
border interaction at the country level or cross-border relationships
between mobile-phone companies.

There is also a clear diffusion gap between the developed and developing regions of the world. Developing economies lag in both the extent of
diffusion and the type of technology in use (Kauffman and Techatassana-

soontorn 2005). Most of Africa, the Asian Pacific, and South America are still in the very early stages of adoption, with penetration rates up to ten times lower than that of developed regions. In terms of technology, most countries in Western Europe and the Asian Pacific have 3G networks in operation.[5] North America is in the early stages of introducing 3G, while the Middle East is transitioning from 2G to 3G. In Latin America, 2G technologies dominate, but a few countries are already using 2.5G or 3G. High-speed networks are largely absent from most developing countries. In 2001, only one low-income economy (Indonesia) had 2.5G or higher in operation, as against 53 high-income countries (ITU 2002b).[6] The first 3G-capable network in Africa was deployed in Mauritius in November 2004, and plans have been announced to launch 3G, using EDGE (Enhanced Data Rates for GSM Evolution) in a few other countries; for example, South Africa and Ghana (Ajao 2005a).

In addition to the existing divide between developed and developing regions of the world, there is also a gap between more- and less-developed economies within each region (table 1.1). In Europe, a 2–3-year lag can be seen in growth levels between the first fifteen (and richer) countries of the European Union and those that joined later. The North and South American regions are also distinguished by the higher rates of mobile-phone diffusion in North America (United States and Canada). Similarly, in Africa, the majority of mobile-phone owners are in South Africa, which also has one of the highest penetration rates in the region. Finally, in Oceania, Australia and New Zealand lead in mobile subscriptions.

Diffusion of Wireless Communication Technologies in Selected Areas and Countries

We have observed so far that globally there has been an explosion of wireless telephony, but that there are differential rates of penetration in different regions. The following section presents a brief overview of the diffusion of mobile voice and data technologies in selected countries.[7]

Mobile Telephony in Europe

Mobile Telephony in the European Union   In 1992, when the GSM system was introduced, only seven European countries (Denmark, Finland, Iceland, Norway, Sweden, Switzerland, and the United Kingdom) had mobile-phone teledensities above the OECD average of 2.44. By 2003, the picture had transformed completely, with all but one European country exceeding the OECD

Table 1.1
Mobile phones per 100 inhabitants by region, 2000–2004

		2000	2004
Europe	European Union	56.9[a]	85.8[b]
	Non-European Union	7.6	53.0
	Europe as a whole	36.6	71.5
Americas	United States and Canada	37.9	66.0
	Rest of America	12.1	30.2
	America as a whole	21.9	42.4
Oceania	Australia and New Zealand	44.1	81.8
	Rest of Oceania	2.2	4.6
	Oceania as a whole	33.9	62.7
Africa	South Africa	19.1	43.1
	Rest of Africa	0.98	7.1
	Africa as a whole	2.0	9.0
Asia	Hong Kong, Japan, and Korea (Rep.)	55.2	74.4
	Rest of Asia	4.2	16.1
	Asia as a whole	6.8	18.9

[a] In 2000, there were 15 members of the European Union.
[b] In 2004, there were 25 members of the European Union.
See online statistical annex (appendixes 5–8) for growth trends since 1992.
Source: ITU statistics (www.itu.int).

average (69.93), and Europe leading the world in diffusion of the technology.[8] On average, the mobile teledensity for European Union (EU) countries exceeds 80 percent. This growth can be linked to the concerted efforts of some Scandinavian countries to take the technology created in the United States and jointly develop and promote a unified standard – first the NMT (Nordic Mobile Telephone) and later the GSM (Group Spécial Mobile) – within their borders, as well as to the decision of the European Commission to mandate the implementation of a single standard (GSM) in the EU (Gruber 1999; Huag 2002; Agar 2003). Among the factors contributing to the success of this standard were its fortuitous timing (occurring during a period of tremendous growth in first-generation systems), its facilitation of international roaming, and economies of scale, leading to lower prices for manufacturers and users (Huag 2002). The European Commission also ensured a competitive environment by directing that there be at least two GSM service providers in the 900 MHz band, and that the 1800 MHz band be opened up to at least one additional operator (Gruber 1999).

Mobile Telephony in Non-EU Europe Diffusion patterns in most of the non-EU countries are similar to those in the whole continent and the European Union, although with a slight lag. Mobile-phone subscriptions overtook fixed lines in 2003, making up 52.2 percent of total telephone subscriptions. Significant divides exist, however: there are the wealthier countries (specifically Iceland, Norway, and Switzerland), which have more than eight handsets per capita, and mobile teledensities above 80 percent; and there are the poorer countries, such as Belarus, Ukraine, Moldova, and TFYR Macedonia, which have less than 20 percent teledensity (see online appendix 13). A majority of countries have mobile teledensities below 40 percent, and some have fairly high mobile-to-fixed ratios (for example, in Albania 81.2 percent of total subscriptions are mobile). The Russian Federation is a case of special interest.

The adoption of mobile telephony in Russia has been slow relative to levels in Europe as a whole. It was not until 2003 that mobile subscriptions reached the same level as fixed lines.[9] The adoption of mobile telephony began a steady climb in 1999 and has been especially steep since 2001. A prediction that the growth rate would decline after the higher-income population had been served (United Nations 2002) has not materialized, as adoption continued to accelerate through 2004. Thus, the compound annual growth rate for 1997–2002 equaled 105.1 percent, while for 2000–2004 this figure rose to 118.5 percent.

At the beginning of 2002, mobile communication networks covered 74 of Russia's 89 regions (United Nations 2002), and by 2003 only two regions of the Russian Federation where still not covered (Chechnya and Tuva; Vershinskaya 2002). Despite this coverage, service remains concentrated in the main urban areas. For example, more that 70 percent of mobile-phone subscribers are from Moscow and St Petersburg (Vershinskaya 2002). Mobile-phone penetration in the Moscow region increased from 47.7 mobile handsets per 100 inhabitants in 2002 to 70 at the end of 2003 (Monitoring n.d.), and is said to stand at 103 in 2005 (Softpedia 2005). Mobile-phone operators are reported to be making attempts to reach the under-served populations, however, by launching simplified terminals designed for emerging markets such as China and India (Softpedia 2005).

Mobile Telephony in the Americas

Mobile Telephony in the United States In 1976 there were 44,000 mobile-phone owners in the United States (Lasen 2002b). With the introduction of cellular phones, this number increased to 5 million by 1990 (2 percent penetration rate; Paetsch 1993). A period of rapid growth began in the early to mid-1990s, driven most likely by falling prices as a result of increased competition among operators (Council of Economic Advisors 2000), as well as the introduction of digital technology. About two-thirds (64.3 million) of US households now own at least one wireless phone (FCC 2002a), and 58 percent of Americans aged 12 and above own a wireless phone (Genwireless 2001). Mobile-phone teledensity was 60.97 in 2004 as against 59.91 for fixed lines.

Cellular telephony was invented in the United States, and the United States currently has one of the most competitive systems, with regional markets having up to seven mobile network operators. Some analysts also argue that the United States has some of the best consumer prices among OECD countries (OECD 2000; Beaubrun and Pierre 2001; Alden 2002). Despite these apparent advantages, the United States has exhibited low awareness and uptake of wireless communication technology relative to other industrialized countries (Lynch 2000; Felto 2001; TNS 2001a; *The Economist* 2002; *Revolution* 2003; *TWICE* 2004). This is primarily the result of delays in allocating spectrum for mobile telephony, as well as other policy and industry decisions about how to structure the market (Gruber 1999; King and West 2002; Lyytinen and Fomin 2002).

The first commercial mobile-phone system in the United States was set up in Missouri in 1946, a "precellular" mobile system providing a limited service owing to capacity constraints in the early infrastructure (Paetsch 1993: 23). Cellular mobile systems were delayed until 1983, owing, first, to the refusal of the Federal Communications Commission (FCC) to grant spectrum for mobile telephony (in favor of giving the spectrum to television broadcasters) and then to delays in processing applications from operators (Paetsch 1993: 150; Lasen 2002b: 32; Steinbock 2003). Once the system was in place, the cost of handsets fell rapidly, but service charges remained high, thus inhibiting the development of a consumer market for mobile phones.

Developments in the AT&T Bell system also affected the progress of mobile telephony. AT&T was broken up at around the same time as mobile telephony was being introduced, an event that probably affected the institutional direction of the company (King and West 2002; Lyytinen and Fomin 2002). In addition, although AT&T had the capacity to develop and deploy cellular telephony on a large scale, it remained focused on fulfilling its universal service obligation via fixed lines, providing cell phones only as a service to commercial entities (King and West 2002). Thus, the development of mobile telephony for the mass market was left to individual entrepreneurs.

Unlike Europe, where a common technical standard was adopted, in the United States there are several incompatible standards (for example, CDMA, TDMA, and GSM) competing for the market (Gruber 1999). This, in addition to aggressive competition in general, has made cross-network communication difficult. Only recently have service providers started to facilitate communication between subscribers on different networks.

Mobile Telephony in Canada Mobile telephones were launched in Canada in the mid-1980s. The use of options such as no-contract, flat-rate pricing, and per-second billing, as well as steady price reductions, has encouraged the growth of the subscriber base. However, Canada is one of the few countries where total fixed lines (20 million) still exceed mobile-phone lines (14 million). Mobile-phone teledensity in 2004 was 47.21, while that for fixed lines was 63.21. Nevertheless, fixed-line operators are experiencing declining shares of total telecom revenues as mobile subscriptions rise. In 2003, only the wireless industry saw revenue increases (Strategis 2005). Fixed residential lines fell from 60 percent of total lines in 1994 to 40 percent of total lines in 2003. Business lines also fell from 28 percent to 22 percent over the same time period. Conversely, wireless lines increased from 10 percent of total lines in 1994 to 40 percent in 2003. Although mobile-phone subscriber rates are rising across the country, some cities have significantly higher penetration; for example, in 2000 Alberta had the highest (60 percent of households) and Quebec was among the lowest (35 percent of households). Pager use also remains high: 20 percent of households use a pager (Choma and Robinson 2000).

Mobile Telephony in Latin America Mobile-phone penetration rates in the Latin American region ranged from less than 1 percent to about 52 percent in 2004. In 2003, the number of mobile subscriptions was equal to the number of fixed lines. Coverage is, however, unevenly distributed, with most users to be found in the large capital cities (Hilbert and Katz 2003).[10] Subscriptions have continued to grow despite economic recession in the region (NECG 2004). The fast diffusion of wireless communication in the Latin American region can be explained by three factors (following MacDermot 2005). First was the inadequacy of the fixed-line segment in the region: the fixed sector was plagued by corruption and inefficiency, and in most countries subscribers faced a waiting list of months before a new line would be installed, if ever. Consequently, the launch of mobile networks around the region at the end of the past decade allowed many consumers to obtain a phone for the first time. The exception however is

Argentina, where, following financial crisis in the economy, mobile-phone subscriptions dropped by 6.8 percent between 2001 and 2002, the greatest decrease in the region. On the other hand, fixed lines also dropped, in this case only by 1.2 percent, leading to an overall 3.8 percent reduction in total telephone subscriptions.[11] Yet, even in such a dire situation, "in Argentina's 2001 crisis ... consumers hung determinedly on to their mobiles (even though they used them less) and the total subscriber base was only temporarily reduced" (MacDermot 2005: 16). Despite this optimistic view, 2001 subscription figures did not recover until 2004.[12]

The Call Party Pays (CPP) and prepaid systems have also been key to the development of mobile telephony in this region (NECG 2004). Prepaid subscription was available in almost all countries of the region by 1999 (Arias Pando 2004), and, in 2003, accounted for slightly more that 80 percent of total subscriptions in Latin America,[13] reaching to over 90 percent in some countries (for example, Mexico, Panama, and Venezuela).

Finally, in all markets, the announcement that a third or even a fourth operator would enter the industry has galvanized the expansion of the subscriber base. Two-player markets, meanwhile, have tended to grow at below-average rates. Nevertheless, competition is, in the view of many observers, the growth driver that is most under threat, considering that two operators – Spanish's Telefonica Moviles and Mexico's America Movil – account for 80 percent of total Latin American subscribers.

Mobile Telephony in Brazil   Brazil is the sixth largest mobile-phone market in the world and the largest in Latin America. Mobile telephony has been an important contributor to increasing total teledensity in Brazil. Poverty, accentuated by unemployment and economic stagnation, has been a significant barrier to the growth and maintenance of fixed-line telephony in Brazil (Lobato 2004). The high cost of fixed-line subscriptions (on average R\$ 33 per month, i.e. US\$ 14.36) resulted in social exclusion for poor people.[14] However, with the introduction of prepaid mobile telephony, the number of households with at least one telephone rose from 23.2 to 27.4 percent between 2001 and 2003, while total fixed lines decreased less than one percentage point and total mobile lines increased from 31 to 38.6 percent (IBGE 2003). By 2003, there were more mobile than fixed lines (ITU 2004). Fixed-line operators are reported to be losing, on average, 2,200 customers per day (Lobato 2004), and the number of households that have only a mobile phone has been increasing since 2001 (11.2 percent in 2003; IBGE 2003). Lower-income households tend to depend more on mobile phones alone (12.1 percent compared to 3.1

percent for higher-income households), whereas higher-income households have higher levels of access to both fixed and mobile systems (89.8 percent versus 19.8 percent for lower-income households; IBGE 2003). As Lobato (2004) observes, "In most cases, the subscriber changed to a prepaid mobile phone because she was unable to maintain the fixed phone despite the fact that it was fundamental for the economic survival of the family."

The national penetration rate in 2004 was 36.32 mobile telephones per 100 inhabitants. In the ten-year period between 1994 and 2004, this means of communication grew by over 8,000 percent (ANATEL 2004).[15] As in other developing regions, mobile telephony is mostly an urban phenomenon. For example, almost all residents of the capital city (Brasilia) have access to mobile phones (93.51 subscriptions per 100 inhabitants). Other major cities such as Rio de Janeiro (51.04) and São Paulo (40.23) also have subscription rates higher than the average.

As tends to be the case with prepaid subscribers, prepaid Brazilian subscribers generally talk less on the telephone (over 50 minutes, on average, per month) than post-paid subscribers (almost 200 minutes per month), an indication of the limited resources they have to spend on communication needs.[16] The importance of prepaid is such that, in July 2003, the Brazilian government created a compulsory register of prepaid mobile handsets in the country in order to prevent unauthorized "cloning" (or replication) of mobile numbers.

Mobile Telephony in Mexico The spread of mobile telephony in Mexico has occurred quite rapidly. In early 2000, both overall mobile subscriptions and prepaid subscriptions alone surpassed fixed lines. By 2004, the penetration rate was almost 37 mobile subscriptions per 100 inhabitants, with more than 90 percent being prepaid. There is still a significant digital divide: in 2002, there was more than one mobile handset per two persons in Nuevo Leon, while in Chiapas there were only four handsets per 100 inhabitants (data from COFETEL quoted by Mariscal 2004).

Mobile Telephony in Asia

The past decade has witnessed major growth in wireless communication in all countries of the Asian region. The total number of mobile-phone subscribers in 2004 was 710,528,100, nearly triple the number in 2000 (240,624,400). There were 18.94 mobile phones per 100 inhabitants (up from 9.46 in 2001), and mobile-phone subscriptions made up more than half (56.9 percent) of all telephone subscriptions.

Mobile Telephony in Japan   The penetration of mobile phones in Japan has grown from 21.3 subscribers per 100 inhabitants in 1993 (Japan Ministry of Internal Affairs 2005) to 71.5 subscribers per 100 inhabitants in 2004. This growth has been achieved at the same time as stagnant or declining fixed-line subscriptions, in part because the fixed-line service is relatively expensive in Japan (Ito 2004). The growth of mobile services, however, has been slowing down for both cellular phones and Personal Handyphone Systems (PHS, a less-expensive but limited mobile service). The year-to-year change in cell-phone penetration was 51 percent growth in 1997; it slowed to 23.1 percent in 1999, and then to 9.5 percent in 2002. In the PHS market, growth rates rose quickly from 0.3 percent in 1995 to 13.1 percent in 1998, but then dropped to 9.1 percent in 2001. Meanwhile, the total number of PHS subscriptions declined from the high point of 6.7 million in 1997 to 5.5 million in 2002.[17]

Mobile Telephony in Korea   South Korea is an important actor in the wireless market and was home to the world's first commercial CDMA service. Since January 1996, the country has played a major role in handset R&D, manufacture, and the launching of new mobile services. New wireless technologies are spreading as a result of the Korean government's "u-Korea" (or ubiquitous Korea) project (S-J. Yang 2003). As of November 2003, 78 percent of Koreans above the age of 15 subscribed to mobile-phone services (KISDI Report 2003: 24). According to the Korea National Statistical Office, the total mobile subscriber population was 33.2 million by the end of 2003.

Despite market liberalization, the Korean government remains a central player in the mobile industry. In order to establish the nation's leadership in information technology, the Korean government chose mobile telecommunications as a key strategic industry for systematic internal capacity-building, which would in turn contribute to the country's future exports and its competitiveness in the global IT market (Yang et al. 2003). Since then, a series of special policies, ranging from handset subsidy to preferential regulation, has been implemented (see Lee et al. 2000; D-Y. Kim 2002; H-J. Kim et al. 2004). In December 1994, the Ministry of Post and Telecommunications (MPT) was expanded in size and administrative function to become the Ministry of Information and Communication (MIC). Since then, under the auspices of the state, Korean mobile providers have played a leading role in the world in launching new services, testing out different standards (for example, w-CDMA and CDMA2000; MIC Report 2003: 20), and in experimentation with 3G services (for example,

IMT-2000; Park and Chang 2004). Most of these transformations occurred during or after the Asian financial crisis of the late 1990s, which expedited the restructuring of the Korean economy, while further enhancing the role of the state (Chang 2003).

Mobile Telephony in China   Modern wireless technologies began in the 1980s in China with the introduction of the pager in 1984 and cellular phones in 1987. While initial adoption was slow and restricted to a very small circle of high-end business users, the speed of growth has been extraordinary since 1990. Pager subscription took off throughout the 1990s to peak with 49 million users in 2000. It then started to decline, becoming a technology used largely by migrant workers.[18] The change happened at the same time as mobile-phone penetration began to surge strongly, together with the rapid spread of short message systems.

The phenomenal boom in mobile-phone subscription began in the mid-1990s, especially after the establishment of China Unicom, which to a great extent undermined the monopoly of the incumbent China Telecom in mobile communications (Mueller and Tan 1997). By the end of 2004, subscriptions had reached 334 million, giving the country by far the largest population of mobile-phone users in the world. According to China's Ministry of Information Industry (MII), mobile phones now account for more than half of the telephone sets in China. This growth is impressive given that China's overall teledensity was only 1.1 percent in 1990, comparable to the teledensity rate of 1.3 percent in the United States in 1899 (US Department of Commerce 1989; *China Statistical Yearbook* 1990). And there were only 20,000 Chinese mobile-phone subscribers in 1990, compared to America's 5.3 million at the time (FCC 2002b; *China Statistical Yearbook* 2003). Yet, despite the speed of growth, 280 million subscribers only account for 21.8 percent of China's total population. A large number of users have a rather limited budget for cell-phone consumption, which is reflected in the popularity of prepaid phone cards. In January 2004, China Mobile, the country's largest mobile-phone operator, had more prepaid that fixed-term contract subscribers, and the popularity of the prepaid plan is even more overwhelming among new subscribers (Liu 2004: 19).

The rapid diffusion of mobile communications runs counter to the conventional wisdom that mobile services are too expensive to be commercially viable in developing nations. In China, the boom in mobile technologies is certainly related to a strong market demand produced under the following conditions: (a) historical inadequacies in the telecom infrastructure; (b) continual economic growth, with GDP rising 7–9 percent per year; (c) massive urbanization, leading to the emergence of large, mobile workforces which migrate within and between the urban areas; and (d) the integration of China into the global market, the arrival of global capital, and the subsequent increase in demand for just-in-time information.

Mobile Telephony in the Philippines   Mobile telephony has been rapidly diffusing in the Philippines since the turn of this century. The subscriber population was less than half a million in 1996 (Kaihla 2001). The number jumped to 6 million in 2000 (Friginal 2003), and then to almost 33 million in 2004. In 2001, there were about 11 million mobile-phone users nationwide (Toral 2003).[19]

In order to secure market share, service and equipment providers have to engage in fierce price competition. As a result, the cost of ordinary handsets has been lowered to about US$ 50 in the open market and half this amount in secondary markets (Rafael 2003: 402). A great majority (70–90 percent) of subscribers use prepaid phone cards instead of fixed-term contracts (Toral 2003: 173–4), which "allowed those without credit history, a permanent address, or a stable source of income to purchase cell phones" (Uy-Tioco 2003: 5).

Mobile Telephony in India   The growth of mobile-phone subscription in India took place in the midst of wider telecom reform when a more liberalized structure was set up to replace the traditional monopoly of the Department of Telecommunication. The liberalization measures significantly lowered entry barriers, bringing great vitality to the mobile market since many new entrants chose to focus on mobile-service provision. This was particularly important as the state-owned companies of MTNL and BSNL have been criticized for making little progress toward their universal service obligations (Dossani 2002).

As a result, the current mobile-phone market in India is characterized by a fairly decentralized structure. The country is divided into 23 telecom "circles" (or "service areas"), in which a maximum of four mobile licenses are allocated to public and private companies. In total, 78 licenses had been awarded to 20 companies by 2004 (Department of Telecommunication 2003–4). Compared to the traditional nationwide licensing scheme, this "regionalized" structure has been effective in driving down prices. For example, following the promulgation of the New Telecom Policy, per-minute costs for mobile telephony dropped by about 90 percent from an average of 16.8 rupees in 1999 to an average of 2 rupees in 2001. The cost of mobile handsets also fell to less than one-tenth of the mid-1990 prices. A major policy change in May 2003, when the "calling party pays" regime replaced the previous two-way charging system, further accelerated wireless diffusion (McDowell and Lee 2003).

Following these reforms, total teledensity rose rapidly from 1.28 in March 1996 (Department of Telecommunication 2003–4) to 8.24 in November 2004,[20] largely as a result of the rapid growth in adoption of mobile phones. In less than a decade, mobile-phone subscriptions grew by a factor of about 400 (26.15 million subscribers in March 2004, up from 76,700 in 1995).[21] This was in addition to the country's 7.55 million wireless local loop (WLL) customers (Department of Telecommunication 2003–4). During this period, the total number of fixed-line telephones (excluding WLL) also increased from 14.54 million in 1997 to 42.84 million in 2004, a significant though slower pace compared with mobile telephony (Department of Telecommunication 2003–4: 92). From March 2003 to March 2004, the annual growth rate of fixed-line phones was 2.9 percent as compared to 159.6 percent in the mobile sector.[22] GSM is the dominant standard (78 percent of total subscribers and nearly 80 percent of new subscribers; *Business Line* 2004). Prepaid services constitute 70 percent of total subscriptions, usually in smaller towns in less wealthy regions (McDowell and Lee 2003).

It is, however, important not to take official statistics at face value, as some analysts maintain that figures are often over-estimated. For instance, as a result of growing market competition, "GSM operators are adopting different cut-off periods to check if the existing pre-paid card subscriber has left their service" (Rambabu 2004). Some operators keep customers' accounts in their records for up to 30 days after the due date for recharging the account; others for as much as 60 days. This is especially the case with smaller operators, which seek to maintain a higher valuation for their enterprise (Rambabu 2004).

Thus, although the mobile subscriber base surpassed the number of fixed-line consumers in October 2004, this was true in only ten of the 23 telecom circles (Thomas 2004). The most concentrated growth in mobile diffusion was in the four major cities – Delhi, Mumbai, Chennai, and Kolkata – which accounted for almost two-fifths of the wireless subscribers in the country in December 2001 (McDowell and Lee 2003). The dominant share of these urban centers has continued (O'Neill 2003: 86). In Delhi and Mumbai, mobile subscription is about twice that of fixed-line users: for example, 4.9 million mobile phones and 2 million fixed lines in Dehli (Thomas 2004).

Mobile Telephony in the Middle East and North Africa (MENA)

The MENA region, as defined by the World Bank, includes the following countries: Algeria, Bahrain, Egypt, Iran, Iraq, Israel, Jordan, Kuwait, Lebanon, Libya, Malta, Morocco, Oman, Qatar, Saudi Arabia, Syria, Tunisia, the United Arab Emirates, the West Bank and Gaza, and Yemen. As in other areas of the world previously dominated by state-owned enterprises, new licenses are increasingly being awarded to new operators to provide telecom services in this region (Ford 2003). For example, there are currently three main mobile-phone operators in Israel: Cellcom, Pelephone, and Partner. These and the other mobile-phone providers cover the largest share of the telecom market (both subscriber lines and revenue; Gandal 2002: 7).

However, telecom markets in the MENA region still remain relatively closed to competition from both private and foreign sources, and the regulatory regimes do not support fair competition (Varoudakis and Rossotto 2004: 59). For example, the Saudi government began the process of privatizing the telecom industry in the late 1990s; however, the Saudi regulator, the Communication and Information Technology Commission (CITC), recently decided to issue only one 25-year license for the operation of a GSM network in the country (*Country Monitor* 2004).

There has been a significant increase in mobile-phone subscriptions in the past decade, and mobile lines outstrip fixed lines in most countries. Mobile-phone penetration in the Middle East region was 16.59 percent in 2002, with an annual growth rate of about 30 percent (Ford 2003). There are, however, large disparities between the Gulf states and the rest of the region (Dutta and Coury 2003). For example, in some states, such as Israel, the United Arab Emirates, and Kuwait, mobile-phone penetration levels are approaching saturation (104.74 percent, 84.71 percent, and 77.07 percent respectively in 2004), and operators are already shifting their attention from simple voice and text messaging to data and multimedia services (Ford 2003; *AME Info* 2004). In others, such as Iran and Libya, the penetration of mobile communication technologies is fairly low (6.16 percent and 2.3 percent respectively).

Mobile Telephony in Africa

Mobile-phone networks are a fairly recent development in Africa, driven primarily by the inability of fixed-line providers to meet the demand for telephone lines, as well as changes in telecommunications policy which allow for the growth of private telephone companies across the continent. The earliest subscriptions began to appear around 1987 in Egypt, Morocco, and Tunisia, followed by South Africa in 1989, and Mauritius and Algeria in 1990. In 1998, there were 4,156,900 cellular-phone subscribers on the continent; by 2004, there were 75,891,900, representing an increase of over 1,800 percent in five years.

A majority (74.6 percent) of telephone subscribers are mobile-phone subscribers. In most countries, over 60 percent of telephone subscriptions are for mobile phones, with the Democratic Republic of Congo having the highest proportion (94.2 percent) and Guinea Bissau the lowest (10.8 percent). Between 2000 and 2004, mobile-phone subscriptions grew at rates ranging from 17.3 percent (Seychelles) to 364.2 percent (Djibouti). Fixed lines in the same period grew slowly at rates of between 0.1 percent (Cameroon) and 28 percent (Sudan) and, in some cases, declined (Congo, Côte d'Ivoire, Gabon, Guinea Bissau, Morocco, Mozambique, South Africa, Tanzania).[23] Mobile subscriptions first surpassed fixed lines in 2001. Not surprisingly, mobile phones have been hailed as the answer to improving telecommunications access in Africa (Kelly et al. 2002).

Uganda was the first country in which mobile phones exceeded fixed-line connections in 1999 (*Business in Africa* 2004). This is largely attributable to the decision by the second national telecommunications operator to focus on cellular technology (McKemey et al. 2003; Panos 2004). It is also notable that Uganda has had one of the lowest fixed-line growth rates on the continent. The majority of mobile-phone users, however, are in South Africa (19.5 million or almost one-quarter of Africa's 75.9 million subscribers), and the island community of Seychelles has the highest penetration rate (60.78 percent).

Despite the high levels of mobile-phone subscription, the penetration of telecommunications services in Africa is low.[24] The continent has only nine mobile phones per 100 inhabitants and there are sharp differences in the levels of mobile teledensity between and within countries in the region. Only seven countries had more than 20 mobile phones per 100 inhabitants in 2004; 17 countries had mobile teledensities between six and 20; and all other countries had mobile teledensities lower than six. It is also a fact that, as happens in all cases of growth, after the spurt of growth in 2001, the annual mobile-phone growth rate for Africa has started to decline, down from 64 percent in 2001 to 47 percent in 2002 and 36 percent in 2003 (ITU 2004a).

Mobile Telephony in Oceania

Of the countries in Oceania, Australia and New Zealand were the first to have access to mobile telephony in 1987, and in 2004 had the largest number of subscribers, as well as the highest penetration rates (82.6 percent and 77.5 percent respectively). It was not until 1994 that any other country introduced mobile telephony. Samoa was the last to have mobile telephony (in 1997). In Australia, the number of mobile-phone service subscribers was 16,449,000 in 2004, representing an average increase of 17.7 percent from 2000 figures. Contract service (57 percent) constituted the largest segment of the Australian mobile-phone services market, with prepaid having 43 percent of market share (Snapshots International Ltd 2005).

Trends in Mobile Data Diffusion

Assessment of the diffusion of mobile data is problematic in many ways, most of them related to global variations in the collection, definition, and measurement of mobile data usage.[25] In this section, we present some indicators of the level of mobile data use in some countries and regions, bearing in mind that the data may not always be directly comparable.

Text Messaging

Text messaging (SMS) has become a critical aspect of mobile communication, adding significant value by expanding the amount of interaction possible in any given amount of airtime. Trends in this use of mobile phones largely mirror trends in mobile telephony, with European and Asian Pacific countries leading the market, and evidence of uneven regional penetration.[26] In 2001, a survey by A. T. Kearney found that 72 percent of respondents in Europe, 39 percent in Japan, and 12 percent in the United States used SMS. Both Canada and the United States fell in the bottom quintile of mobile-phone owners who used their mobile phones to receive text messages, while South Korea, urban Turkey, the Czech Republic, Hungary, and Finland fell in the top quintile, according to Taylor Nelson Sofres research (TNS 2002b).

Subscribers in Europe, especially Western Europe, are avid users of text messaging: 186 billion messages[27] were sent in Western Europe in 2002 (CellularOnline 2004), 16.8 billion of them from the UK (Mobile Data Association 2005). In the UK alone, 26.2 billion messages were sent in 2004, and an average of 85 million messages per day were sent in May 2005 (Mobile Data Association 2005). However, texting is not popular in all of Western Europe; for example, in France adoption is low (Gartner 2003). In Central and Eastern Europe there were about 45 million SMS users in 2003, with usage being more common in Slovenia and the Czech Republic and less common in Poland and Romania (IDC Spain 2004).[28]

SMS is also the primary mobile-phone application in Asia. The Philippines is reportedly the world's highest texting nation, with the average user sending over 2,000 messages a year. Singapore, the Republic of Korea, and Japan have particularly high statistics for text message per subscriber per month (184, 167, and 111 messages respectively; Minges 2005).[29] Two hundred and twenty billion messages were sent in China in 2003 (Adelman 2004). Text messaging in Asia is facilitated by very attractive pricing packages, sometimes including hundreds of free text messages.

In North America, on the other hand, there is relatively limited (but steadily rising) use of SMS (Harter n.d.; *Revolution* 2003). Although SMS is now the leading mobile data service (according to In-Stat/MDR 2004b), as late as 2004, *New Media Age* stated that the average American cell-phone user had little idea what SMS was. In 2003, there were 27 million SMS users in the United States, up from 18 million in 2002 (eMarketer 2003), and, according to Greenspan (2004b), more than one-third of US wireless phone owners use SMS. SMS traffic is also increasing: for example, in June 2001, US wireless phone users sent 30 million text messages, in June 2002 they sent nearly 1 billion (Trujillo 2003), and by mid-2004, they were sending about 2.5 billion per month (IT Facts 2005). In Canada, business organizations have been using SMS since 1997 to facilitate cross-border communication between units and workers who operate in Canada and the United States (Lahey 2003). Among the general population, usage is slightly lower than in the United States: numbers have risen from 174 million messages in 2002 to 352 million in 2003 and 710 million in 2004. Canadians are now sending over 3.4 million messages a day (CWTA 2005b). In both countries, a significant barrier has been incompatibility between the major carriers' interface technologies, a situation that was finally resolved around 2002 (*Communication Today* 2001; CWTA 2005a; Fitchard 2005).

Wireless Internet

Although there are a variety of ways to access the Internet wirelessly (for example, mobile phones, pagers, laptop computers, PDAs), access via mobile phone has received the most attention in the research community, probably because it is the primary mode used in countries where wireless data access is popular. Again, Asian and European countries surpass most others in the diffusion of Internet-capable wireless phones, and there are intra-regional variations. Japan, on its own, has consistently had wireless-phone Internet access levels exceeding those of North America and all of Europe (A. T. Kearney 2002, 2004). Japan also has the highest percentage of mobile Internet users to total mobile-phone subscribers (Srivastava

2004a; Minges 2005). In addition to Japan, Korea, Singapore, and Hong Kong have relatively high levels of wireless phone Internet access, but other countries in the same region (for example, China and Thailand) have low uptake (TNS 2002b; Minges 2005).

Data from selected European countries show that the percentage of Internet users who access the Internet using mobile phones is very low: for example, less than 18 percent in Germany, and just over 2 percent in Bulgaria, in 2001 (Eurescom 2004). Furthermore, in 2002, usage levels were generally stagnant (staying at around 12 percent in the UK and 2 percent in Bulgaria, for instance) and, in some cases, even fell (from about 10 percent to less than 8 percent in Norway, and under 18 percent to about 12 percent in Germany). The main cause of this pattern is the high cost of the service, although low data-transmission capacity may also be a factor.

Mobile phone-based Internet access is equally low in the United States (Harter n.d.; DeJong 2001; ITU 2002a). Surveys report low visibility and usage of the wireless Internet in the United States. For example, 89 percent of respondents in a 2001 survey said that they were "unaware" or "poorly informed" about wireless Internet technology (Felto 2001), and, in another survey, only 30 percent of respondents who were aware of Internet access on their wireless devices said that they used it (Genwireless 2001). While 28 percent of Americans are "wireless ready" (that is, they own a wireless device that can be used to access the Internet), only 17 percent have actually gone online wirelessly (Pew Internet and American Life Project 2004a). For the US population, the wired personal computer (PC) is still the primary means of accessing the Internet for information and communicative purposes (US Department of Commerce 2002). PDAs, Blackberries (especially in the corporate world), and laptops are the most popular devices for mobile data access. In 2002, ComScore Media Metrix reported that, although Internet users own more cell phones than PDAs, more PDA owners than cell-phone owners use their devices to go online (Reuters 2002). Still, the use of mobile data applications is growing: 3.4 million users in 2000, 9.5 million users in 2001, and an estimated 23.4 million users in 2002 (Entner 2003). A survey by In-Stat/MDR (2004a) found that 54 percent of respondents were using some form of wireless data service and that there was little difference demographically between users and non-users, indicating that wireless data usage is becoming more mainstream.

Wireless Internet delivery services are relatively undeveloped in Canada. Providers began rolling out services in the early 2000s, mostly low-speed connections on specialized handsets. Some providers are now offering high-speed fixed wireless access in a few urban areas (mainly Ontario, Quebec, Manitoba, Saskatchewan, British Columbia), and several companies have announced plans to implement high-speed wireless data systems such as EDGE and EVDO (Evolution Data Optimized; Strategis 2005). Wireless data services are available via phone, pager, or PDA. However, only 2 percent of mobile-phone users used their phones for Internet access in 2000 (Strategis 2005).

Other parts of the world have not seen much growth in the use of the wireless Internet. In developing economies especially, where such use exists it tends to be limited to large national and multinational companies, and even there, dial-up and cable are utilized more. Such trends have been observed in Australia (Department of Communications 2004; Paul Budde Communication 2004) and India (Srivastava and Sinha 2001), for example. Africa, the Middle East, and Latin America also exhibit relatively low usage of mobile data in general and wireless Internet in particular, mainly because the focus there is still on promoting voice and text messaging. In any case, there is currently limited bandwidth and limited market size to support data applications in these regions, although some services are being targeted at the international business community.

Wi-Fi

Wi-Fi technology as a means of organizing wireless access to digital data is gaining popularity in several countries. It is also considered to hold immense potential for bringing Internet access to poor communities. The trend here differs significantly from that for mobile telephone uses. While Europe leads overall in the uptake of wireless-phone technology, and while diffusion of the wireless Internet via the mobile phone has been most significant in the Asian Pacific, the establishment of Wi-Fi systems, based on laptop usage, seems most prominent in North America, particularly in the United States (Sharma and Nakamura 2003). As of July 2005, there were at least 68,643 Wi-Fi locations worldwide, a majority of them in the United States, followed by the UK and Germany (Intel 2005).[30] In 2003, 47 percent of the 42 million Wi-Fi users worldwide were in the United States, followed by Western Europe with 35 percent and the Asian Pacific with 17 percent (Maddox 2003; *Wired* 2003).

The set-up of Wi-Fi systems provides the possibility of free access to digital information wherever such access is not deliberately blocked. For example, the warchalking phenomenon[31] illustrates how attempts are being made to use the availability of Wi-Fi zones to bypass economic organizations and enhance free access to information. In addition, plans are underway in a number of communities in the United States to provide public Wi-Fi coverage as a free service. According to research at the University of Georgia, there are 38 Wi-Fi clouds and 16 Wi-Fi zones throughout the United States, most of which have been set up to enhance the value of communities rather than to generate revenue (New Media Institute 2004).[32]

It is not surprising that the United States is the leading Wi-Fi user since Wi-Fi usage is largely laptop driven and the United States has the greatest penetration of laptop computers. Hotspots in the United States are generally located in airports and cafés, as well as some parks. Wi-Fi connections also exist to a limited extent in some homes (Pew Internet and American Life Project 2004b). Nevertheless, it is important to emphasize that the technology is diffusing and attracting attention from people who find it interesting and potentially significant. However, the extent of actual Wi-Fi usage is lower than expected. Studies by a number of research firms, such as In-Stat/MDR and Jupiter Research, indicate that visitors to various hotspot locations use them infrequently – less than six times a year (Biddlecombe 2003). Jupiter Research has found that, although a large number of people are aware of public Wi-Fi availability (70 percent of online consumers), only 15 percent have used it and only 6 percent have done so in a public place (Vilano 2003). Only about 11 percent of visitors to Wi-Fi-enabled US Starbucks locations are taking advantage of the service (Biddlecombe 2003).

Factors Accounting for Differences in Penetration Rates of Wireless Communication Technology

The variation in diffusion of wireless communication technologies in different countries can be explained by a combination of factors that confound the process of adoption. We present here a summary of elements that have been identified by analysts and researchers, as well as our own observations on factors affecting the adoption of wireless communication technology around the world.

Level of Economic Development Measured by GDP
The global diffusion of mobile telephony reflects the traditional digital divide between wealthy and poor countries, as most of the countries in the low range of diffusion are poorer, developing countries (although the telephone divide is smaller than the Internet divide; Kelly et al. 2002). Thus, the economic status of a country appears to have some impact on the speed

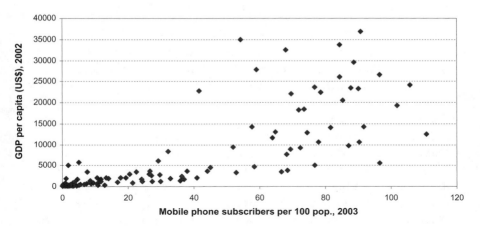

Figure 1.4
Mobile-phone penetration versus GDP per capita (US$), 2002–3. *Source*: ITU statistics (www.itu.int).

and reach of mobile-phone subscription.[33] More than half of the mobile subscribers worldwide are in the high-income group (ITU 2004b). To explore this further, we looked at the relationship between mobile-phone penetration and GDP (see figure 1.4), which confirms that, generally, mobile-phone penetration rises with GDP (although this does not imply a causal relationship).

GDP alone, however, only explains about half of the variation in mobile-phone penetration rates (TNS 2002b). While the data suggest that initially a country's GDP affects its ability to adopt wireless technology, this must be considered in relative terms. Poorer countries may in fact be more enthusiastic adopters of mobile technology, as has been seen in several developing countries. For example, mobile-phone subscriptions continue to rise in Latin America, despite economic recession in the region (NECG 2004), and Africa is registering the highest growth rates worldwide (ITU 2004a). The ability to diffuse the technology to a majority of the population (i.e. the penetration rate) is, however, still limited by economic constraints, resulting in asymmetric diffusion of mobile telephony to urban areas. This has been the case in Africa, Latin America, and the poorer regions of Europe and Asia.

Existing Fixed-Line Infrastructure and Service
Wireless communication systems, being relatively less expensive to install and easier to distribute, tend to be a good alternative to fixed lines. Poorer

countries generally also have inadequate fixed-line infrastructure, which makes the use of mobile telephony more attractive to telecom operators seeking to enter these markets. Furthermore, faced with long waiting periods and unreliable service from fixed-line operators, citizens in poor countries are more likely to turn to wireless telephony when it becomes available. This to a large extent has led to the remarkable growth of mobile-phone use in emerging economies.

A majority of countries with poor fixed-line infrastructure now have more mobile phones than fixed lines. Still, bearing in mind the high growth but low penetration rates in poor countries, some evidence has been found that higher numbers of fixed lines are associated with greater mobile-phone penetration (for example, ITU 2004b; Varoudakis and Rossotto 2004; Kauffman and Techatassanasoontorn 2005). The ITU analysis of the global dispersion of fixed and mobile telephone lines shows that, while most countries now have more mobile than fixed lines, developed and newly industrialized countries have high levels of both – that is, more than ten fixed and ten mobile lines per 100 inhabitants (ITU 2004b). Most developing countries have less than ten fixed and ten mobile lines per 100 inhabitants; and a mix of Latin American, Asian Pacific, and a few African countries have less than ten fixed and more than ten mobile lines per 100 inhabitants. Finally, there are still a few countries, mainly from the former Soviet Union and the Middle East, with more fixed than mobile lines. The existence of a positive relationship between fixed- and mobile-phone penetration levels, if significant, could have important implications for efforts directed at leapfrogging fixed-line technology in the developing world.

The problems of low income and inadequate fixed-line infrastructure lead to the type of growth trends observable in most developing countries: higher than average growth rates, lower than average penetration rates, high mobile to fixed-line ratios, and urban concentration of coverage. Clearly, while mobile telephony enables developing countries to do some leapfrogging of communication technologies, they still lag in overall uptake, compared to more developed countries. Furthermore, diffusion is occurring faster in some developing economies than in others.

Geographic Factors
Countries with small land mass (for example, most European countries) and more densely populated residential settlements (for example, Japan) are able to speed up the adoption of wireless communication because it is easier to set up wireless infrastructure. More effort, expense, and collabo-

ration are needed to establish such systems in wide areas such as the United States (Ling 2004). On the other hand, it could be argued that some types of wireless systems are more easily installed in wide, uninterrupted areas, as some Wi-Fi operators have found. For example, the builders of a 1,500 square mile Wi-Fi network in a rural part of Columbia state found it easier to build that network than to build citywide networks because there was less interference from buildings and radio signals (Cook 2004).

Industry Factors

Conditions within the telecommunications industry have been particularly conducive to adoption in some countries and may have inhibited growth in others. It is not clear which of these factors is most important, and few causal links have been established. However, all have played a part in producing different adoption rates around the world. We discuss below the contribution of pricing systems, billing systems, technological standards, amount of competition, and types of service.

Pricing Systems In most countries, the emergence of flexible tariff packages (Min 2000: 42), brought about by competition between operators, has resulted in falling prices for consumers, and subsequently higher overall adoption. However, differential adoption rates have been attributed to the choice between the two primary pricing systems: calling party pays (CPP) and receiving party pays (RPP).

In most countries with the CPP pricing system, consumers are more willing to use mobile telephones because they are only responsible for the calls that they make (in other words, it is free to receive calls). This is the case in most of Europe. On the other hand, consumers in countries with RPP pricing tend to be slower adopters since this system makes them partially responsible for calls other people make to them (as has been the case in the United States). The ultimate impact of the RPP system remains unclear[34] since the US market, for example, is so price competitive (most packages have a large allowance of free minutes) that the cost of an incoming call is no longer a big issue for most consumers (Lynch 2000). There are also some countries that have the RPP system in which there has been high adoption of mobile telephony (for example, Hong Kong and Singapore, which have 70–80 percent penetration rates). In general, however, in the absence of other mitigating factors, the introduction of the CPP system positively influences the penetration rate of a given country. For instance, monthly subscription growth hit record figures in Mexico just after the introduction of CPP in May 1999 (Passerini 2004).

The system of free local calls and un-metered fixed-line packages may also have reduced the incentive to use mobile telephones for both voice and data purposes in the United States. It should be noted, though, that other countries with similar market features such as un-metered, local, fixed-line calls (for example, Australia) have still shown greater levels of adoption than the United States (OECD 2003). On the other hand, high pricing of local, fixed-line calls may have facilitated the success of mobile telephony in the UK (Banks 2001). High mobile-phone charges have also been a factor behind the popularity of text messaging in Europe (Zhang and Prybutok 2005). This only emphasizes the point that no single factor determines how mobile communication technologies are adopted.

Billing Systems: Prepaid vs Contract   It has become obvious that the availability of prepaid systems has made it possible for the global telephone divide to be narrowed more rapidly than the Internet divide. For poorer countries and people with limited access to credit, prepaid billing systems have been the primary factor facilitating adoption of mobile telephony. This phenomenon is observable in both developing and advanced economies, an issue we discuss in chapter 2.

Three main groups of countries can be defined with respect to the prepaid billing system and its relationship with income level and mobile-phone penetration. Table 1.2 shows the results of a cluster analysis that sheds more light on the development of this phenomenon. The first type corresponds to high-income countries[35] in which mobile penetration is high (more than 55 handsets per 100 inhabitants) and the presence of prepaid subscriptions is moderate to high (averaging 55.5 percent of total subscriptions). Type 2 countries are low-income economies with low mobile penetration (less than 55 handsets per 100 inhabitants) but a very

Table 1.2
Cluster analysis of mobile-phone penetration and prepaid subscriptions

	Type 1	Type 2	Type 3
Mobile penetration (subscriptions per 100 population)	High	Low	Low to moderate
Prepaid subscription (% of total mobile subscription)	Moderate to high	High	Low
Country income level	High	Low	Moderate

Details of this analysis are available in the online statistical annex (appendix 62).
Source: ITU statistics (www.itu.int).

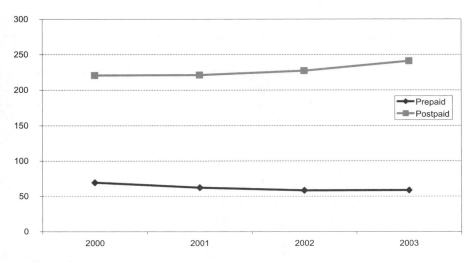

Figure 1.5
Average monthly voice minutes of mobile-phone use in Latin America, 2000–2003.
Source: Pyramid Research, 2005: Latin America Mobile Forecast (http://www.
pyramidresearch.com/store/fcmbla_t.htm).

high incidence of prepaid subscriptions (averaging 82.6 percent of total
subscriptions) with higher presence of this billing system when the penetration of mobile telephony is lower. Finally, type 3 countries are those
moderate-income economies in which mobile subscribers per 100 inhabitants are low or moderate (averaging 10.4 percent of total subscriptions)
with very low presence of prepaid subscriptions (20 percent of total
subscriptions).

The prepaid billing system is, however, associated with lower average
monthly minutes of use (MOU). Statistical analysis of the relationship
between prepaid and average monthly MOU shows a negative correlation
(58 percent).[36] This is only to be expected since prepaid is most used by
cost-conscious consumers whose objective is to control their expenditure
on mobile telephony (as illustrated for Latin America in figure 1.5).[37]
However, since, compared to fixed-line telephones, mobile operators
assume a much lower sunk cost when adding a new subscriber, they can
still benefit from this system (NECG 2004: 4), although they would prefer
higher usage levels.

The combination of the calling party pays system and prepaid billing
appears to create the ideal conditions for the growth of mobile telephony
(NECG 2004). Thus the OECD found that the introduction of prepaid cards

is strongly correlated with faster growth of mobile-phone subscriptions in countries with CPP than in countries with RPP (NECG 2004: 26).

Technological Standards   The existence of uniform technological standards promotes interconnectivity between telecommunications and information systems, and reduces uncertainty in the market (Gruber 1999). This enables manufacturers to enjoy economies of scale in equipment production, service providers to provide smooth and reliable service, and makes it more convenient for consumers to use the technology. Analyses by Kauffman and Techatassanasoontorn (2005) have shown that the existence of multiple technical standards slows adoption rates, while Sarker and Wells (2003) also found that lack of inter-network operability was a significant factor inhibiting the adoption and use of mobile communication devices.

Currently, as an outcome of the free-market orientation, there are several incompatible standards operating in the United States (for example, CDMA, TDMA, and GSM). This, most observers agree, has been one of the critical factors slowing down the rate of adoption of wireless technology in the United States, in contrast to European countries, which have the uniform GSM standard (Banks 2001; Zhang and Prybutok 2005). As already noted, it is only recently that cell-phone owners in the United States and Canada have been able to send text messages across different networks. However, as a counterpoint, it is worth noting that the existence of multiple standards in the Chinese mobile-phone market has not inhibited the growth of adoption in that country.

Amount of Competition   Related to this last point is the level of competition in the wireless market. In general, there is a perception that higher levels of competition should lead to lower prices and greater growth. However, because the extension of wireless communication services depends so much on the integration of different systems and technologies, it is also important for competition to settle down fairly quickly, so the market can gain momentum (Lyytinen and Fomin 2002). The fragmentation of the US cellular market caused by the decision to encourage experimentation and competition in standard formation may have been a significant factor in slowing down the growth of the market. Extreme competition also tends to result in the application of the "walled garden" approach in which operators limit access to their networks from competing networks, leading to the inter-operability problems that have persisted in North America. On the other hand, policy and regulatory provisions can moderate some competitive tendencies: in Europe, policies were implemented to encourage cooperation in standards formation and competition only in service provision.

A slightly different dynamic seems to exist in developing regions. On the one hand, Mureithi (2003) suggests that, in African countries, fewer operators may be better than many at the initial stages to allow operators the benefits of economies of scale. Analysis of the industry in Kenya (two operators), Ghana (four operators), and Tanzania (five operators) showed that a higher number of competitors was associated with higher tariffs and lower penetration rates. On the other hand, in the MENA region, it appears that lack of market competition is inhibiting the penetration of mobile telephony in some countries (for example, Tunisia, Iran, Syria, and Libya; Varoudakis and Rossotto 2004). Thus, the overall impact of competition may depend on other contextual issues. A study of global trends in mobile-phone diffusion concluded that the number of competitors was positively related to initial mobile-phone penetration, but had no impact on subsequent growth rates (Dekimpe et al. 1998).

Types of Service It has also been suggested that, especially with value-added wireless services, adoption has been higher where industry operators have offered consumers appropriate applications. Technological devices and services cannot succeed, no matter how innovative, if they do not meet the motivations and goals of consumers. For example, the services provided by NT Docomo were attuned to what its Japanese subscribers wanted, hence its tremendous success. Wireless service providers in the United States, on the other hand, have been successful in offering valued services to the business community, but less effective in finding killer applications for the personal market.

Government Policy
Institutional environments, while not fully determining the social uses of communication technology, can cause developments to occur at different speeds because of their bureaucratic ability to control access to required technical resources, such as spectrum. In some countries (for example, China), the government has been particularly active in promoting wireless technology, leading to high adoption rates. In others, such as the United States, this has been left largely to the market, as a policy decision. In the early days of wireless, US regulatory bodies actively inhibited the growth of the market by limiting spectrum allocation to wireless telephone companies. The United States was also slow to adopt digital cellular

technology, unlike other countries (for example, in Europe) which made the switch fairly rapidly to digital and to a common standard. The introduction of policies to liberalize the telecommunications industry in most African countries was probably the first and most important action that made the flow of mobile telephony into that region possible.

Sociocultural Factors

Population Size and Characteristics   Most research has not focused on these aspects, but two studies report that population size and features may influence mobile-phone adoption rates (Dekimpe et al. 1998; Varoudakis and Rossotto 2004). Dekimpe and colleagues (1998) found that initial penetration levels were positively related to population growth rates, and negatively related to the existence of larger numbers of major population centers and ethnic heterogeneity. Illustrating the influence of population centers, for instance, they note that the whole Belgian population was covered in the first year of mobile-phone service, whereas this took longer in the United States. However, when considering penetration growth, it was found that ethnic heterogeneity slowed growth rates, but higher numbers of major population centers led to faster growth. Dekimpe et al. conclude that:

the variances in global diffusion patterns are explained by variances in social system characteristics that affect long-run ceilings (which vary between .001 and .99) and social system sizes (which vary between 2,000 and 1.1 billion), variances in the initial penetration level (which varies between .00001 and .033), and variances in the growth rate coefficient (from .001 to .705). (Dekimpe et al. 1998: 126)

Similarly, Varoudakis and Rossotto's (2004) analysis showed that penetration of telecommunications in general, including mobile phones, tended to be lower in MENA countries with larger and more dispersed populations.

PC Penetration   The astounding uptake of wireless Internet technology in some Asian countries, such as Japan, has been attributed to their low levels of PC penetration (Banks 2001; Zhang and Prybutok 2005). It seems, however, that in order for this to be a promoting factor, the country also needs to be fairly wealthy, since other developing countries with low PC penetration levels have not so far been able to develop similar wireless Internet devices and systems. Conversely, the high penetration of Internet-enabled PCs in the United States is seen as a reason for low interest in mobile phones and the mobile Internet (Sharma and Nakamura 2003). On the other hand, there is also evidence that high PC penetration does not necessarily lead to low mobile-phone adoption: Korea, for example, has high levels of broadband Internet access (70 percent of households in 2003; MIC Report 2003: 3), high mobile-phone adoption (78 percent of adults; KISDI Report 2003), and high levels of text messaging (Minges 2005).

Dominant Transportation System Another critical difference between national systems is related to the predominant transportation method: in the United States, for example, where most people drive their own cars, certain types of mobile communication activities (such as SMS) are less viable. In contrast, where public transport is the main means of movement (as in parts of Asia and Europe) people have a greater ability to use wireless technologies on-the-go and consequently develop expertise faster. When obstacles to PC use also exist, the motivation to use text messaging may be even higher.

Communication Preferences Preferences for different styles of communication have also been cited as a reason for the different adoption rates of wireless technology, especially cell phones. For example, Moschella (1999) suggests that Americans prefer asynchronous communication methods, such as e-mail and voice-mail, because they are considered more efficient, keep things brief, leave users in greater control, and are more formal and guarded. Other researchers have identified similar differences in cultural traits. Mante's (2002) survey of people from the United States and the Netherlands found that both groups used mobile devices to enhance their reachability, but that Americans had a preference for devices that allow control over flow, such as pagers and caller ID. Plant (n.d.) also concluded that Americans tend to place a high value on their privacy; hence their reluctance to be fully available through ownership of mobile communication devices. Such national generalizations have to be handled with caution, however. "National" characteristics combine in different ways to influence citizens' behavior. For example, there has been immense adoption in Finland, where citizens are generally perceived to be more reticent than Americans. The influence of cultural factors is discussed further in chapter 2.

Summary

Globally, there has been an explosion of wireless communication, especially of mobile phones. Wireless-phone subscription is growing faster than fixed-line subscription, but both are growing. Rates of growth vary widely between countries, based initially on economic wealth, but also beyond that, even among both poorer and richer countries. European countries have exhibited the highest and most rapid uptake of mobile telephony, followed by some Asian Pacific countries, and then North America. Mobile-phone Internet access, on the other hand, is more prevalent in the Asian Pacific region, while access via Wi-Fi appears to be dominant in North America.

It is clear that the adoption of wireless communication technologies occurs for different reasons in different types of economies. For example, in countries with poor fixed-line infrastructure, wireless becomes a tech-nological substitute for fixed lines; in countries with adequate fixed-line infrastructure but competitive rates for wireless, wireless is becoming an economic substitute for fixed lines. Per capita figures also show that coun-tries with extremely high growth rates may still have low proportions of their population with access to mobile technology, as is the case in most developing countries (for example, China, Uganda).

Overall, the differential rate and level of diffusion in different areas of the world, as well as in different regions of countries, results from the inter-play of a number of factors, among which level of development, industry structure and strategies, and government policies seem to be paramount. This is to say that technological diffusion is the result of the interests and values underlying the social context of their adoption. On the other hand, the most important trend is the extraordinary rate of growth of mobile telephony around the world, particularly in developing countries. The fact that most of the world's population is entering the electronic communi-cation age in a wireless mode has social and cultural consequences, the importance of which we are only beginning to perceive. Let us examine what we already know about the social processes being constructed around these communication technologies.

2 The Social Differentiation of Wireless Communication Users: Age, Gender, Ethnicity, and Socioeconomic Status

The adoption of a given new technology is influenced by the possibilities and limitations associated with it. These, combined with end-user needs, generally lead to different processes of appropriation. Social groups, often defined in demographic terms, are likely to adopt wireless technologies in different ways, as long as the utilities they seek are different. When accounting for utility, costs and needs should not be considered only from a monetary point of view because the adoption process has two stages: first, the acquisition of the device, and, secondly, the appropriation of the technology. Once barriers to acquisition of the device are overcome, further differentiation can still take place. For example, a voice call is easily made and does not require special (new) skills, but text messaging (SMS), while being cheaper, requires some – physical and literacy-related – abilities and some free time to become an efficient user. From this point of view, it is more likely that young adults who have to pay their own mobile-phone expenses will develop text-messaging skills than those of similar age who are introduced to the technology in their working lives where their employers pay the bill. For the first group, the utility of the preferred mobile-phone function is in saving money; for the second group, it is in saving time.

In this chapter, we examine evidence of the socially differentiated patterns of diffusion of wireless communication technologies with regard to four basic social variables: age, gender, ethnicity, and socioeconomic status. It is important to note that, with the exception of gender, researchers around the world often use different conceptualizations to define demographic groups.[1] We do not adopt any particular framework here, but accept categorizations presented in the reviewed research as representative of the general meaning of any specific demographic. The focus of our analysis is to distill general patterns for the social differentiation of wireless diffusion in different societies of the world, and then consider two questions: (a) Are there global trends, or points of convergence, in the patterns of social differentiation among adopters of wireless technology? (b) What are the distinctive demographic patterns for each region or each type of country, if any?

As in chapter 1, there will naturally be some imbalance in the areas covered by our observation (in terms of countries and topics). There is a general dearth of data on the interaction between most demographic factors and patterns of mobile communication adoption and use (Buckingham 2004; Livingstone et al. 2004). So, notwithstanding some observable contemporary trends, it must be remembered that these patterns are all emerging rather than set, and that their long-term continuance is by no means certain.

Is There an Age Divide?

In almost all the societies we have studied, wireless technologies were initially marketed to attract adult members of the business community. Business users worldwide remain a leading group in the use of high-end mobile services, and this is particularly true in developing countries such as China and the Philippines. But, across the globe, adolescents and young adults are emerging as playing a very active role in adopting and appropriating mobile services such as SMS. The trend is particularly noteworthy in the OECD countries of Europe, America, and the Asian Pacific because the younger generation has more free time, can live on a loose budget, and most of them do not own a landline. Indeed, the increasing importance of mobile diffusion among young people has made the subject a popular one among industry analysts and academics, and analysis of the practices of young users constitutes the bulk of existing evidence with regard to age in the public domain. For this reason, we leave discussion of the mobile youth culture to chapter 4, and here consider only the specific case of older users.

The lowering of the price of mobile services has made the technology increasingly affordable and therefore more likely to attract adopters in all age groups. With the youth groups of certain developed markets reaching near saturation, there have been more deliberate corporate efforts to target older generations. In the European context, middle-aged and elderly users are still using mobile voice telephony more frequently than the younger generation (Ling 2002). Since ease of use is a more important issue for the elderly market, two initiatives for this age group from different mobile phone operators should be highlighted. First, Japan's main mobile operator, Docomo, released the *raku-raku* (or "easy-easy") handset in September 2001, which has a bigger keypad and an easier-to-read screen specially designed for the elderly. More than 200,000 units of the *raku-raku* cell phone were sold in the first two months (ITU 2002a: 132). Secondly, in 2005, Vodafone launched a similar handset called "Vodafone Simply" in a number of European countries, including Spain and the UK.

Thus, the age factor plays differently regarding the uses of mobile telephony. In the first stage of diffusion of the technology, the middle-aged group was the one targeted for the adoption of the technology. As soon as commercial and technological barriers to entry were lowered, however, young people became the drivers of mobile communication. Yet, even in this stage of diffusion, adults are still the most frequent users of voice communication, while the younger groups use more SMS, features, and services, within the limits of their budget. The group that remains less inclined to use wireless communication is the older segment of the population, but even this is changing as new devices and services adapt to the needs of the elderly. It looks as if the trend is toward the general diffusion of mobile communication within the whole population, with age continuing to specify the type of use rather than the use itself.

The Gendering of Mobile Communication: Myth and Reality

As far as mobile communication technologies are concerned, available data and reports indicate that, in general, where diffusion levels are high, the gender differential is growing smaller.[2] Following the trend of a decreasing gender gap among Internet users (Rice and Katz 2003), the diffusion of mobile technologies among the female population has been catching up with or surpassing the level of adoption among males. This is hardly surprising: as mobile phones become more accessible, there is a greater likelihood that people of all social demographic characteristics will be drawn into the market. However, a report by TGI Global (2004) states that there is still a significant divergence in mobile-phone ownership between men and women, with men being more likely to be owners than women. This conclusion is supported by Huyer et al.'s (2005) analysis of gender disaggregated data on the use of information and communication technologies. We can gain a more precise view of diffusion by gender by focusing on several countries in Europe, North America, and Asia, for which there are now some reliable sources of data.

Across Europe, male workers were the early adopters of wireless technology (Ling 1999a; Frissen 2000; Fortunati and Manganelli 2002; Agar

Table 2.1

Access to mobile telephones by gender: selected European countries (% of users over total population of a given gender), 2002

	Female	Male	Ratio female : male
Finland[a]	92	95	1 : 1.03
Norway	91	94	1 : 1.03
Sweden	87	92	1 : 1.06
Denmark	77	84	1 : 1.09
Italy	71	88	1 : 1.24
UK	71	80	1 : 1.13
Germany	61	78	1 : 1.28
Catalonia	56	64	1 : 1.14
Spain	51	60	1 : 1.18
Bulgaria	13	21	1 : 1.62

[a] 2001

Sources: own elaboration and (in alphabetical order): Bulgaria, Germany, Italy, and UK: Eurescom (2004).

Denmark: StatBank Denmark, *Statistics Denmark* (www.dst.dk).

Finland: Household Budgets Survey, *Statistics Finland* (www.stat.fi).

Norway: Survey on Media Use, *Statistics Norway* (www.ssb.no).

Spain and Catalonia: Household Information Technology Survey, National Institute of Statistics, Spain (www.ine.es).

Sweden: *Statistical Yearbook of Sweden 2004* (www.scb.se), *Statistics Sweden*.

2003; Lacohée et al. 2003), particularly young men in Belgium (Lobet-Maris and Henin 2002: 103), Russia (Vershinskaya 2002: 141), and Norway (Ling 2002; Skog 2002: 261). Research in the UK showed that mobile phones were initially owned predominantly by men (60 percent of owners), while women were twice as likely to use someone else's mobile phone occasionally (*Marketing Week* 1998). But studies in recent years have found narrowing gender differences in adoption rates. As table 2.1 shows, in all cases male adoption of mobile telephony is greater, but the gaps are not very wide. The ratio of female to male users indicates that gender differences tend to disappear as mobile penetration rates increase. In the countries with high mobile teledensities, there is an almost one-to-one ratio, while in those countries with lower mobile teledensities there are slightly more male subscribers (for example, three males for every two females).

In contrast to Europe, data for the United States show a higher proportion of women currently using mobile phones. Between 1993 and 2000, the percentage of female users rose from 39 to 52 percent, as the percent-

Table 2.2
Mobile phone use by gender in the United States (% of users), 1993–2000

	US population	1993	1995	1997	1998	2000
Women	53	39	48.5	52	55	52
Men	47	61	51.5	48	45	48

Source: Robbins and Turner (2002).

Table 2.3
Wireless data use by gender in the United States (% of users), 2004

	Wireless data users	Non-users of wireless data	All wireless subscribers
Women	45	48	46
Men	55	52	54

Source: adapted from Smith (2004a).

Table 2.4
Visits to text-messaging sites by gender in the United States (% of users), 2003

Sites	Predominant gender
Sprint PCS – Messaging	Female (59%)
Verizon TXT Messaging	Female (55%)
SMS.ac	Male (57%)
Yahoo! Mobile SMS	Female (57%)

Source: Hitwise, cited in Greenspan (2003b).

age of male users dropped from 61 to 48 percent (Robbins and Turner 2002; table 2.2). More recent reports confirm that women slightly exceed men in mobile-phone ownership (TGI Global 2004). One of the few statistical analyses of the relationship between gender and mobile-phone use (Rice and Katz 2003) found that females exceeded males in all categories: non-users (55.3 percent), current users (53.4 percent), veteran users (1997 or earlier, 51.5 percent), recent users (1998 or after, 55.5 percent), and dropouts (58.5 percent).[3] Information on the use of mobile data is unclear. Some estimates indicate that men use mobile data more than women do (table 2.3) and that men are also significantly more interested in using Wi-Fi (*Wireless Week* 2003). Other estimates have found that it is women who are more interested in Wi-Fi.[4] However, women tend to be more frequent users of text-messaging websites (table 2.4).

Relative to Europe and America, the Asian Pacific exhibits a traditional patriarchal gendered pattern of diffusion. This is unsurprising in the newly developed markets, such as China and the Philippines, because of the persistence of the traditional marketing approach, which mainly targets the – mostly male – business community. But even in Japan and South Korea, the male gender continues to dominate. For example, by December 2003, male users accounted for 57 percent of Docomo's 2G subscriptions and 64 percent of its 3G subscriptions (NTT Docomo 2003). In South Korea, by November 2003, 88 percent of adult males subscribed to mobile services, as against 69 percent of adult females (KISDI Report 2003; see also S-D. Kim 2002: 63–64). In the case of SMS in China, while one survey in ten cities and nine towns showed that the penetration of SMS tended to be higher among male users (Xinhuanet 2003), another by BDA China (2002) found that the majority of frequent daily SMS users were female.

The limited data available for the Middle East and North Africa region indicate varying levels of gender disparity. Statistics from Turkey show that men are twice as likely as women to own a mobile phone: roughly 68 percent of men and 30 percent of women (Öczan and Koçak 2003; TGI Global 2004). In Israel, which has an extremely high penetration of mobile phones (95 percent), this presumably translates into a fairly equitable distribution between the genders.

There is limited demographic breakdown of data on the diffusion of mobile phones in Africa. Considering the situation of gender inequities in most of these countries, it is likely that ownership and use among women is currently low relative to men. However, there appear to be some important contextual variations, the reasons for which are not immediately clear. For example, Huyer et al. (2005) report that in Ethiopia, Uganda, South Africa, Rwanda, and Cameroon, males have much greater access to mobile phones than females, while in Botswana, Namibia, and Zambia the gap is smaller. This does not seem to be related to overall mobile-phone penetration rates since South Africa and Rwanda, for example, have both high mobile-phone teledensities and large gender distribution gaps (Huyer et al. 2005). Contrary to the above findings, however, another survey of users in South Africa found that females were the predominant owners (56.8 percent of owners) and users (60% of users) of mobile phones (Samuel et al. 2005).[5] The same survey found a more balanced state of affairs in Tanzania where women comprised 48.4 percent of owners and 52.7 percent of users.

Gendered Uses

Research in developed economies has long shown that women are heavy users of the domestic landline, having "'appropriated' a practical, supposedly masculine technology for distinctively feminine ends" (Fischer 1988: 212). It has been suggested that the socioeconomic conditions of women led to the growth of this trend: using the telephone enabled women to deal with isolation and to fulfill their socially prescribed role of network maintenance, as well as express their natural enjoyment of social interaction (Fischer 1988: 226). Thus, in addition to its liberating and enjoyable impacts, the domestic line has also been seen as reinforcing gender differences and roles (Fischer 1988; Rakow 1992), turning it into a gendered technology, at least in use.

Mobile communication technologies have added a new dimension to this argument. While some view mobile communication technologies (in particular, the mobile phone) as gender-neutral, researchers around the world are finding both instances of gendered uses and evidence of a leveling out of gender differences in the use of the technology. At least three trends are identifiable in the literature: clear gender differences in acquisition and use, a blurring of traditional gender lines in usage or observations of no gender differences, and culturally specific gendered usage patterns. Within these processes, some scholars see mobile communication technologies as liberating users from gender limitations. Others see them as only adding to already existing mechanisms that maintain traditional gender divisions.

One of the most consistent gender-related findings across countries is that females tend to prioritize safety and security as reasons for acquiring a mobile phone (Scott et al. n.d.; Johnson 2003; Plant 2003a; Gergen 2005; IDC 2005; Lemish and Cohen 2005b). Men, in a sense, also fulfill their gender role as protectors through the tendency to act upon a perceived need for women to have mobile phones for safety; for example, by being the one to buy the phone for a wife or daughter. It appears that, at the broadest level, mobile phones provide females with a sense of security that is considered less necessary for men. In this sense, mobile communication technologies are becoming a tool associated with protecting "vulnerable" groups such as women, children, and elderly people, notwithstanding other potential and actual uses. This claim has, however, been described by Rakow and Navarro (1993: 145) as "dubious" and mainly serving to perpetuate gender inequities.

Another generally consistent finding is that women are more likely to use mobile phones for maintaining social networks and coordinating family activities (Rakow and Navarro 1993; Ling 2002; Plant 2003a; Gergen 2005; Lemish and Cohen 2005b). For example, women (in the United States) reportedly make more personal calls: 80 percent of airtime versus 67 percent for men (Cingular Wireless 2003) or 82 percent for women versus 62 percent for men (Forbes 2005). Women are more likely to call family and friends (making 40 percent more of these calls than men; O'Connell 1999). Researchers in Europe have also observed differentiated mobile-phone usage between the genders. Commenting on three surveys that examined the evolution of mobile-phone usage in Norway, Ling (2002: 44) states that "the mobile telephone has changed from being a gadget for the guys into being more of a social networking tool for girls." Moreover:

[w]omen often have a central position in this activity and, thus, the adoption and use of the device, particularly for social communication can be seen as a type of pre-socialization of adolescent girls and their role as keepers of the social network. While during the recent past much of this activity is often carried out via the fixed telephone, the newer technology has opened a new possibility here. (Ling 2002: 44)

Similarly, Skog (2002: 268) points out that "for Norwegian boys, the importance of the mobile phone relates to its functional, practical and instrumental qualities, whereas girls stress the symbolic and expressive aspects, particularly in terms of social relationships and interpersonal ties."[6]

Rakow and Navarro (1993) also note that middle-class North American women's uses of the mobile phone are similar to their uses of the fixed line in terms of the goals they seek to achieve, although some different meanings are attached to the device. Ultimately, mobile phones help women, in particular mothers, to practice "remote mothering," whereby they are able to manage their traditional female roles in the face of the increasing mobility of family members. Rakow and Navarro's in-depth interviews of 19 women led them to conclude that "cellular telephone technology may appear to provide a solution to two important problems faced by middle class, suburban women: the problem of safety and security in a violent and mobile society, and the problem of carrying out family responsibilities across barriers of time and space" (1993: 155). They conclude that use of the mobile phone is "likely to reproduce gender inequities, albeit with some shifting of public and private ground, under the guise of *solving* those very inequities" (Rakow and Navarro 1993: 145). In essence, as used, mobile phones enable men to extend the reach of the public world into their personal lives, while women use it to extend the reach of their private lives into the public world (1993: 155).

Lemish and Cohen (2005b), conversely, are less critical of this association, suggesting that the apparent gendering of the mobile phone is a
remnant of traditional gender roles still manifest in discourse about mobile
communications rather than being valid in actual practice. In the Israeli
case described by Lemish and Cohen (2005b: 520), "Both men and women
discussed their perceptions of its role in their lives in quite a traditional
gendered manner – activity and technological appropriation for men, and
dependency and domesticity for women," although examination of their
practices showed limited differences. They note, therefore, "the discrep-
ancy between the conventional construction of gender in discourse about
the mobile phone versus the actual practices associated with it that indicate a process of feminization . . ." (2005: 520). For example, in the quan-
titative section of the study, very few significant gender differences were
found in terms of whom people call, where they call from, and attitudes
toward the phone. However, the interview section revealed striking differences: men saw the phone as an extension of themselves without which
they could not function. They also regarded mere ownership of the device
as a sign of social inclusion. Women, on the other hand, measured social
inclusion by the number of calls they received. Men also saw the advantages of the mobile phone in terms of their ability to get access to others,
while women saw the benefits mainly in terms of other people having
access to them.

Similar observations were made in Norway, where little gender difference was observed in the culture of mobile-phone use, but clear differences
emerged in how men and women described their use and expertise (Nordli
and Sørensen 2003). For instance, both genders appeared to use SMS in
similar ways, but men emphasized texting for practical purposes, although
they equally used it to send "sweet notes" to wives and girlfriends, and
"goodnight notes" to children (Nordli and Sørensen 2003: 4–5). Women
saw their male counterparts as more skilled in handling the mobile phone,
but the researchers' observations did not support this either. This tendency
is attributed to a possible "issue of gender differences in the way one
accounts for one's own use of a mobile" (Nordli and Sørensen 2003: 19).

A more extreme version of the gendering of the mobile phone has been
observed among South Asian families in the United States and in India
(David 2005). Here, exploratory interviews revealed that women saw the
mobile phone as a medium of control. After an initial appreciation of the
liberating effects of the mobile phone, they began to resent their perpetual accessibility to others – to the extent that some would deliberately leave
it at home when they went out. Additionally, it was observed that care of

the phone translated into an extension of traditional gender roles, especially among older married couples. Generally, women were in charge of carrying and taking care of the phone (for example, cleaning and charging the battery), as illustrated by this respondent: "I always look after things at home. This new toy . . . his new toy . . . is another thing for me to take care of. I charge it and when we go out together, I'm so scared that it will get scratched when he puts it in his pocket that I carry it in my purse" (David 2003).

With regard to attitudes toward mobile communication technologies, there has been a general perception that females have a low affinity for technological gadgets and services. Findings in this regard seem to depend on what type of communication technology or device is in question and are generally inconclusive when taken together, at this stage. One study found that, along with other individual characteristics, gender had no significant impact on perceptions about mobile phones (Kwon and Chidambaram 2000). Another study found support for the proposition that females are more technophobic than males. Focusing on psychological rather than biological gender, the study of attitudes toward a mobile Internet device that looked either like a computer or a mobile phone showed that people with feminine psychological gender experienced higher levels of computer anxiety and negativity than those with masculine gender. However, these tendencies were moderated by perceptions of self-efficacy and previous experience with the technology (Gilbert et al. 2003). Min and Yan (2005) also conclude that when making decisions about mobile Internet access, men are more concerned with perceived usefulness, while women are more concerned with ease of use, an indication that women are more wary than men of the wireless Internet. On mobile commerce services, another study concluded that there was little difference in the attitudes of females and males toward such services, although women exhibited higher interest in more services than men did (Anckar and D'Incau 2002).

Blurring of Gender Lines

Some research has found that gender is not an important determinant of mobile-phone usage. Most gender differences observed are found to be statistically insignificant. For example, in Spain, an online survey by Valor and Sieber (2003),[7] concluded that there is no significant gender difference in uses among teenagers. Although boys were more likely to use the mobile phone for fun, the researchers declared that this was not a very important aspect. Girls were also found to use the mobile phone because it would make them more available for others to contact, whereas boys tended to have a more technical and autodidactic interest in the phone. In Turkey, Israel, and Norway researchers have identified little or no differences in how men and women use mobile phones (Nordli and Sørensen 2003; Öczan and Koçak 2003; Lemish and Cohen 2005b).

Recent data suggest that some gender stereotypes are also being challenged by wireless communication uses. According to Plant (2003a), in the early days mobile phones tended to be "deployed as symbols of sexual prowess by men," and although there are still instances of males using the mobile phone as a status symbol, as mobile phones proliferate, gender differences tend to disappear. This includes supposed differences in the amount of time spent on the phone, the tendency to send frivolous text messages, such as rumors and gossip, and the use of secret codes in text messages. In all these cases, it has been found in global surveys that males are, after all, equally likely to engage in such activities (Plant 2003a).

This is certainly evident in surveys in the United States, which indicate that, since 2001, men have been the greater users of mobile phones in terms of actual talk time.[8] In 2003, they talked 14 percent more than women on their mobile phones (Cingular Wireless 2003). The following year, men used 455 minutes per month versus 391 minutes for women (Ankeny 2004), and the latest report found that male subscribers used 571 minutes per month as against 424 for women (Forbes 2005). This trend is partly linked to the finding that men make more business-related calls (twice as many as women do), which is also found to be consistent over the years.[9] It has also been documented that, at least in 1999, women were 25 percent less likely to talk unnecessarily on the mobile phone, and more likely to make calls out of necessity (O'Connell 1999). Similarly, in Israel, it appears that men make significantly more mobile-phone calls in absolute terms as well as relative to fixed-line calls. For example, 12.5 percent of men made 21 or more calls per day as against only 2.9 percent of women (Lemish and Cohen 2005b). The same pertains in parts of Europe where, for instance, 31 percent of Norwegian men make daily use of mobile phones as against 15 percent of women.[10]

It would appear that, in this respect, *the mobile phone is not following the same trend noted for domestic landlines*. It is possible that this is an effect of the emergence of mobile telephony as a ubiquitous communication medium, thus erasing the ease with which one could distinguish between calls made at home (which made it possible for women to be identified as the dominant users of domestic landlines) and calls made outside the home or at work. Lower-income levels for females, and the tendency for men to have work-related mobile phones for which their employers pay the bill, may be additional reasons why women seem to talk less on mobile phones. In Norway, three times as many men as women have their mobile-phone bills paid by their employer (Nordli and Sørensen 2003).

A similar transcending of traditional gender attributions can be seen in the adoption of mobile games. Although gaming is generally considered a male domain, it turns out that women are playing mobile games as much as or sometimes more than men. For example, in the United States, a 2003 study stated that women (6 percent) were using the gaming feature on their mobile phones twice as much as men (3 percent; Cingular Wireless 2003). A 2004 study put the figure higher: at 28 percent of women as compared to 17 percent of men (Dano 2004), while a third reported that the difference between men and women playing mobile games was only 6 percent (Ankeny 2004). This is despite the fact that men are more likely to download mobile games (67 percent of respondents who had downloaded a game; McAteer 2005). This may be because mobile-game offerings do not serve the interests of the female market with almost half of the titles offered being action or sports oriented, whereas women reportedly tend to download puzzles and card games (comprising about 10 percent of mobile games on offer; McAteer 2005). Women download 46 percent of mobile content. A survey by Enpocket found that men download more games (58 percent of mobile game players) and women more ring-tones (over 50 percent of ring-tones; BBC 2004a; Telecomworldwire 2004).[11]

There are also indications that men are motivated by social status and women by use value when acquiring and using mobile communications. Contrary to claims that gender use of new technology is becoming similar, among young Finns (16–20 years) males tended toward "trendy use" (a focus on design and technical functions), while females tended toward "addictive use" (a focus on use value; Wilska 2003). Interestingly also, the study found that trendy and impulsive use of mobile phones (stereotypically female characteristics) was linked to technological enthusiasm, which was more associated with males. Wilska suggests that this may be an indication of changing attitudes toward consumption by males, in a society where efforts have been under way to develop products for the male market. Thus a situation has developed where "the 'cool' consumption styles include both 'masculine' technology and 'feminine' trend-consciousness" (Wilska 2003: 459).

A recent global survey came to similar conclusions: young adult males were more likely than females to be "style- and status-conscious" when it came to selecting mobile communication devices (IDC 2005). On the other hand, a study of college students in the United States found that male respondents were generally not particularly concerned about their phones being "up-to-date": less than half the sample considered this very important or important, compared to just over half of the female sample (Gergen 2005). A surprising discovery in China was that subjective norms, "the degree to which an individual believed that people who were important to her/him thought she/he would perform the behavior in question" (Min and Yan 2005: 151), influenced men more than women's intention to adopt the mobile Internet. Yet, in light of other findings on social pressures, this is actually not surprising. It would appear that men associate ownership of and dexterity in the use of mobile communication devices with social standing to some degree.[12]

Gender and Culture

Clear cultural trends are particularly identifiable in the Asian region and in Europe. In these areas, users have developed unique and innovative patterns of usage, some of which have distinct gender characteristics. Referring to the Norwegian case, Skog observes, on the basis of his fieldwork during 1999–2000, that "[t]he gendering of mobile phones may be described via the use of mobile phones, as well as in how gender leaves its imprint on mobile phones. The mobile phone companies seem to design phones to match the traditional female and male cultures" (Skog 2002: 268–269).[13] Although mobile-phone companies actively promote this gendering, certain aspects also emerge autonomously from within the female population. So far, it appears that it is the female population who are appropriating mobile phones for their own purposes. Less is documented about the development of any male-oriented practices.

Concerning the relationship between gender and culture, one of the most relevant fields of observation is the process of gender differentiation in the culture of *kawaii* in Japan. Japan is particularly well known for its culture of *kawaii* or "cute culture," which has embraced mobile phones as the latest female fashion item, using flashy colors and cute characters as decorations (McGray 2002; Hjorth 2003; Richie 2003). Among the most essential social uses of *keitai* (the mobile phone) is the manifestation and celebration of the female gender in the culture of *kawaii*. Among early-adopting young users, the *keitai* culture built on the previous pager culture that was primarily led by girls. Japan already has a long tradition of "intimate, personal, and portable media technologies" (Ito et al. 2005), such as the Walkman, Tamagotchi, and Pokemon cards, which are cute and have a strong feminine appeal. Mobile phones are lightweight, portable, and easily customized as a wearable item to suit different lifestyles and fashions. As a result, *keitai* decorations and their associated cultural expressions have become the latest epitome of the culture of *kawaii*. According to Hjorth (2003: 57), "the implications for women in Japan, who have been both key consumers and producers of *keitai* technology, are considerable." "The colonization of high-tech spaces such as the Internet by the cute characters usually associated with the female realm in Japan is an important signifier of the power afforded women by this new technology" (Hjorth 2003: 52). The best symbolic illustration of this new-found power is probably Ms Mari Matsunaga. Given her 20-year editorship of *Recruit* before she became head of Docomo's marketing team, many have attributed Docomo's success in part to Matsunaga's female perspective as both consumer specialist and media producer (Lightman and Rojas 2002; Rheingold 2002).

The changing power dynamic is most manifest among cell-phone-equipped *kogyaru* (high-school girls), as explained in chapter 4.[14] Nevertheless, the image of a trendy, female, mobile-phone user is not without its critics. Kogawa (n.d.), for instance, has argued that Japan's "independent" woman has become no more than "a new consumer." This line of thinking suggests that the culture of *kawaii*, further promoted by *keitai* usage, does not really empower women. In fact, what it does is to further subdue females to the dominance of technological consumerism.

Gendered Mobile-Phone Culture in Asian Countries The *keitai* culture occurring in Japan has spread to females in other Asian countries, particularly those of younger age. In South Korea, mobile-service providers have started to concentrate on the female market by introducing distinct handsets, rate plans, and special service packages appealing to women. These include SK Telecom's "Cara," KTF's "Drama," and LG Telecom's "i woman." Meanwhile, Chinese female users, especially those in the white-collar class, are known for their preference for red clamshell designs with ornaments made of synthetic or real diamonds. A number of handset manufacturers consequently began to produce such cell phones to meet the needs of this market segment.[15]

Female users in the Asian Pacific region not only appropriate the mobile phone as a fashion item but, more importantly, also as a key channel to maintain intimate personal relationships, as opposed to men who tend to use mobile phones for instrumental purposes.[16] Although, overall, more males are using SMS, the intensity of usage is higher among female texters because they use the technology to communicate more with their close friends and family members.

Gender and Mobile Phones in Africa   Two studies that have investigated new communication technologies in Africa found little difference between males and females in terms of usage patterns, attitudes toward the technology, and perceived social pressure to use mobile phones. There were some differences, though, in motivations to use telecommunications in general, including mobile phones (Scott et al. n.d.; Samuel et al. 2005). In Tanzania, researchers found that female respondents, in particular, saw the mobile phone as a household asset, unlike South Africa where most people saw it as an individual possession (Samuel et al. 2005). In Ghana, Uganda, and Botswana, men tended to prioritize business and livelihood-related uses for the mobile phone, while women tended to prioritize security-related issues, such as the risks of traveling and the ability to contact others in an emergency (Scott et al. n.d.). Likewise, studies conducted in Benin, Burkina Faso, Cameroon, Mali, Mauritania, and Senegal found that women use the mobile phone more for personal and social purposes, while men use it more for professional activities (Huyer et al. 2005). These findings are significant in light of the tendency to link mobile phones with economic activity, especially for women, in developing countries. The development discourse that tends to surround discussion of mobile communication technologies in developing economies has led to a strong belief that putting mobile phones in particular into the hands of poor women will enable them and their families to rise out of poverty. The success of the village-phone system pioneered by the Grameen Bank among women in Bangladesh has lent support to this belief. In the cases reported above, at least, women were not so focused on the potential of mobile phones to create or enhance employment.

A distinctive gender-related practice emerging in African countries is the gendering of "beeping" (the practice of dialing a number but hanging up before the call is answered with the objective of getting the recipient to call back). While "beeps" are mainly used to redistribute financial resources, such that people perceived to be wealthier end up paying for calls, there is a clear gender dimension as well, especially when romantic associations are involved. It is generally considered improper for a man to "beep" a woman, regardless of his financial situation (Chango 2005). Chango reports having "seen an unemployed man receive 'beeps' from a lady with a rather good job who was aware of his situation. Traditionally, because the man controls the material resources, he is expected to pay"

(2005: 80). But it is absolutely forbidden for a boyfriend or suitor to make such a call-back request to a romantic interest. An investigation of the rules of "beeping" in Rwanda illustrates this point:

If you are chasing after a lady, you cannot beep. You have to call. Beeping is for friends. When a girl you do not know well beeps you, you have to call back if you are interested. You cannot even text. She has to see that the effort is being made. Borrow a friend's phone if you do not have airtime. Text is not two-way. (survey respondent in Donner 2005a: 9)

According to Donner, this reflects traditional norms and gender roles already existing in society, which have been transferred to the new technology. This does not mean the practice is accepted without question. In Uganda, men complain about "beeps" from girlfriends that turn out to be for no important reason (BBC 2001).

The Interplay between Gender and Communication Technology across Cultures Looking at the interaction of gender and mobile communication technologies, we can observe some general trends such as an initial bias toward male ownership of mobile phones, followed by increasing levels of adoption among women. There is limited evidence to draw conclusions about mobile data, except that women appear to be more intense users of text messaging. Only in Asia, and especially Japan, is there sufficient literature to describe the clear development of a mobile communication subculture defined by gender. In other parts of the world, current research does not conclusively support any particular gender trend, but rather reveals multiple dynamics at work. There is an emerging tendency for people to view and use mobile communications in both gendered and non-gendered ways. Thus Nordli and Sørensen (2003: 19) aptly conclude that mobile-phone use "could be characterized as trans-gender due to the observations of flexible co-constructions of gender and mobile phone use. It is not gender neutral. The mobile phone seems to facilitate a broad set of symbolisms and practices, without any one of them being allowed to or able to dominate."

At this point, it appears that gender is just one factor among others in determining mobile-phone adoption and usage patterns. Exactly what these variables are and how they drive, moderate, or mediate different usage patterns remains to be clarified. Variations in gender use of mobile communication technologies might, for example, be explained by factors such as work status, location of workplace, family status, and lifestyle. The complex relationships that people of different genders are developing with mobile communication technologies signify the flexibility of the technologies and their ability to promote both gendered and non-gendered behavior, depending on cultural contexts, while at the same time blurring the lines between gendered practices. There is also an important difference between gendered discourse and actual usage patterns, which should be taken into account when examining the uses of mobile communication.

Haves and Have-Nots in the Mobile Universe: The Class Dimension of Wireless Communication

Is mobile communication a privilege for people of higher socioeconomic status? Not all that long ago, in the mid- to late 1990s, a colloquial Chinese expression referred to the mobile phone as *dageda*, meaning literally "big-brother-big." This phrase, spread via Hong Kong movies, indicated the exceptional wealth and power associated with the owner of the device at the time. To varying degrees, the distinct socioeconomic status of mobile-phone users is still observed in large parts of Asia, Africa, and the Americas. In South Korea, for example, although the country has a fairly advanced mobile market, differential mobile-phone penetration rates persisted in high- and low-income populations in 2003. As the KISDI Report (2003) shows, 84.3 percent of Koreans with a monthly income above KRW 3.5 million (slightly less than US$ 3,000) had adopted the technology. But for those who earned less than KRW 2 million per month (slightly less than US$ 1,700), the percentage was only 69.9. This gap is noteworthy because there was very little difference between the two income groups with regard to their landline subscription rates (KISDI Report 2003). The same is observed, for instance, in Lima (Peru), where despite the fact that penetration of mobile telephony is higher for high-income and middle-income groups, even low-income, very-low-income, and extreme poverty groups have significant take-up of mobile services (NECG 2004: 18). Indeed, the mobile penetration for each of those groups equaled, in 2001, 78, 53, 22, 10, and 7 handsets per 100 people (Fernández-Maldonado 2001; also quoted in NECG 2004). In the same sense, observations from Mexico presented by Mariscal and Rivera (2005), following Telecom CIDE (2005), demonstrated the same pattern. In 2003, lower-income groups (levels D and E) showed a penetration of nine mobile phones per 100 people, while higher ones (levels A/B and C+) reached 85, and middle-income groups (levels C and D+) accounted for 43. Moreover, preliminary results for 2005 showed that in the lower socioeconomic levels penetration had tripled to 27.

In developing countries with lower average income and more serious social disparity, access to mobile communication technology is more limited to people of higher income, education, and social status. Although, as will be discussed in the following pages and in much more detail in chapter 8, the mobile communication society brings fresh opportunities for upward social mobility and for bridging the digital divide, the general pattern holds that, at the current stage, technological diffusion is greater among those of higher socioeconomic status. This is true not only for developing countries but also for the United States, where the median income of mobile users was found to be much higher than the national average, and level of income was shown as a significant predictor for mobile-phone adoption (Rice and Katz 2003).

There are, however, some exceptions to be noted, which suggest that socioeconomic differentiation in mobile adoption patterns is not a permanent phenomenon. It is instead a function of the stage of technological diffusion, which means that the influence of existing socioeconomic inequalities decreases, or even evaporates, when penetration gets close to saturation in a given society. The most obvious case is Europe, where average penetration is 71.5 percent for the continent and over 90 percent in a number of countries (ITU 2004b; see chapter 1). For instance, in the European Union we can see a trend indicating that income is now less important as a predictor for mobile-phone adoption as long as mobile telephony reaches almost all the population. The same is happening in major metropolitan areas in the rest of the world. From Tokyo to Kuwait, from New York to Sydney, because mobile telephony has become so affordable, and so easy to use, there is a growing trend for the device to become a routine and habitual object for people from all walks of life, rich or poor, educated or not. This is particularly important for voice telephony because even the illiterate can talk on a mobile phone.

On the other hand, if we look at the latest mobile gadgets and applications, like 3G or Wi-Fi services, more often than not the bulk of subscribers still tend to be wealthier and better educated because these more advanced, non-voice services are usually more expensive and require a higher level of literacy. More important, people of higher class are often targeted in the R&D and promotion of newly developed mobile technologies, which are more expensive and usually specifically adapted to the cultural norms and practices of the upper and middle classes. This is why the business community still remains the primary user group for high-end services, such as Blackberry and other wireless data applications in the United States (Fitchard 2002). This is also why, in many cases, we see the transformation of the mobile phone from its basic communication functions into a key fashion object, a signifier of modern urban lifestyle, and a major component of contemporary consumerism, which we will explicate in chapter 3.

Because income often closely correlates with educational attainment, it is not surprising that, on average, users of wireless data services are often better educated than those who only use voice telephony, who, in turn, are better educated than those who do not use any mobile service. By wireless data services, we mean not only advanced wireless Internet applications and Bluetooth but also more mundane, non-voice applications including SMS. A study in China, for example, shows that college graduates are significantly more likely to adopt SMS than those without a college education (Xinhuanet 2003). This is also the case in Africa, where a significant proportion of owners and users have only primary education or no schooling at all. In South Africa, for example, most owners (56.2 percent) and users (56 percent) have secondary education, while a majority of non-users (44.8 percent) have primary education only. In Tanzania, a majority of owners (37.7 percent) have secondary education, while most users (62.4 percent) and nonusers (71.4 percent) have been educated only to primary level (Samuel et al. 2005). McKemey et al.'s (2003) studies in Africa indicate that people with higher levels of education have already incorporated the optimal level of telephone usage into their lives. They also found that most people with low education and income tend to live in areas with little or no access to telecommunication services.

When we examine the case of Africa, higher income is clearly associated with higher levels of mobile-phone ownership. For example, there is a higher level of mobile-phone use in Botswana than in Ghana because higher average income levels enable more people in Botswana to own and use mobile phones (McKemey et al. 2003). However, studies showing that over 50 percent of users fall in the lowest income bracket indicate that income may be a barrier to ownership but not to access or usage (Samuel et al. 2005). Not surprisingly, in areas where text messaging is feasible, there is some evidence to suggest that the proportion of text messaging to voice calls also increases as income decreases (for example, in South Africa; Samuel et al. 2005). Of particular importance is the tendency for mobile-phone ownership to spread more evenly across income groups than has been the case with other consumer durables (Samuel et al. 2005). People in developing countries also spend large amounts of their income on their communication needs. For example, in South Africa, 10–15 percent of income is spent on mobile phones. Typical tariffs in the Southern African region range from US$ 19 (Zimbabwe) to US$ 71 (Mozambique) for 100 minutes of use per month (Minges 1999). It appears, however, that spending on mobile phones is not among the top priority uses of income, as national statistics indicate that food, fuel, and energy top the expenditure list for poor black South Africans (Minges 1999).

Low penetration in Africa also has to do with the fact that rural dwellers, who make up the bulk of the population, have limited access to mobile telephony (Shanmugavelan 2004). This suggests that there may be a high percentage of urban-based mobile-phone users who possibly have both fixed and mobile-phone subscriptions, consistent with Hamilton's (2003) conclusion that mobile phones act both as complements to and substitutes for fixed lines in African countries, but that complementary use dominates. The differential distribution of mobile phones within countries is often hidden or overlooked because most national statistics are not broken down to show the rural–urban divide (Panos 2004), which is also a problem in other parts of the world. In Africa, indications of this situation can be seen in a few cases. For example, over 70 percent of mobile phones in Senegal are located in the capital city, Dakar (African Connection Center for Strategic Planning 2002). Forty-five percent of Botswanans, 20 percent of Ghanaians, and 76 percent of Ugandans use mobile phones "regularly";[17] however, the usage levels fall when only rural areas are considered: Botswana 29 percent, Ghana 6 percent, and Uganda 66 percent. For many of these low-income Africans, the postal system is still their main means of communication across distances (McKemey et al. 2003).

Yet the correlation between socioeconomic status and mobile adoption does not always have to be a linear relationship. This is what China Mobile discovered in its large-scale survey of wireless Internet users in 2004, which, despite the limitation of its online sampling procedure, offers valuable insights into adoption patterns at different social strata (*Guangzhou Daily* 2004).[18] The survey shows that the majority of wireless Internet users in the country are of middle income and education. Forty-five percent of the users have a monthly income of 800–3,000 yuan (or US$ 96.7–362.5), whereas the high-income bracket of more than 3,000 yuan only account for 10.6 percent of total subscribers. Moreover, the educational breakdown of users is 44 percent with middle-school education, 31 percent with junior college education, 24.5 percent with college education, while those with masters and doctoral degrees only account for 0.44 percent. Apart from the early stage of diffusion, the popularity of wireless Internet services among people of middle income and education levels probably has to do with the fact that most people in the high-income and high-education bracket tend already to have computer-based broadband Internet access at home and at work.

There are other structural constraints to be noted. For instance, yearly mobile-phone usage in very low-income countries could cost more than twice the GDP per capita (Pigato 2001). Service providers also tend to limit their coverage to high-population cities and towns, and usually only incidentally include rural communities. Thus, even with access and the means to acquire mobile phones, it is still impossible for some potential subscribers to use the service. There are reports of people in villages who have been provided with mobile phones by their urban-based relatives. However, these phones are mere artifacts in this setting because no mobile-phone operator provides coverage in their community.

It is also important to note that, because of their association with higher income and education, mobile communication gadgets serve as indicators of social status in the context of developing countries, thus adding a peculiar dimension to the processes of social appropriation. The "mobile mania" in the Philippines, for example, has resulted in both adoption among the middle class and the spread of the technology to the more impoverished urban populations (Rafael 2003: 404–405; see also Arnold 2000; Strom 2002). As Vincent Rafael (2003: 404) points out, several factors account for the rapid technological diffusion among the poor, including the use of prepaid phone cards, the availability of used handsets, the deterioration of public infrastructures, the offering of "free and, later on, low-cost messaging" by private telecom operators, and the display effect of the gadget in the middle of the Manila crowd.

Two lessons can be drawn from the case of the Philippines that have wider implications for the entire developing world. One is that the spread of mobile technology is taking place in tandem with an unprecedented wave of urbanization around the globe, when billions of rural-to-urban migrants use the mobile phone as both a functional tool to cope with a new social environment and an instrument to shape their recently acquired urban identity. The second, perhaps more important, observation is that when wireless technologies diffuse to groups with lower income and education, less-expensive services emerge to meet the particular needs of the new adopters, which is often inevitable as diffusion reaches saturation point among those of higher socioeconomic status. The trend can be observed in developing countries, as well as some of the more developed ones, in several wireless services and user practices, including pager, Little Smart (or PHS), SMS, prepaid phone cards, handset sharing, and the "beep" call. When these inexpensive uses materialize, they serve the interests of the underclass by providing not only new ways of accessing information and of social networking but also employment opportunities and chances for upward social mobility, hence giving rise to a particular social category of the "information have-less" as we have identified in the Chinese context to include migrants, laid-off workers, and pensioners (Cartier et al. 2005). In the following, we will give a brief account of these less-expensive services and how they reflect the existing cultural norms of people of lower socioeconomic status in China, India, Africa, and elsewhere.

Wireless Communication for Lower-Income Groups: When Class Shapes Technology

Let us start with an almost forgotten technology that is still relevant: the *pager*. The pager is a mobile service that often gets ignored in scholarly analysis, yet it plays a significant role in both the history and reality of mobile communication. Despite the challenge posed by the cell phone, the pager service still has a customer base among those with lower income. In the United States, there is a rather loyal pager market because the technology is cheaper, less conspicuous, has better coverage, and allows users greater control over whom they communicate with. As a result, revenues from pager sales in the United States increased by 17.2 percent between 1998 and 2002 (Euromonitor 2003).

The appropriation pattern is, however, very different in China, which had over 50 million pager subscribers in 2000 (China Ministry of Information Industry 2001). As the conspicuous consumption of mobile phones surges, the pager has been increasingly stigmatized among urbanites as "outdated," "unreliable," and suited only for culturally "unsophisticated" migrant workers (Qiu 2004). Given this discourse, even the new migrants in Chinese cities feel the need to disassociate themselves from this technology, leading to the undermining of the customer base. The trend was exacerbated when many pager operators took flight, often in an irresponsible manner, into other, more profitable businesses, leaving low-income subscribers ill informed or completely uncatered for (Qiu 2004). The country thus quickly lost almost 30 million pager subscriptions during 2000–2002 (*China Statistical Yearbook* 2000–2002). The number dropped to just 3.6 million by January 2005 (China Ministry of Information Industry 2005).

Crossing boundaries and cultures, we should now consider a very significant mobile communication technology used by millions of Chinese: *Little Smart*. While the Chinese have been quick to abandon pagers, they have been even quicker to adopt the low-end service of Little Smart (or *xiaolingtong*), a limited mobility service that allows subscribers to use a mobile phone but only pay the price of a landline. Similar services, such as the corDECT in India, have also been growing in recent years (Jain and Sridhar 2003; McDowell and Lee 2003; O'Neill 2003), although China remains the most successful case of the adoption of limited mobility services worldwide. The Chinese Little Smart system is based on the PHS (personal handyphone system) standard from Japan, where this technology was designed for low-income consumers but did not take off in this more affluent market. It proved to be a major success, however, in China, setting a sales record of US$ 2 billion in 2003, when 25 million subscribers were added to the Little Smart market in a single year. By January 2005, the total number of Little Smart users in the country had surpassed 67 million (China Ministry of Information Industry 2005). This reflects an extraordinarily strong demand for inexpensive mobile technologies, especially among the lower classes. In chapter 8, we will discuss the Little Smart phenomenon in more detail because of its great relevance for the rest of the developing world.

A third manifestation of the appeal of low-budget wireless services has been the popularity of *short message systems* (SMS) from the beginning of this century. In China, SMS is priced at the inexpensive rate of RMB 0.1 (a little more than one US cent) per message or eight text messages for the cost of a one-minute mobile-phone call (Turchetti 2003). A survey thus shows that 40 percent of Chinese mobile subscribers between the ages of 18 and 60 have used SMS (Xinhuanet 2003). The popularity of SMS was confirmed by BDA China, whose report finds that 70 percent of urban mobile subscribers in the country have used some form of mobile data services (BDA China 2002). As previously mentioned, the surge in SMS usage is nearly a universal phenomenon among lower-income groups of students and teenagers in countries as diverse as Japan, the Philippines, Bahrain, and the Nordic nations. While there are other factors leading to the high adoption rates among the younger generation, such as more free time and the ability to work intensively on a small keypad, the much lower price of SMS as compared to voice telephony is also recognized as a major consideration for young adopters under various circumstances.

Fourth, the *prepaid service* is arguably the most important form of appropriation that caters for the needs of those with lower income and education. In the Philippines, for instance, a great majority (70–90 percent) of mobile subscribers choose to use prepaid phone cards instead of fixed-term contracts (Toral 2003: 173–174), which "allow[s] those without credit history, a permanent address, or a stable source of income to purchase cell phones" (Uy-Tioco 2003: 5). According to China Mobile, the largest mobile operator in China, the company had 144 million subscribers in January 2004, including 93 million prepaid subscribers as opposed to its 51 million fixed-term contract subscribers; among the newly added subscribers in January 2004, only 15,200 signed contracts, whereas 233,000 subscribers chose the prepaid plan (Liu 2004: 19). The prepaid service is also important for less wealthy Europeans and Americans. Data from 2002 show that about 70 percent of Norwegians in their early teens use prepaid subscription. This percentage decreases for those in their middle twenties (those with a more stable income) and rises again for those close to the retirement age of 60 (again, those with lower income) (Ling 2004: 113). The case is similar in the UK among its low-income populations (Oftel 2003b: 11), and in the United States, where for a long time young people and low-income groups could not afford cell phones because of their limited access to credit (Robbins and Turner 2002).

Prepaid services are also highly popular in Africa. Despite being several times more expensive than the mobile contract and fixed-line systems, prepaid services have become the only viable means of access for most people. Prepaid systems dominate in Africa, making up over 80 percent of mobile subscriptions in most countries,[19] and are particularly suited to the environment for several reasons. These include consumers' low income levels, inability to meet credit requirements, and the difficulties operators would have in keeping track of customers for billing purposes.

Prepaid phone cards have become so important that they are reportedly being used as an alternative to cash for the payment of goods and services (Wright 2004). In East Africa, streetside kiosks, shops, and salons feature entrepreneurs reselling prepaid mobile-phone cards which cost about US$4.80, almost a week's wages in some countries (Cronin 2005). This provides an important source of income for those engaged in the trade and makes it easier to access phone cards in the city outskirts. However, the high cost means that profitability is limited by the number of customers who can afford it.

The mobile society in Africa appears then to contain different classes of users, which makes it possible for the existing cost structure to persist. At one extreme, there are the heavy users who have a high need for telephony, as well as the means to meet contractual obligations. These may be organizations or entrepreneurs who expect to receive high pay-offs from their communication activities, or individuals who are simply wealthy enough to sign up for the post-paid service. At the other extreme are the light users, who may make only occasional use of the phone or have restricted budgets. Though they are sensitive to per-unit cost, the prepaid system enables them to use and pay for airtime only as needed. In this sense it is more cost-effective for them. Also, those who use text messaging more than voice calls can derive greater communicative power from the same number of units.

In some countries, however (especially in Southern Africa), the prepaid market is already reaching saturation, although not all market segments have been served. This indicates that for the rest of the population, even the access, cost, and convenience benefits of the prepaid system are beyond their reach. Nevertheless, the fact that South Africa has extensive mobile coverage (80 percent of the population; Minges 1999) shows that these tariffs are not too high for its citizens. Other obstacles to rollout include high tax rates (for example, in Kenya, operators pay up to 26 percent of revenue in taxes) and lack of cooperation among operators in such areas as the sharing of infrastructure and support services at cell sites (Cronin 2005).

In Latin America, there is a similar situation with respect to prepaid systems, as can be seen in figure 2.1. The popularization of the prepayment system is due, primarily, to the ability to control expenditure and to avoid credit checking and contracts (Oestmann 2003). Indeed, this kind of contract between final users and operators has led to a growth in mobile

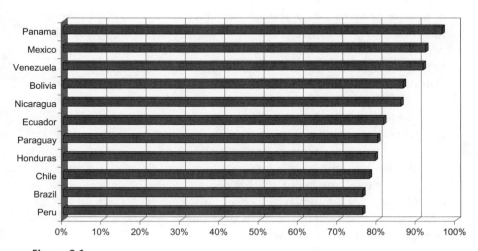

Figure 2.1
Mobile prepaid subscriptions as a percentage of total subscriptions in selected Latin American countries, 2002. *Source*: our elaboration based on ITU (2004b).

penetration increasingly driven by social groups who were previously not served by any form of telephony (NECG 2004). In this sense, it is worth highlighting the fact that fixed operators have already introduced the prepaid system in different countries such as Brazil, Gabon, the Philippines, and Mexico (Oestmann 2003).

Community out of Necessity: Collective Sharing

Another often overlooked factor is the existence of high levels of *mobile-phone sharing* in developing countries, which implies greater access levels than teledensity figures suggest. This was the case in India when researchers found a collectivistic tendency by which "group communication, trust, and emotional bonding are highly valued" (Konkka 2003: 106). As Konkka maintains, in India "experiences with mobile communication tools are shared between friends and family members" (2003: 102). Thus, "a phone call is a social event. When one family phones another family, such as relatives or friends, each family member is expected to talk to each member of the other family . . . Sons and daughters ask to borrow dad's mobile when going out." Mobile-phone calls are also commonly shared among friends and colleagues to the extent that it is "considered unfriendly to omit someone among friends or colleagues from the distribution" (Konkka 2003: 104–105). At the present stage, one may conclude that the mobile phone in India is used more as a *collective* communication instrument rather than the *personal* communication device observed in more developed markets. Although this situation may change in the long run with further diffusion of the technology, collectivistic mobile-phone practices are likely to remain a central defining feature of the social uses of wireless communication in the subcontinent.

In Africa, the sharing of mobile phones exists to varying degrees, which may occur for a fee or free of charge. As Wachira (2003) notes, community-access models have been on the rise in Kenya, where individuals who own a mobile phone become a commercial communication node for friends and family. However, surveys conducted in Uganda, Botswana, and Ghana find that little of this type of community sharing is happening. Rather, people see sharing of another person's cell phone as "a gesture of goodwill, and not a commercial service" (McKemey et al. 2003: 3). Increasingly, however, commercial models of wireless-phone sharing are being established in different countries (see chapter 8). Payphone-style mobile-phone access is also being used to provide a slightly cheaper alternative to owning a mobile phone. For example, the *Simu ya watu* (People's Phone) in Tanzania takes the form of a container where patrons can make metered calls on Vodacom's mobile network for about 29 cents per minute (Cronin 2005). Set-up costs are still high: about US$ 13,000 for the container in which the phones are housed, which is purchased from Vodacom.

Another form of sharing that happens between friends is the sharing of handsets, which has been observed in a form that eliminates accounting issues.[20] In this case, a person can remove the SIM card from his or her mobile phone (for example, if the battery is running low or the handset is faulty), place it in a friend's mobile phone, make a call, and then remove the SIM card. In this way, the cost of the call goes to the owner of the SIM card used, as if they had made the call from their own handset.[21]

The existence of sharing mechanisms means that in African countries there are at least three categories of people in relation to mobile phones: owners, users, and non-users. In countries where there is a high degree of sharing, one can find large numbers of users: people who do not own a mobile phone personally, but have access to one. For example, a study in Tanzania (which has a low mobile-phone teledensity) found that 43 percent of respondents were owners, 42 percent were users, and 16 percent were nonusers; while in South Africa the percentages were 67 percent, 8 percent, and 25 percent, respectively (Samuel et al. 2005).

The network effects of sharing depend on the type of communication required by the phone user. As Samuel and colleagues (2005) point out, where outgoing calls are more important (for example, to search for information), the shared system is adequate enough to meet the user's needs. On the other hand, where incoming calls are more important (for example, to be contacted by potential clients, employers, and family), the system may be less adequate since "a non-owning user can make calls out but cannot receive spontaneous inbound calls" (Samuel et al. 2005: 46).

Another important sharing practice is a collective response to the lack of electricity in some parts of Africa. In Tanzania, for instance, ownership of a mobile phone was found to be positively related to access to electricity, and people without access to electricity were more likely to be mobile-phone users than owners (Samuel at el. 2005). In some communities, a system has been developed whereby all the phones in the community are taken to a nearby town once a week to be charged.

Finally, two studies from Latin America show two further instances of the collective use of mobile telephones among low-income populations. Ureta (2004) found that mobile-phone handsets act as the family telephone, essentially functioning as a fixed line, managed by the mother, which is not surprising since women are the central point of communication among family members. The handset is usually kept at home, unless the spouse carries it out with her. Ureta's ethnographic study, which focuses on low-income families living in a peripheral neighborhood of the metropolitan area of Santiago de Chile, found that the most common uses of the mobile phone are communication with members of the extended family, who often live far away, and the receiving of work-related calls. While in the cases reported there was no fixed line in the household, in other countries such as Brazil (IBGE 2004; Lobato 2004), some families decided to change from fixed-line to prepaid wireless telephone because of the lower commitment and low cost. Our hypothesis, which is yet to be tested, is that the mobile phone in those households would play a similar role to that which Ureta (2004) describes where families with two or more members are involved.

Second, the mobile telephone is used, in rural areas, as a community means of communication, as showed by Carrasco (2001) in her study of the rural area of Los Ranchillos in Cordón de Chacabuco, Chile. There, a single mobile phone was introduced into the community for safety reasons and was shared and maintained by a number of families.

The Last Resort: "Beeping"

Finally, there is also the growing popularity of "beeping" or "flashing" in sub-Saharan Africa (McKemey et al. 2003), a practice also known as "menancing" or "fishing" in Indonesia (Barentdregt 2005: 56) or "boom call" (see chapter 4). This is where a person dials a number but hangs up before the call is answered with the objective (usually) of the recipient calling back. This practice is used for different reasons, including requesting the recipient to call back when the sender has almost run out of prepaid units but needs to talk, or simply to say "I'm back," which does not require a response and enables essentially free communication.[22] This practice has been developed into an art in some countries (for example, Rwanda), with rules and systems of filtering "beeps" (Donner 2005a). As a means of transferring the cost of communication to another person, however, "beeping" can be an inconvenience – to the extent that, in some areas, people are choosing to ignore "beeps" (McKemey et al. 2003; Donner 2005a).

Ethnicity, Culture, and the Uses of Mobile Communication

Issues relating to ethnic or cultural factors as facilitators, inhibitors, or shapers of the uses of technology can be controversial since representations or explanations may be more stereotypical than real, especially in the absence of empirical data. Nonetheless, some trends have been observed that seem to distinguish users of different ethnicities and cultures. What the data so far tell us is that ethnicity and/or culture do not act as barriers to acquisition of mobile communication devices, but may limit the range of applications and services that users have access to, and/or are interested in using. Furthermore, some types of communities may be disadvantaged in that their use patterns compel them to devote large proportions of their resources to paying for communication technology. Owing to limitations of data, we do not present a comprehensive section on the diffusion of mobile communication technologies among different ethnicities, but refer to available statistics throughout the general discussion.

Limiting ethnicity to tribal or cultural social groups within a country, the question of its influence on mobile communication can be addressed from two perspectives: social exclusion and usage levels/patterns. Generally, when communication technologies are discussed, ethnicity is included as a potential barrier to use. In other words, the focus tends to be on what are considered to be ethnic minorities, and how their specific location in society (for example, language differences, low educational levels, unattractiveness to commercial interests) puts them at a disadvantage in terms of access to technology and skill in using it. This has been shown to be the case with Internet use in the United States, for example (NTIA 1999, 2000), but is not necessarily the same with mobile communication technologies, especially mobile phones. Indeed, the limited evidence so far suggests that ethnicity is a barrier only insofar as it interacts with other factors, such as residence in under-served areas, language differences, the need to make expensive international calls, and low income levels.

In this respect, there may be parallel trends occurring in the United States. On the one hand, it has become necessary for special programs to be developed to provide affordable mobile phones (Tohtsoni 2001; Nokia 2002) and Wi-Fi access (Nokia 2002; BBC 2004d) to Native Americans living on some reservations, where telephony in general is limited. For example, the Vision One program, undertaken by Cellular One in conjunction with the federal government and the Universal Service Fund, provides mobile phones for US$ 1 a month to Native Americans living in under-served areas. The program had provided mobile phones to over 25,000 residents by the end of 2002.

On the other hand, African American and Hispanic populations appear to be making significant strides in the acquisition and use of mobile telephony without the need for external assistance. As a result of the introduction of prepaid systems, the pattern of wireless-phone ownership among whites, African Americans, and Hispanics changed over the period 1999–2002. While African Americans were significantly behind in adoption in 1999, they surpassed whites and have been in the lead percentage-wise since 2001 (see table 2.5). Both African American and Hispanic populations (particularly teens) have been found to have higher mobile-phone ownership and usage levels than the Caucasian population. They use text messaging more, are more likely to buy new phones, and are early adopters of new services (Charski 2004; Silho 2004). For example, Scarborough Research found that Hispanics spend more on wireless bills: US\$ 67 versus the national average of US\$ 62 (Charski 2004). Penetration of the mobile phone into African American culture can be seen in the frequent appearance of new mobile-phone models in rap music videos (Silho 2004).

A survey of Norwegian youth (13–15 years and 16–19 years) to find out if ethnicity, among other factors, affected ICT-usage patterns found that mobile-phone ownership was less common among children with parents born outside Norway than among those with Norwegian-born parents (Torgensen 2005). In the UK, a higher proportion of the non-white population depends solely on mobile phones (table 2.6). This is probably related to income rather than ethnicity, as a qualitative study of low-income groups, including Turkish, Bangladeshi, Chinese, Afro-Caribbean, asylum-seekers, and white British, found that ethnicity was not an important factor shaping telecommunication use, including mobile phones (Oftel 2003b). In fact, none of the respondents felt excluded from society, or had a reduced sense of security, mainly because most had mobile phones.

Language may act to some extent as an ethnicity-based barrier to the use of mobile communication technologies, even voice services. For instance, in Britain, non-English speakers had more problems than English speakers

Table 2.5

Mobile-phone ownership by ethnic group in the United States, 1999–2004 (%)

	1999	2001[a]	2002	2004
African Americans	37	74	65	74
Whites	42	56	62	62
Hispanics	n/a	56	n/a	58

[a] Age range 12–34 years.

Source: data adapted from Genwireless (2001) for 2001; Fattah (2003) for 1999 and 2002; Silho (2004) for 2004.

Table 2.6
Use of mobile phone only, by ethnic group, in the United Kingdom, 2003 (%)

Ethnicity	Only mobile	Neither mobile nor fixed
White	7	1
Non-white	12	5

Source: Oftel (2003b: 14).

in matters related to Directory Enquiries, accessing information, and inter-
acting with sales personnel (Oftel 2003b). Conversely, provision of multi-
language services can promote the use of services by non-English speakers.
The existence of Spanish-language operating platforms on most phone
systems in the United States has made it possible for Spanish speakers
to be able to navigate mobile-phone technology and gain a fairly good
understanding of the capabilities and limitations of the mobile phone
(Leonardi 2003).

From another angle, as observed in Bosnia, technology can be used to
reinforce ethnic divisions. In a situation that is possibly unique to Bosnia,
although there is little difference in the pricing plans of the mobile-phone
operators there, "almost without exception, Bosniaks use the company
from Sarajevo, Serbs the one from Banja Luka, and Croats the service
provider based in Mostar" (Slatina 2005: 92). A news-magazine writer suggests that this is a remnant of the war: "People do not want to contribute
to the economic prosperity of those who were on the other side during the
war" (Slatina 2005: 92). Further evidence of continuing ethnic division
in Bosnia can be seen in the fact that Muslims and Croats live apart to
the extent that they reportedly have different mobile-phone area codes
(BBC 2004b). Furthermore, in the current security-conscious environment
caused by the terrorist attacks in the United States and Europe, some ethnic
groups may see the location-tracking capabilities of mobile phones as facilitating surveillance of their particular communities, as evidenced in comments by respondents in Green and Singleton's (n.d.) study of young black
and ethnic minority groups in Britain.

A particular manifestation of ethnicity in mobile-phone use that is
appearing in some regions is the tendency for ethnic minorities to be
heavier users of mobile telephony than the general population. One British
mobile-phone company reports that ethnic minorities make more international calls and use mobile phones and text messaging more than the
average mobile-phone customer (Varney 2003). And, as already men-

tioned, ethnic minorities in the United States are also more engaged with mobile telephony than other users.[23] In the case of African Americans and Latinos in the United States, the reasons for this heavy use are not obvious. It has been suggested, though, that this may be culturally based. For example, there is some evidence that Hispanics place high value on familial communication (Leonardi 2003; Charski 2004). Thus, Leonardi (2003) found that Latinos in the United States view the mobile phone positively, in large part because it suits culturally valued behaviors of interpersonal communication. This positive attitude was particularly significant when set against their attitudes toward computers and the Internet. Despite their undisputed benefits, the latter two technologies were considered to impede interpersonal communication and were consequently viewed in relatively negative terms. Mobile phones, on the other hand, were considered to be a necessity rather than a luxury in contemporary society. Furthermore, most Hispanic immigrants are used to using mobile phones since fixed lines are luxuries in their countries of origin (Charski 2004).

Where ethnic minorities are immigrants, or descendants of immigrants, heavy mobile-phone use is largely explained by their need to stay in touch with family in their home countries. This appears to be the case with asylum-seekers in Britain (Oftel 2003b) and children of Chinese immigrants in Canada (Tsang et al. 2003), as well as Sudanese resettlers in Canada and their families in Kenya (Thomson 2005). It makes sense that this is probably the case wherever there are significant immigrant or refugee communities. The advantages of mobile telephony to immigrants are, however, also associated with significant expenses for this population. Since they mostly have to make international calls, they end up spending larger proportions of their household budget on communication. For example, in Britain, non-white groups had more problems and generally incurred more costs than white groups because of their need to make international calls and their inability to avoid peak-time charges owing to timezone differences (Oftel 2003b). Thus, while expressing appreciation of the benefits of mobile telephony (especially easier monitoring and control of costs), all respondents who did not have fixed lines aspired to have them. Asylum-seekers and immigrants, in particular, face a conundrum since they not only have to make expensive international calls via mobile phones, but also often have to call their relatives' mobile phones, this usually being the only form of telephony available in their home countries. Thus, in the Oftel study, most asylum-seekers reported using payphones rather than mobile phones when making international calls. Similarly, although they had access to mobile phones, Sudanese refugees in a Kenyan camp were heavy patrons of an Internet café that offered satellite phone calling because it is the cheaper option (Thomson 2005).

International differences, more than intra-national variations in mobile communication usage, have attracted more attention from researchers and journalists. As mentioned in chapter 1, in addition to differences in mobile communication diffusion rates, certain behavior patterns distinguish countries from each other and have been attributed in part to national cultural characteristics. First, there are arguments that cultural tendencies are to some extent responsible for the rapid diffusion of mobile telephony and data in some countries and the slow diffusion in others. For example, it has been suggested that people in collective, sociable cultures accept certain use behaviors that would be frowned upon in more individualistic societies. For example, in Thailand or Italy, where there is high use of mobile phones, there is little concern about the volume, location, or nature of public conversations. Conversely, in the United States, a tendency to be extremely concerned about privacy and personal space has inhibited the adoption of mobile telephony (Mante 2002; Plant 2003b).

Adoption of particular mobile applications can also be linked to cultural differences. The popularity of text messaging among young Japanese has been attributed to the situation of small living spaces and hence limited privacy, as reported by anthropologist Geneviève Bell (Erard 2004). This can again be contrasted with the United States and some European countries, where children often have their own rooms. While demand for mobile content is significantly related to age, cultural factors also come into play: for example, British residents apparently tend toward material that can be consumed in short bursts, such as music videos and one-minute comedy clips, as they are only seeking content to while away small periods of free time (McCartney 2004).

Comparing Japan and Korea, Lee et al. (2002) conclude that there are significant value-structure differences between the two that affect mobile Internet adoption and use patterns. In Korea, users prefer to download entertainment content, such as live TV, music videos, or games, and this is linked to the higher premium placed on emotional values in this culture, as well as high tolerance of public mobile-phone use (see also McCartney 2004). Consequently, the less-intrusive mobile e-mail is less popular. Conversely, in Japan, "Although mobile Internet is a convenient communication device, Japanese users may not want to use it for socializing with others" (Lee et al. 2002: 236). Here, functionality is more highly valued, and the society is fairly introverted; hence e-mail is the most used mobile Internet application, and information services in general are popular. An observed preference for self-generated picture and video mail in Japan has also been linked to the culture, in which it is considered impolite to make phone calls in public places – thus, the alternative is to use picture mail (McCartney 2004).

These trends have certainly been observed and commented on in the popular literature, but there is yet to be a preponderance of data to support cultural explanations. Preliminary findings of a study exploring factors affecting the use of handheld mobile devices among participants from the United States, Norway, China, Korea, and Thailand indicated that individual characteristics, technological characteristics, communication task characteristics, context, and modalities of mobility all contributed to different patterns of adoption and use (Sarker and Wells 2003). The influence of cultural origin was observed particularly with text messaging: in "high power distance" cultures,[24] such as Korea, it was considered unacceptable to send text messages to a superior; in "low power distance" cultures, such as Norway, this was not considered offensive, although potentially inappropriate for strangers who may not be familiar with abbreviations and slang (Sarker and Wells 2003).

One study that has recently attempted to systematically examine this issue on a global scale is an analysis of diffusion data for 64 countries between 1981 and 2000 (Sundqvist et al. 2005). The study sought to discover whether and how country characteristics, cultural factors, and time of adoption explain the differences in mobile-phone diffusion processes across countries. Ultimately, the researchers found support for the hypothesis that culturally similar countries (using Hofstede's 1991 framework) have similar adoption behaviors. They identified four adoption clusters. Cluster 1 contained the late adopters, mainly from South America and Eastern Europe. These consisted of collectivist cultures with "high power distance" and "uncertainty avoidance." Cluster 2 was made up of early adopting wealthy countries with penetration rates of around 70 percent. These were mainly from Western Europe and North America, and consisted of individualistic, masculine cultures with "low power distance." Cluster 3, containing rapidly developing Asian countries, was similar to cluster 1, except for "low uncertainty avoidance." Results for this cluster were, however, inconclusive. Finally, the earliest adopters fell into cluster 4. They had penetration rates of around 100 percent, and were mainly Scandinavian and Central European countries with high individualism, femininity, and "extremely low power distance."

This would seem to show that individualism, "low power distance," and risk-taking facilitate the adoption of mobile telephony, since these were defining characteristics of the early adopting nations in clusters 2 and 4. It is interesting to observe, however, the inability to explain adoption patterns in the newly industrializing Asian countries, which arguably tend toward collectivism and "high power distance," but are apparently not risk averse. Thus the cultural explanation for adoption rates does not hold in all situations.

Apart from the general trends discussed above, there are a few uniquely cultural mobile-phone-related practices that have been observed across the globe. These provide evidence of interesting appropriations of mobile communication technologies and structures to match local needs, interests, and beliefs. For example, in an Indian village, people use farm produce such as milk, wheat, or sugar cane to pay for a mobile payphone service provided by another farmer. The farmer receives produce in lieu of cash but pays cash to the mobile payphone company (Bajpai 2002).

Religion has also found a place for mobile communication technology. This can be seen in the development in Korea of a mobile phone with an embedded compass (using GPS technology) to enable Muslim users to locate the direction of Mecca for their prayers (Erard 2004). Furthermore, the practice of using SMS messages to conduct the *talaq* – the process by which a man can divorce his wife by pronouncing the word *talaq* three times – is now recognized as a valid means of divorce in a number of Muslim communities, for example, in India, the United Arab Emirates, and Malaysia (Jyothi 2003). The Sharia recognizes technology as a means of communication (including fixed-line telephony, postal mail, and telegram which are already used to declare divorce) and, thus, "digital divorce" is possible, as long as the message can be authenticated (two witnesses need to be present during the divorce proceedings). This practice is viewed differently across countries and even within them. For example, while in the United Arab Emirates it seems generally acceptable, a woman in Malaysia has contested her divorce communicated in this manner, and the Malaysian government reportedly does not recognize SMS divorce (Jyothi 2003).

In Asia, families have been observed burning paper cell-phone effigies for the use of departed family members in the afterlife, while Chinese mobile-phone owners take their phones to the temple to be blessed (Rae-Dupree 2004; *The Economist* 2005e; Maddox 2005). In Germany, a religious sermon was transmitted by mobile phone in May 2001 in an attempt to get more young Germans interested in compressed spiritual messages delivered to their phones: 1,300 signed up for the sermon (Kettmann and Kettmann 2001). It is not clear how successful this trial was. A Finnish service, now shut down, went so far as to claim to deliver text messages from Jesus (*The Economist* 2005e; Maddox 2005). Religious authorities have also reportedly curtailed the emerging trend whereby people in the Philippines were making confession and receiving absolution by text message (*The Economist* 2005e; Maddox 2005). The supernatural emerges in mobile telephony when cultural beliefs about the power of numbers make contact with specific phone numbers potentially life-changing. Thus, a Beijing resident is said to have paid over US$ 200,000 in order to get a "lucky" phone number (*The Economist* 2005e). And in Nigeria, mobile phones are becoming a source of "fear, panic and anxiety . . . due to the current belief that people are dropping dead after receiving calls from certain mysterious numbers" (Agbu 2004).

Summary

In sum, it is reasonable to assume that mobile telephony has opened up access to communication and its advantages for ethnic minorities in different countries, enabling them to overcome obstacles that may have prevented their access to other types of communication technology. It is also a critical tool for immigrant or displaced populations, making it possible for them to stay close to their cultural origins. The only note of caution is that we cannot extend this impact to all possible applications of mobile communication technology, in particular non-voice uses (apart perhaps from SMS). Just as there is a technological gap in global mobile communication, even as the diffusion gap is closing, a similar quality-of-use gap may exist between consumers who can afford and/or are literate in the use of a variety of mobile applications, and other consumers who may use only basic services.

There is also evidence of variations in usage patterns and preferences in different countries that appear to be based on cultural characteristics. Thus, alongside the development of trends in mobile communication that could be considered global, other trends unique to individual ethnic, cultural, or national characteristics are also found. The extent to which these trends are rooted in cultural backgrounds, belief systems, norms, and values is arguable, as is the question whether these tendencies will persist over time. Nonetheless, while avoiding stereotypical characterizations, and based on the limited data, one can surmise that culture plays a part in how mobile communication technologies are incorporated into the routine lives of users. Indeed, what we observe throughout this chapter is the interplay of the different factors and processes defining the social structure with the diffusion and uses of wireless communication.

Technology follows differential paths of diffusion and use according to age, gender, class, ethnicity, race, and culture, and it is appropriated by people in terms of their values and needs. And because these needs and these values evolve over time, and are specified by their social context, so the uses of communication technology vary. This is to say that the social differentiation of technology closely reproduces the social differentiation of society, including the cultural diversity manifested within countries and between countries. Technology, as it is practiced, is society, and it embodies society. And society, which is made of communication, reproduces its cleavages and diverse models of existence in the expansion of its communication mode into the realm of mobile communication.

3 Communication and Mobility in Everyday Life

Everyday life is made up of a broad range of social practices that recur in the routine of people's experience. They include work, family, sociability, consumption, health, social services, security, entertainment, and the construction of meaning through perceptions of the sociocultural environment. Because mobile communication is pervasive and reaches all domains of human activity, its mediating effects can be observed in these different dimensions. And from the observation of its specific role in each one of these realms, a pattern emerges that may yield some clues about the actual transformative effects of wireless communication technologies. While the specific patterns of mobile diffusion differ across geographic regions and social groups, it is clear that mobile technologies are becoming an integral part of people's everyday activities. The ubiquitous influence of mobile technology has been signified in recent years by the emergence of a series of "m-" neologisms, such as m-commerce, m-learning, m-government, m-literature, m-entertainment, m-gaming, m-etiquette, mobil-ization, and moblog. The list goes on, and the spread of mobile technology will continue to change the ways in which people conduct their lives.

As mobile devices are "personal, portable, pedestrian" (Ito et al. 2005), they have been quickly adopted – attached to the body like watches (Ling 2001; Fortunati and Manganelli 2002; Oksman and Rautiainen 2002; Kasesniemi 2003) – for a wide range of social practices, in addition to the main function of communication (Harrington and Mayhew 2001; Varbanov 2002; Lacohée et al. 2003). Hence, as elements of daily routine, wireless technologies, especially the mobile phone, are perceived as essential instruments of contemporary life. When they fail, users tend to feel lost because of the dependency relationship that has developed with the technology (Ling 2004). Consider, for instance, the address book that is often stored only in the mobile, and the fact that nowadays there is no need to memorize phone numbers because they are always available in the handset. If there is a failure in the address-book function or the mobile phone is lost, the owner of the handset will be seriously disabled, if not totally isolated, in his or her social networks.

The purpose of this chapter is to summarize a number of studies on mobile technology and the transformation of everyday life, highlight key developments that are central to the rise of the mobile network society, and make analytical sense of the observed trends. To understand this vast field, we will first examine changes in work and work processes, including not only high-level professionals and middle-level employees, but also lower-ranking migrant workers. Then we will focus on the family, the domestication of mobile technology at home, and its effect on private relationships among family members. This will be followed by a discussion of the transformation of sociability, and then issues of personal safety and security. The next section will deal with new modes of public service provision, including m-government, alternative modes of access, and services for people with disabilities. We will then consider how mobile devices transform consumption patterns, giving rise to new types of entertainment and fashion, while users are enabled to personalize the technology and be more actively engaged in the construction of their identity. Finally, the chapter will examine social concerns related to the perceived risks of the new communication technology, including issues of health, spam, the inappropriate use of camera phones, surveillance, sexuality, and media representations of these real or perceived dangers in the mobile network society.

Mobility at Work

A profound transformation is under way in the processes of work because mobile communication allows for coordination among co-workers at a distance. This is important because of the increasing tendency for businesses to operate in diffused networks, globally or locally. It is therefore not surprising that businesses "heavily reliant on field personnel" are among the first to adopt mobile phones (Netsize Group 2005: 70). In Scandinavian countries, for example, the early adopters included truckers, construction workers, maintenance engineers, and the police (Agar 2003: 52; Lundin 2005). These are employees who may or may not have an office at all, but need to coordinate work across long or short distances. With the help of wireless technology, a mobile worker can keep in constant contact with his or her headquarters, while working anywhere that has communication coverage. This is a key development because, while members of staff are away from the office, contextual constraints become unpredictable (Perry et al. 2001). Yet the use of mobile technology connects the different contexts into an extensive work environment which shares a common network logic.

Professionals on the Move

As Watson and Lightfoot (2003: 348) write: "The aim of mobile working is to allow staff to access a range of systems and services whilst they are away from the office – but without the restrictions of wire." Hardly can there be a more fundamental change for mobile workers and professionals, who account for an increasing share of today's labor force. According to O'Hara et al. (2002), those who travel long distances, and need to be away from the office for at least one night, tend to check their e-mail in the evenings in their hotel room, while they use the mobile phone more frequently, and sometimes as a way of solving disconnection problems experienced during a trip. In the case of mobile workers who routinely travel short distances, it has been reported that the mobile phone is often deployed in automobiles, when people can check voice-mail and reevaluate traveling time (Laurier 2002). Hence, mobility is organized "on the ground" (Laurier 2002: 47), allowing work processes to become flexible and responsive to circumstances in the field. Workers can go beyond the strict objectives of the trip to take advantage of specific situations and perform other tasks as need be. They can make calls or continue working on various jobs that are pending when there is time available between two programmed activities (Perry et al. 2001; Laurier 2002). All these activities can now be done despite the limited access to documents and information as compared to work in the office (O'Hara et al. 2002), which explains why the use of laptops fell by 45 percent when Goldman Sachs employees were given Blackberry pagers.

There are, however, still notable differences between mobile work and work in a fixed location. Although wireless technologies, such as the mobile phone, Wi-Fi, and PDAs, offer an array of flexible ways to access e-mail and other documents (O'Hara et al. 2002), all these services have to operate within particular physical contexts that guarantee different levels of privacy. This is why certain activities are not performed, for example, at an airport lounge, despite the growing importance of the airport as a central node of coordination for the professional world (Breure and van Meel 2003). When it comes to task performance and information processing, mobile professionals cannot always accomplish what they routinely do with their desktops (Sherry and Salvador 2002: 114). Given the contextual limitations, a major function of mobile devices is to "check in." They are usually used, not as the equivalent of office computers, but as a tool for short conversations to make sure that both parties are "okay," along with some brief discussion of work progress (Sherry and Salvador 2002: 116). In many cases, co-workers call each other at the end of the day using the mobile phone in order to offer social support and avoid feelings of isolation (Perry et al. 2001; O'Hara et al. 2002; Sherry and Salvador 2002), which is very important for productivity and cohesion among colleagues.

But perpetual availability can also be a far less benign development for mobile workers. For one thing, it often serves the needs of companies by providing new channels of surveillance. Thanks to the ubiquity of mobile communication, supervisors are now not only able to monitor mobile workers constantly during working hours but can also exert control around the clock; for example, by ordering their employees to turn on their mobile phones twenty-four hours a day, seven days a week.[1] Moreover, they can instantly control task performance in diffused space by using a location awareness device; for instance, using wireless tools designed for the tracking of distribution fleets. Already some researchers see the return of the "assembly-line" model in which workers' every activity is directly controlled and timed by management (Laurier 2002: 50) using mobile communication as a wireless leash.

There is, therefore, no simple conclusion about the actual social implications of permanent connectivity at work. As observed during working hours in South Korea, "[m]anagers can constantly check if their salespersons are working properly outside the company, while employees find less opportunity to slacken off" (S-D. Kim 2002: 73). One specific example is the n-Zone service in operation at Samsung Electronics (Ha 2002), where workers get fixed-line phone calls automatically forwarded to their mobiles when they are away from their desks. To reach their colleagues, they only need to dial the last four digits on their handsets as if they were using traditional wired intra-organizational networks. Samsung Electronics and KTF jointly developed this mobile phone system. Though still in an initial stage of development, such a service is becoming popular among corporations because of its promise to improve work efficiency at an inexpensive price. Workers subscribing to n-Zone can call their co-workers and have unlimited use of wireless Internet at a cost of only US$ 1 per month. When calls are placed to people outside the corporation, only the price of the wired phone service is charged, which is cheaper than the cost of wireless phone calls. By 2003, Samsung Electronics had launched this service in their headquarters at Seoul and Suwon. It was reported that they planned to extend n-Zone to many other plants (Ha 2002). In South Korea, other institutions using similar services include the Korea Advanced Institute of Science and Technology (KAIST) and Daewoo Ship Construction. In 2002, about a thousand members of staff of KAIST are subscribers, and the service is to be extended to about seven thousand KAIST students (Ha 2002). Another service in the workplace is Bizfree from KT, which connects offices in other locations (i.e., branches) as if they were in the same building (Han 2001).

What, then, is the overall relationship between mobile communication and productivity? Despite the conventional wisdom that mobile technologies tremendously increase work productivity – for example, by improving economies of scale (UMTS Forum 2004) – the evidence suggests a much more complicated picture. As discussed above, wireless devices allow permanent availability, which means "off time," such as time spent traveling, can become productive time, when mobile workers can respond more directly to local problems in a diffused network of information, coordination, and social support. While "always ready" workers may yield a greater load of working time, it does not mean that they increase productivity. In fact, they may become less productive, because they put in more working time without necessarily producing significantly more. At the same time, employers may also use the technology to locate employees and control them, anytime and anywhere, which is why "the mobile phone is blamed for the loss of leisure" (Katz and Aakhus 2002: 8) and for increasing the tension between work and life (Sherry and Salvador 2002: 118) which may undermine productivity in the long run.

In the context of developing countries, the mobile phone is thought to increase productivity by reducing the need for small entrepreneurs to be tied to specific locations where family members can be in contact (Donner 2004). But regression analysis in Africa does not show any correlation between the use of mobile phones and productivity (Matambalya and Wolf 2001). The only significant relationship is between productivity and fax machines, for which the researchers offer several reasons, including the time lag between investment and visible outcome and the fact that fax machines are used more exclusively for business purposes. However, people *perceive* mobile phones to be helping them expand their markets, more so than fixed phones or fax machines. The researchers conclude, however, that "ICT investment is not more productive than other investment in the short run" (Matambalya and Wolf 2001: 24).

Blurring Boundaries of Social Space/Time

Although it is too early to draw any conclusion on the issue of productivity, most researchers would agree that work and work processes are fundamentally transformed with the rise of mobile communication, and that a most notable change is the blurring of the boundary between work and the private sphere. While permanent connectivity allows work to spill over into homes and friendship networks, it is also likely that personal communication will penetrate the formal boundaries of work. In Rwanda, for example, only one-third of the calls made by local entrepreneurs were found to be for business, with the other two-thirds being of a personal nature, leading Donner to question "popular images in development articles and in the mainstream press, which present the microentrepreneur's typical call as an instrumental, market-oriented action, made to reach customers or check on prices" (2004: 7). Among the calls made in Rwanda, friends were the most frequent call recipients, followed by family and customers. Chatting was the most common purpose of calls, followed by arranging meetings, and other business matters. The results held for outgoing and incoming voice calls as well as SMS (Donner 2004).

In the United States, a study by Grant and Kiesler (2002) found that workers at Carnegie Mellon University became attached to their mobile phones a few months after acquiring them for work purposes and began to see them as personal possessions. The "wearability" and interactive capabilities of wireless devices give them a different character and meaning in people's lives from other types of communication equipment, such as the fixed-line phone or desktop computer. Another notable trend is people's eventual attachment to their cell-phone number, as shown by the ruling of the Federal Communications Commission (FCC) that made it possible for people to migrate to new wireless telephone service providers without changing their existing phone numbers. By 2004, 54 million people had taken advantage of this option (Pelofsky 2004).

This is not to deny the importance of wireless communication for business. Numerous studies show that the spread of mobile phones leads to more employment opportunities and expands the range of business activities, for example, through the resale of airtime in Africa (Wright 2004) and the refurbishing of mobile handsets in Indonesia (Barendregt 2005). It is now routine for small shop-owners in Chinese cities or individuals throughout Europe to distribute business cards that show their mobile phone numbers, a practice seldom observed in the United States.[2] As a survey demonstrates, in Egypt and South Africa there was a significant rise in the use of mobile telephony among small businesses between 1999 and

Table 3.1

Increase in mobile-phone and fixed-line subscriptions among small businesses in South Africa and Egypt, 1999–2004 (%)

	1999		2004	
	Mobile	Fixed	Mobile	Fixed
South Africa	34	52	89	60
Egypt	11	45	85	80

Source: Samuel et al. (2005).

2004 (Samuel et al. 2005). Not only has the use of mobile phones increased at a much faster rate but the percentage of mobile subscriptions has also surpassed that of fixed-line telephony in both cases (table 3.1). Another study in South Africa found that more than 85 percent of small, black-owned businesses have only a mobile phone (Donner 2004).

While observing that micro-entrepreneurs in Rwanda made two-thirds of their mobile calls for personal reasons, Donner (2004) nevertheless argues that the mobile phone brings two major benefits to the work processes of local businesses: an enhanced facility to reach and be reached by business partners, suppliers, and customers, and an improved ability to communicate with family and friends who also play important roles in business practices. Donner concludes that the use of mobile phones by micro-entrepreneurs in Rwanda amplifies certain social networks (family and friends) and enables others (new business contacts). By measuring how often respondents communicated with call recipients before and after buying a mobile phone, this study revealed that new entrants into networks were more likely to make business-related contacts in the form of weak ties. Amplified relationships were generally those with family and friends, a significant proportion of whom lived far away. Mobile phones were used to establish new business contacts but less frequently new social contacts, especially with people outside Kigali.

Mobile Usage among Migrant Workers

It is worth noting that migrant workers often constitute a particular user group, whose mobile communication patterns differ from upper- and middle-class professionals, as well as from small entrepreneurs who work at their place of permanent residence. Overall, contrary to popular perception, most migrant workers do actively engage in telecom services of all kinds, including the mobile phone (Qiu 2004; Barendregt 2005; Cartier et al. 2005; Thompson 2005). In Singapore, for example, migrant workers from the Philippines, Indonesia, India, Sri Lanka, Thailand, and China were all found to use mobile phones on a frequent basis, using, for example, prepaid mobile phone cards to call home (Thompson 2005). The only exception was Bangladeshi construction workers, who, compared to others, tended to have lower income and fewer people to call in their home country because of the extremely low teledensity in Bangladesh (Thompson 2005). In the case of China, where Qiu conducted fieldwork in 2002–2005, it is commonplace for migrant workers to own mobile phones, especially in large cities such as Beijing, Shanghai, Guangzhou, and Shenzhen. Moreover, they often spend a larger proportion of their income and time on mobile phones in comparison with ordinary, long-term, urban residents. Yet the services they receive are usually unsatisfactory, and they often feel that they are badly served by mobile operators (see Qiu 2004).

There are approximately 150 million migrant workers in China who have left the countryside to seek jobs in the cities (Zhang 2001). A major component of the emerging "information have-less" (Cartier et al. 2005), these workers constitute a critical labor force in the processes of urbanization and modernization in general, and of the telecommunications industry in particular. They represent a sizable user group with distinct patterns of communication (Cartier et al. 2005).[3]

In the summer of 2002, Qiu conducted fieldwork by living with migrant workers in an industrial zone of South China for two weeks. He held three focus groups among migrant workers in Shenzhen, Guangzhou, and Zhuhai regarding their telecom usage patterns, including those of wireless technologies. A survey was conducted which included 272 migrant workers, purposively sampled from ten different dialect regions, covering most provinces in southern, eastern, and central China. Although, admittedly, these data were collected on a small scale, they nonetheless give a preliminary snapshot of how the mobile phone is playing a key part in the everyday life of migrant workers. Qiu resided with two migrant workers who were junior college graduates from the mountainous regions of northern Guangdong and southern Jiangxi Provinces. Working for a fertilizer company, they lived in an apartment-style dormitory equipped with a fixed-line telephone. There was also a phone that they could use at their workplace, which was 20 minutes away by bike. Yet, they both still had their own mobile phones.

At the start of the project, one of the workers had just bought a new cell phone with a color screen, using his savings for an entire year. He told Qiu that he had spent many hours in the past few months searching for the best deal: "Although it's pricey, I feel pretty good about it because everyone says that's a good deal." For this migrant worker, the purchase of a mobile phone was less a rational choice for communication purposes than a social and psychological need to achieve a status symbol which could be shown off in front of friends or even strangers in public. Just as industrial workers in the early twentieth century would save for years to buy a personal timepiece, these migrant workers also attach much non-instrumental value to the cell phone. For them, obtaining a mobile is a milestone that indicates success, not only financially but also culturally in terms of their integration within the city.

But there is a critical difference between a mobile phone and a mechanical watch: whereas both entail a sizable initial cost, the mobile phone requires further regular expenditure. Even if you are using prepaid cards, the expenses continue to accumulate as long as the phone is in use. The more you call, the more you spend. Or if you do not use it much, the purchase of the handset would not be justified. Thus, it was not surprising that migrant workers in the survey spent an average of US$ 37.4 (RMB 309.5) a month, which accounted for 20.2 percent of their total income.[4] In contrast, while long-term urban residents had a higher average income, they only spent 13.8 percent of their monthly income on the mobile phone. Furthermore, as new cell phones come out, handsets become outdated in a couple of years, and you are compelled to update your handset, which, according to the survey, costs approximately US$ 231.1 (RMB 1,912.4) each. So, it starts again – the vicious circle of conspicuous consumption, which, conversely, can be seen as a covert means of labor control because, in order to purchase mobile devices and pay phone bills, migrant workers have to discipline themselves to work overtime, and work hard.

Moreover, although migrant workers spend a high proportion of their income on mobile-phone-related expenses, they are often subject to poor service. All participants in the three focus groups conducted by Qiu in Shenzhen, Zhuhai, and Guangzhou agreed that the mobile services they received were unsatisfactory. Voice telephony was over-priced. Customer service was poor, with billing disputes often unresolved and their complaints neglected. Because of the lack of security in places where migrant workers congregate, their cell phones were easily stolen. There are also unscrupulous organizations or individuals who use cell phones and SMS to deceive migrant workers, such as offering them fake jobs after collecting contact information in local labor markets. The list of pitfalls and ordeals goes on. But the point is clear: although the technical infrastructure for mobile phones is established, the social infrastructure supporting mobile-phone usage among migrant workers is nearly non-existent. This is because, from service providers to local government, attention is paid exclusively to the urban middle class and to multinational corporations, and the fact that migrants are becoming an increasingly important group of consumers in the mobile market is ignored.

The absence of a larger social infrastructure, on the other hand, offers a more profound explanation for the boom in mobile phones among migrant workers. In the focus groups, participants from different cities reported widespread problems in using fixed-line telephones. While Qiu's junior-college graduate roommates were lucky to live in an apartment-style dormitory equipped with a landline telephone, the majority of the participants in the focus groups had only high-school or junior high-school diplomas and had to live in much worse conditions with no shared domestic phone. In addition, several of them reported restrictions on the use of an office phone, especially in restaurants where they worked as waitresses. It is common practice to bar the migrants from using the work phone during business hours. One boss required employees to pay a fee every time they used the phone at work. In two other instances, migrant workers were allowed to receive but not to make calls. This leaves public phone booths as the only option for migrants if they need to be connected with friends and families for both emotional attachment and job-related information. Yet the phone booths were notorious for security reasons because they were often designed with little protection for privacy. Migrant workers often had the pass code or security number of their prepaid phone cards stolen while using public pay phones.

With all these problems in accessing domestic, office, and public pay phones, having one's own cell phone has obvious advantages. In this sense, part of the reason for the high demand for mobile phones among migrant workers has to do with the constraints of the larger social structure. Left with no choice but the cell phone in these circumstances, those migrant workers who have adopted the mobile phone still have only limited power to control when, where, and how they communicate with others. Sometimes, they make rational decisions, such as using SMS to save money. But, at a higher level, this is not a rational system for migrant workers, who have to pay more and get less, who have to go through a series of hardships which ordinary, permanent urbanites do not face while using the new technology. While the migrant workers are active in pursuing their urban dream, in which the cell phone is now the centerpiece, this new technological condition of mobile use among the "information have-less" only operates at the micro level, and cannot solve the macro problems, whose solution depends on innovative adjustments and combinations of public policy, corporate strategy, and working-class culture.

Micro-coordinated Families

As contemporary families often exist as micro-distributive networks across multiple sites with translocal and sometimes transnational reach, mobile technologies have been widely adopted in the family setting throughout the world. It should be emphasized that the demand for mobile communication has long existed, as family members always want to stay in touch and adjust their activities to ensure the functioning of the family unit. Thus, while the new technologies bring new means of coordination and of social support, they are appropriated in a way that strengthens existing family relationships instead of causing any revolutionary change. As chapter 4 will focus on young people and wireless communication, including the transformation of intergenerational relations, the following discussion examines mobility and family communication in more general terms.

As can be seen in figure 3.1, which looks at household ownership of mobile telephones in selected European countries, penetration of mobile telephony in the private sphere is very high.[5] These figures include mobile telephones provided at work to family members, which are actually used for private purposes and do not focus on the number of devices available in a given household. The data, thus, differ from ITU data, but are also of interest as they reveal, among other things, the practice of handset borrowing in the private sphere. Although in more wealthy societies individual family members will eventually have their own device (Ling 1999b), the practice of borrowing within the family is a common phenomenon in both developed and developing countries, meaning that some family members can be active users of mobile telephony even though they do not own a mobile device.[6]

Unsurprisingly, Nordic countries are in the lead, with Finland achieving a penetration rate of 92 percent of households in 2003. What is most relevant is that the gap between Scandinavian and other European countries has decreased over time, leading to a situation, in 2003, in which seven out of every ten European households had a mobile telephone, except for France (66 percent). The German case is interesting because, in 2000, only 30 percent of households had at least one mobile telephone, while in 2003 this penetration rate had reached 73 percent. In other parts of the world,

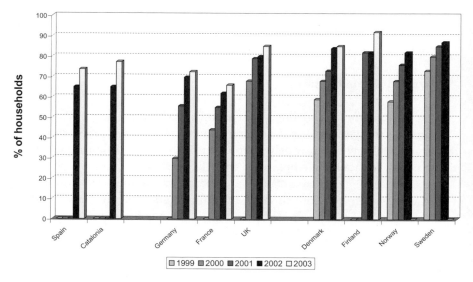

Figure 3.1

Household ownership of mobile telephones (% of households) in selected European countries. *Sources*: Spain and Catalonia: Household Information Technology Survey, National Institute of Statistics, Spain (www.ine.es), and own elaboration. Germany: Survey of Income and Expenditure, Federal Statistical Office, Germany (www. destatis.de). France: Permanent Survey of Household Life Conditions, National Institute of Statistics, France (www.insee.fr). UK: Consumers' Use of Mobile Telephony Survey, Office of Telecommunications (www.ofcom.org.gov). Denmark: StatBank Denmark, Statistics Denmark (www.dst.dk). Finland: Household Budgets Survey, Statistics Finland (www.stat.fi). Norway: Survey on Media Use, Statistics Norway (www.ssb.no). Sweden: Statistical Yearbook of Sweden 2004, Statistics Sweden (www.scb.se), and own elaboration. Source data in online statistical annex (appendix 63).

such as Latin America, the situation is slightly different because penetration is lower, but the growing trend is the same.

Crude as they are, the adoption data at the family level suggest that mobile phones are now pervasive, not only for work purposes or among rich families, but for people from low-income families as well (Oftel 2003b; Ureta 2004). In the United States, the Asian Pacific, Oceania, and urban areas in the developing world, including Africa (McKemey et al. 2003; Samuel et al. 2005), the trend is clear that mobile phones are becoming an increasingly integral part of everyday family life. While the reasons supporting the purchase of a mobile phone can be diverse, the main justification for the household is that the technology is effective in solving communication and coordination problems within the family (Frissen 2000: 65). In almost every part of the world, mobile communication has unequivocally demonstrated its usefulness for the coordination of daily family activities (Ling and Haddon 2001; Ling and Yttri 2002; Ling 2004). Therefore, although in Europe there was a period of unfavorable discourse about the mobile phone in the late 1990s (see Castells et al. 2004: 78–81), this had little observable effect on use during and after this particular period. And the general pattern remains: the busier the parents, the earlier the adoption of the mobile phone (Frissen 2000).

With the increasing use of the mobile phone for family communication, telecom operators around the world began to provide special packages targeted at this market from the beginning of this century. Two frequently used marketing strategies, seen in the United States, are to offer unlimited call minutes among family members or free long-distance calls after 9 p.m. As a result, an unprecedented phenomenon has emerged that almost all family members of a large number of households are now networked at all times. One of the results is that this makes remote care-giving to both children[7] and senior citizens more convenient. In this respect, it would be interesting to have more information about how elderly people use the mobile telephone and, particularly, whether there is any difference between countries in which family solidarity is high (for instance, Mediterranean ones) and those in which family ties are less intense.

For families, the most important function of the mobile phone is to provide micro-coordination. According to Ling (2004: 70):

[m]icro-coordination is the nuanced management of social interactions. [It] can be seen in the redirection of trips that have already started, it can be seen in the iterative agreement as to where and when we can meet friends, and it can be seen, for example, in the ability to call ahead when we are late to an appointment.

This can be done through voice calls or SMS with the same purpose of making adjustments to the activities of family members who are habitually traveling for work or other daily errands, such as going to the supermarket, picking up children from school and driving them to out-of-school activities. These journeys can be made by car, public transport, or even by foot. Although, in some cases, micro-coordination makes some trips unnecessary, in other cases it may generate more need to travel. Hence, as demonstrated in a study by Ling and Haddon (2001), mobile telephony is not significantly changing the number of trips a person makes, but it does allow the redirection of journeys that have already begun.

Thus, the availability of a mobile phone allows increased levels of efficiency in everyday activities thanks to the micro-coordination enabled by
perpetual contact. However, it must be added that the mobile telephone
can also be misused to decrease, at least partially, this efficiency. In this
sense, some activities of coordination which previously could have been
done with lower costs are, nowadays, more expensive in terms of both time
and money. This is because, besides the instrumental uses of information-
sharing and coordination given to the mobile telephone, the technology
is also widely used for expressive purposes; for instance, when calling to
say nothing but "I love you."

Another important aspect to highlight is the fact that, within the new
family relationships, the crisis of the patriarchal family leads to the weakening of traditional forms of parental authority and to the early psychological and social emancipation of young people, as we will explore in
more detail in chapter 4. Nevertheless, there are exceptions in certain social
contexts. Thus Yoon found that the appropriation of the mobile phone
reproduces the authority of the father in Korean families, where the
mother and the children are also attached to certain fixed places in this
structure of intimate relationships. As he writes,

the detailed mobile parenting is likely to be carried out by the mother, while the
father has more fundamental control of all family members. Some fathers do not
call their children directly, but call their wife at home and ask her to track down
their children by calling him/her: "It's unusual to have a call from my dad. If he
has anything to talk to me about, he calls my mom at home and then my mom
calls my mobile phone." This mode of contact is perhaps rooted in traditional patriarchal communication whereby fathers' contact with children is usually mediated
by the "domestic person," the mother. (Yoon 2003: 334)

We are, indeed, pointing not only to the relationship between adult
family members but also between sons/daughters and parents, and this
issue, as it explicitly concerns young people, will be thoroughly developed
in chapter 4. Keeping our focus on the whole range of family relationships,
the cultural tension between individuation and the maintenance of traditional collectivistic values centered on the family unit are described in the
Chinese movie *Shouji* (*Cell Phone*) released in 2004. What the film most
fundamentally reflects is the contradiction between the pursuit of individual pleasure and traditional social values, especially those attached to
the family. It depicts the mobile phone as a menace to society, showing
the device being used to deceive, cover up extramarital affairs, and tear
existing relationships apart. It reflects middle-class anxieties associated

with the new device, which, however, arise from the individual–collective tensions prevailing in Chinese communities before the diffusion of the mobile phone. In this sense, it is useful to highlight the fact that the introduction of new ICT – whether the Internet or mobile telephony – does not necessarily lead to an improvement in the quality of family communication, but reflects its conflicts (Fortunati and Manganelli 2004).

Finally, the family is also a place where invasions of privacy may occur, especially in the contemporary age when the family unit may exist in a variety of settings, from marriage to cohabitation, and include people of different sexual orientations. Thus, while spouses and partners increasingly use mobile devices to check up on the activities of each other, it is not uncommon for the new technological convenience to be abused. Katz (1996) recounts the example of a young man who accessed information on calls his girlfriend received on her wireless device and contacted male callers to warn them off. A Chicago woman reportedly left her boyfriend because she felt he was making too many calls to her cell phone to check up on her (Plant n.d.). On a more serious note, a Californian man was arrested for using a GPS-enabled cell phone to locate and stalk his ex-girlfriend, by attaching the phone to the bottom of her car (Associated Press 2004). Again, like so many technical innovations, the effects of mobile telephony are processed by the patterns of social relationships, often characterized by domination and conflict. Micro-coordination of the family goes hand in hand with the use of mobile communication to deepen the problems of families in crisis.

The Transformation of Sociability

The role of mobile communication in the private sphere extends beyond the family to the networks of friends, peers, and other social relations that together constitute the realm of sociability. Thus, in contrast with the initial perception of mobile telephony as a tool for working professionals, Grant and Kiesler (2002: 129) found that mobiles were used for both work and social purposes and that there was "a clear shift in work and personal communication in behavior settings." For example, there was more sending and receiving of personal calls in the work setting, and vice versa. Similar results were obtained by Palen and co-workers (2000) in their study of new mobile-phone users. As the sociability of young people is treated in depth in chapter 4, this section will deal with the aspects that affect the whole of society, especially the strengthening of intimate social networks and the creation of new social norms.

The "Full-Time Intimate Community"

Essential to the sociability function of mobile technology is the permanent and ubiquitous form of connectivity that allows mobile users to stay in touch anytime and anywhere in a habitual mode of communication. Based on her ethnographic fieldwork, interviews, and analysis of communication diaries in Japan, Ito maintains that this connectivity through *keitai* (handphone) is different from Internet-based connectivity because it is "a seeping membrane between the real and the virtual, here and elsewhere, rather than a portal of high-fidelity connectivity that demands full and sustained engagement" (Ito 2004: 17). The use of SMS adds to the uniqueness of this technology because "[u]nlike voice calls, which are generally point-to-point and engrossing, messaging can be a way of maintaining ongoing background awareness of others, and of keeping multiple channels of communication open" (Ito and Okabe 2005: 264). In this sense, mobile communication is "not so much about a new technical capability or freedom of motion, but about a snug and intimate technosocial tethering, a personal device and communications that are a constant, lightweight, and mundane presence in everyday life" (Ito 2004: 1).

The habitual use of mobile phones keeps alive what Misa Matsuda calls a "full-time intimate community" (cited in Ito 2004: 11). This finding was confirmed in a survey in Japan, which found that those who use the mobile Internet more frequently also spend more time physically with friends; and that "the mobile Internet serves distinctly different social functions from the PC Internet" (Ishii 2004: 57).[8] The key difference is, while high-intensity users of the PC Internet tend to spend less time with friends and families, heavy users of the mobile Internet are actually more active in interpersonal communications and socializing (Ishii 2004: 56). Mobilephone users are also found to have high disclosure of their subjective self because mobile subscribers tend to use the technology for close interpersonal relationships. This is consistent with earlier findings that mobilephone users are more sociable than non-users (Hashimoto et al. 2000), and that the use of e-mail via mobile phones enhances sociability among university students, both for women and men (Tsuji and Mikami 2001).

However, this "full-time intimate community" facilitated by the mobile phone differs from the traditional face-to-face community because of the ease with which one may maintain key interpersonal relationships across spatial distance at any time. While intimate relations may well be strengthened by increased communication (oftentimes ritualistic greetings and repetitive expressions of affection), on other occasions, heavy usage may also lead to the weakening of communal ties beyond the most intimate group of close friends (Ito 2004: 10–11). This is the phenomenon that Ichiyo Habuchi describes as "tele-cocooning," the production of social identities in small, insular social groups through mobile communications (cited in Ito 2004: 11).

To sum up, collectively oriented cultural tendencies play a part in the fast growth of the wireless Internet in Japan. Many i-mode services aim at sustaining and reinforcing existing social relationships. And the majority of e-mails exchanged by mobile phone are between people with intimate relationships (Ito 2004: 10–11). Researchers such as Barnes and Huff (2003: 83) thus believe that the rapid growth of mobile Internet use in Japan owes much to the normative beliefs in Japanese society that attach high value to interpersonal relationships.

Similar patterns are found in China and the United States. Focusing on SMS, a survey shows that this means of communication is used mostly to maintain personal relationships, such as chatting with friends, contacting family and relatives, and communicating with significant others, a pattern quite similar to the "intimate" mobile-based relationships described in Japan and Europe (Sohu-Horizon Survey 2003; see figure 3.2). In the United States, text messaging has also become a useful tool for developing and managing romantic relations, by helping to remove the awkwardness that comes with some face-to-face situations, which is also the case in Indonesia (Barendregt 2005). Even though SMS is in its early stages in the United States, text-message users are already using it in dating, as revealed in a study by AOL and Opinion Research Corporation (Greenspan 2004a).

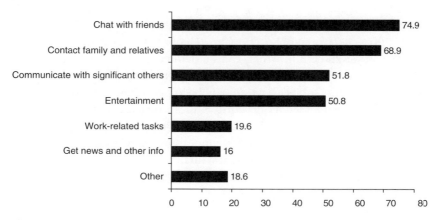

Figure 3.2
Uses of text messaging (SMS) in China. *Source*: Sohu-Horizon Survey (2003: 8).

This development is similar to Japan's online encounter service of *deai*, which allows people to meet virtually or in person, and is "strongly associated with issues of dating, companionship, sexuality and romance" (Holden and Tsuruki 2003: 34) for mostly heterosexual but also homosexual relations. Unlike earlier matchmaking and friends-making settings, "individuals [using *deai*] are able to operate in virtual isolation, freer of weighty social structure and claustrophobic external surveillance" (Holden and Tsuruki 2003: 35). Many *deai* services consequently become a hotbed of fraudulent dating, online pornography, and even open solicitation for prostitution (Holden and Tsuruki 2003: 37). Yet the combination of *keitai* connectivity and commercial operations need not necessarily lead to moral corruption. Holden and Tsuruki point out that *deai* may also afford "some of the advantages of the institutional orbit – namely trust and self-defense. In this way, dual benefits are provided: individually-established and managed social connections, as well as a modicum of security" (Holden and Tsuruki 2003: 35).

m-Etiquette

Not only do the mobile phone and the wireless Internet play a role in the integration and disintegration of communities, they also provide a generic social space in which collective practices become regularized and formalized, giving rise to social norms that shape future development in the social uses of the technologies. In some of these situations, new practices enabled by mobile technology may run into conflict with existing customs. In other cases, new social norms take shape either spontaneously or through the efforts of mobile providers, government authorities, and/or private business owners. This is known as the formation of m-etiquette (Lacohée et al. 2003: 207) or mobile etiquette (Licoppe and Heurtin 2002: 102; Crabtree et al. 2003: 38), which can be defined as the collection of rules that establish the public use of mobile telephony. Even though some handbooks on this subject have already been published (Haddon 2000: 5), we are talking about a set of non-written rules which are, indeed, still under construction (Katz and Aakhus 2002: 10; Puro 2002: 26, referring to Finland). These emergent norms define appropriate protocol in public places, such as libraries, theaters, restaurants, places of worship, public transport systems, classrooms, and even airplanes.[9] In this sense, we may talk about a social learning process about how to deal with permanent availability, and the increasingly inevitable intrusion of wireless communication, which has an important random component because part of the interaction (for example, incoming calls) is unexpected (Licoppe and Heurtin 2002: 99).

This social learning process, which can also be identified as part of the process of social appropriation or domestication, has two sides. The first one refers to manners: people have learned how and when a mobile phone should be used, though some old embarrassments no longer occur nowadays thanks to the generalization of various technical features. The design of new devices has evolved in the direction of etiquette, and vibration or silent modes are now commonly used. The second aspect concerns custom, especially in those countries with an average penetration rate greater than 70 percent in which, nowadays, everybody has become used to hearing a mobile phone ring in almost all situations. For instance, in a typical day in Barcelona, London, Los Angeles, Moscow, Hong Kong, or Tokyo, it is practically impossible not to hear a ring-tone or a third-person conversation via a mobile telephone.

Nevertheless, each country will define what are good manners and what are bad ones, according to cultural differences and the different diffusion patterns of mobile telephony in each country. Some of these norms are self-regulatory effects, while others are initiated by social institutions. And they are needed because, as members of the older generation might maintain, preexisting "manners seem suddenly to have evaporated in this era of perpetual contact" (S-D. Kim 2002: 65). Hence, in the United States, a bill has been considered in Illinois to have separate areas designated in restaurants for diners with cell phones (Georges 2001). Also in the United States, a study by Caporael and Xie (2003) found that respondents tend voluntarily to switch off their phones in certain public spaces, such as churches or concert halls, while they were not so concerned about the phone intruding into their interactions with friends and family.

A comparative study by Lassen (2002), carried out in London, Madrid, and Paris, provides interesting results about differences between countries when considering mobile phone use in public places. The use of a mobile telephone in indoor public places entails the possibility of being overheard and disturbing people around you (for instance, on the train or waiting in line). In the three cities, and also around Europe,[10] a common rule of mobile etiquette is to talk in a low voice as noisy users are considered the main negative aspect of mobile telephony in public.

In Japan, there are more restrictions in this respect. Japanese society has been quick to formulate new norms to regulate the use of the mobile phone when it affects others. For instance, public transport offers a particularly accessible time–space in which individual usage of *keitai* has been subject to strict collective control. According to McVeigh (2003):

In train and subway stations, one can see signs reading "Thank you for not using your portable phone." Other signs are posted warning how cell phones might affect pacemakers, as long as, in public transportation and classrooms it is considered necessary to reduce noise. In addition, many people won't answer their mobile phones in a crowded train, or even if they do so, they will speak in [a] low voice. (McVeigh 2003: 29)[11]

However, when it comes to texting, rules of m-etiquette are often less strict, as revealed in the following description of the situation in the Philippines:

Faye Siytangco, a 23-year-old airline sales representative, was not surprised when at the wake for a friend's father she saw people bowing their heads and gazing toward folded hands. But when their hands started beeping and their thumbs began to move, she realized to her astonishment that they were not in fact praying. "People were actually sitting there and texting," Siytangco said. "Filipinos don't see it as rude anymore." (Arnold 2000)

The school setting is also an important place for the fostering and negotiation of m-etiquette. In their study of a Tokyo school, Ito and Okabe (2003: 15) found that their student respondents would usually set their mobile phones to the "manner mode" (silent mode) in classrooms. Another important norm, at least among college and high-school students, is that "[b]efore initiating a call to a *keitai*, they will, almost without exception, begin with a text message to determine availability; the new social norm is that you should 'knock before entering'" (Ito 2003a).

It is important to emphasize that defining good manners is a dynamic process. According to Kasesniemi (2003: 34), in Finland a total ban on mobiles was issued in schools some years ago, but this has been revoked without exception, and replaced by more specific restrictions. Some schools have included a section on the use of mobile phones in their disciplinary regulations, while others have chosen to draw up a complete mobile phone etiquette. One result across the board is that, in general, mobile usage during breaks has been determined as acceptable but usage during classes is still forbidden. Now, of course, being connected during a class is seen as a symbol of "coolness." This last point about the Finnish case serves as a reminder that what *should* be done may differ from what actually *is* done. The formation of m-etiquette, like social norms in general, is not a fixed entity but an evolving process.

To sum up, it can be said that the transformation of sociability with the diffusion of mobile technologies is constituted by multiple processes that point to different configurations between the individual and the collective. However, all these configurations share one major feature: they strengthen networks of interaction, whether among families, peer groups, friends, or selective personal relationships. Mobile communication deepens and extends the logic of networked individualism that was identified some time ago by Internet researchers (see Wellman 2002; Wellman et al. 2003).

Personal Safety and Security

Consideration for personal safety is among the main factors behind the diffusion of mobile technologies, in both developed and developing countries, especially during the global "war on terrorism" which has heightened security from the US to the Middle East, from the UK to the Philippines.[12] In general, this consideration is more common among adults, especially the elderly, and also among those who were initially reluctant to adopt this technology (Ling 2004). Indeed, the notion of a mobile telephone as a lifeline is one of the central images of the device (Oksman and Rautiainen 2002; Ling 2004: 35).

In the United States, for example, among the main reasons for using a cell phone, according to research conducted for Cingular Wireless (2003), are convenience (60 percent) and safety (21 percent). This is supported by other empirical research, which found that most people had cell phones in order to be able to deal with emergencies (Wei 2004). However, once the cell phone is acquired, other use considerations come into play. Thus, Wei (2004) also found that, for those who use mobile phones primarily for social purposes (twice that of work-related calls), other activities, such as playing games, text messaging, and surfing the web, followed by e-mailing, were also prominent because using the cell phone initially for voice calls often activates other uses.

While communication technologies may become almost seamless aspects of society, sometimes specific events cast light on their critical functions to individuals and groups. This is why, in the United States, one cannot talk about wireless communication without referring to the September 11 attacks on the World Trade Center. Mobile phones played a critical instrumental and emotional role in the unfolding of events at that historical moment and were used to coordinate rescue efforts, report ongoing events on the hijacked planes, discover the status of families and friends, and say goodbye to loved ones (Palen 2002; Dutton and Nainoa 2003; Sharma and Nakamura 2003). In this respect, wireless communication was instrumental in reconfiguring access to information between people on different sides of the tragedy (Dutton and Nainoa 2003), sometimes with lifesaving consequences, as when cell-phone users trapped in the collapsed buildings were able to direct rescuers to their location.

Similar uses of cell phones, along with the Internet, were highlighted in the aftermath of the devastation brought by Hurricane Katrina to the Gulf Coast of the United States in September 2005. In some cases, a call from a destroyed home made it possible to locate survivors and to save lives. Thus, apart from its obvious practical uses for safety and security in normal times, access to wireless communication can now mean the difference between life and death at a time of crisis. In fact, a direct outcome of these events was the prioritization in policy circles of the need to establish effective wireless location capabilities. In the United States, the FCC had already mandated wireless carriers to enhance 911 capabilities by October 2001 (FCC 1999), although this requirement has now been relaxed somewhat to give operators more time.

Dutton and Nainoa (2003) also note how wireless communication facilitated the formation of flexible networks and enabled users to bypass formal or hierarchical channels both inside and outside the hijack situation of 9/11. The flipside of this, the authors note, is the observation that the only location where there appear to have been no calls in or out, was the Pentagon, an illustration perhaps of how social and institutional environments can inhibit people's ability to bypass hierarchy even when the technology to do so is available. On the other hand, cell phones were also used to facilitate the activities of the hijackers, highlighting the "double edged sword of communication," which can be used for both positive and negative ends (Dutton and Nainoa 2003: 242).

Wireless data services were also shown to be particularly useful in times of crisis when voice networks are unavailable. As wireless voice networks became saturated, e-mails, instant messaging, and SMS became more useful, mainly for people to check up on or offer emotional support to friends and loved ones (Sharma and Nakamura 2003). Wireless data and messaging networks were also critical for the continued operation of public and private organizations in the aftermath of the attacks. Blackberry devices, already popular in the New York business district, were particularly prominent (Wolfensberger 2002; Sharma and Nakamura 2003). Apart from September 11, when this capability was demonstrated, mobile Internet users reportedly used moblogs (mobile blogs) extensively during the August 2003 blackouts in the Midwestern and Northeastern United States to record and distribute pictorial information about their experiences (Srivastava 2004b).

The high profile of cellular communication associated with the events of September 11 leads one to expect a noticeable increase in the purchase of cell phones in the ensuing period. Most discussions of the events conclude that cell phones have subsequently acquired increased significance both in private lives and public policy. For example, Dutton and Nainoa (2003: 243–244) state:

In the aftermath of September 11, the wireless industry experienced a major boost. Stocks rose. More phones were sold. More minutes were billed. People and the press began to talk about the cell phone as a "lifeline" in the case of an emergency, for example with some schools in the US giving cell phones to teachers and lifting bans on students having cell phones on their campuses. To some degree, this emergency role was a factor in cell phones' early diffusion, but the rapid expansion of colorful covers, sharp designs and ubiquitous use enabled notions of fashion, conviviality and easy contactability to define the cell phone more as a luxury or necessary everyday social and business aid – until September 11.

After September 11, 43 percent of respondents in a survey said that they now felt safer with a cell phone than without one (Genwireless 2001). Moreover, safety was already one of the main reasons why people get a cell phone for themselves or their children (Katz 1996; Selian 2004).

However, we should refrain from claiming a direct relationship between safety concerns and the rise in demand for cell phones. Data on subscriptions after 2001 indicate that the influence of September 11 may not have been that great. The US wireless-phone market grew by 25.5 percent between 1998 and 2002, but this growth level is still lower than predictions made even before the tragic events of September 11 (Euromonitor 2003). The growth figures show that, after a period of high growth, the rate of growth of cell-phone subscriptions actually slowed drastically after 2001 (for example, subscriptions grew by 14.8 percent between 2000 and 2001 and by only 9.4 percent between 2001 and 2002, although they began to rise somewhat in 2003). This is confirmed by data from Henry Fund Research (2004), which shows that the rate of growth of the mobile-phone market has been decreasing steadily since 2000.

Economic reasons might have contributed to the slowing down of subscription growth in this context. On the other hand, just as the use of telecommunications has been critical in other disasters (for example, earthquakes) without having a lasting effect on perceptions of the technology (Dutton and Nainoa 2003), the initial association of the cell phone with safety and security has taken second place to the values of sociability, business efficiency, and personal expression.

However, real-world events continue to appear in which mobile devices help enhance personal security. In Europe, car-accident victims were saved thanks to the availability of a mobile telephone and people were able to report to their dearest ones the dramatic situations in which they were involved (Agar 2003; Ling 2004). In the United States, a teenage boy was reportedly able to effect the capture of a potential molester by taking a picture of his attacker's license plates and transmitting it to the police (Associated Press 2003a). In China, people trapped under collapsed buildings were also able to use the mobile phone to contact rescue workers (Ji 2005). Although the scale of these incidents is not comparable to September 11 or to Hurricane Katrina, the multiplicity and continuation of such lifesaving uses across the globe demonstrate again the close integration of mobile technologies within our everyday lives.

Public Services via Mobile Devices

Mobile technology is often seen as a personal device operating in the private sphere and a business instrument used by commercial enterprises. Part of the reason is that, comparatively speaking, the public sector has not sufficiently utilized mobile technology despite some progress in m-government, mobile health services, and special applications designed for the disabled. The public service potential of mobile technology is, therefore, far from being fully realized. Of course, the division between public and private should not be taken as impermeable. While public sector entities do play a central role in shaping the technology for the needs of public services, commercial players can be active participants too as long as they are matched with the right market niche. One such example is a Los Angeles company which has launched a service that can send drivers a text message reminding them to move their car from a restricted zone and patients can also get reminders to take their medication (Miller 2003). At the more fundamental level of access provision, nonprofit community organizations may also help enhance wireless accessibility to the general public through such means as the free wireless movement[13] and warchalking (see, for example, New Media Institute 2004).

m-Government
m-Government is an extension of e-government so that citizens can access public information, obtain government services, and/or become involved

in public administration processes using their mobile phones, PDAs, Wi-Fi computers, and other portable devices. This was the case when Korea initiated its "m-government" project in November 2002 and the "u-Korea (ubiquitous Korea)" plan to be completed by 2007 (Yang 2003). In the Philippines, the National Bureau of Customs attempted to use mobile phones in innovative ways to streamline the payment of duties during February 2002 (Hachigian and Wu 2003: 88).

Most m-government experiments, however, are conducted at the local level; for example, in several local states in Japan and Spain. City governments such as Sagamihara in Kanagawa Prefecture, in the southern part of Tokyo, launched an m-government experiment in April 2004 which allows users to report damage or defects they find on the streets and public signs by sending pictures from their camera phones (Suzuki 2004). The same service is offered in the town of Jun (Granada, Spain).[14] Another example is in Nakashibetsu, a farming community on the eastern tip of Hokkaido, Japan's northernmost island (Holden and Tsuruki 2003: 37–38). As a result of agricultural mechanization and modernization, local young people, especially women, have been leaving the area to live independently or to enter college. So the city of Nakashibetsu started to use mobile *deai* to promote matchmaking, which has proved to be quite successful. In this case, the wireless Internet, including chatting via i-mode, was used to select singles to attend "the biannual, three-day pre-marital mixers" (Holden and Tsuruki 2003: 38). In Spain, several municipalities, large and small, provide services through the mobile phone, usually comprising relevant local information and events.[15] The amount of information provided varies from city to city. In some cases, simple but efficient SMS services are provided, for example, by the city council of Barcelona, to keep citizens informed about the processing of their applications, such as the permission needed before undertaking a large house renovation project.[16]

Another notable type of public service has to do with equipping transport vehicles with smart mobile devices, which is a major point of development in Europe. These include the first European vehicle which made its experimental trip in 2004.[17] In Germany, a formidable project, which was impossible without wireless devices, introduced a toll management system via satellite on German motorways.[18] In this respect, it should be noted that the development of mobile communication technology moves more rapidly than the automobile market. As the life cycle of a mobile phone is usually shorter than that of a motor vehicle, auto makers have been hesitant to install state-of-the-art wireless devices in their products. It is also difficult to establish standards for car-to-car communication, which, for example, explains why the adoption of mobile technologies has been delayed in Europe's transport systems.[19]

m-Government experiments are not limited to more developed economies like Japan and Europe. They also exist in less-developed regions such as Africa, where the technology has been used for a range of purposes including public health and voting. A study found that handheld computers were useful for the collection of health data and the dissemination of information in Ghana, Uganda, and Kenya. The researchers conclude that, with appropriate training, technical support, and content, they would be appropriate for the African environment, although the technology might still be too expensive for ordinary people there (Bridges.org 2003). On the other hand, in Senegal, mobile phones were used extensively by poll watchers to report on-the-spot vote counts to FM stations, making it impossible for the ruling government (which ultimately lost the election) to contest the results (Ashurst 2001). Similar events have taken place in the Ghana elections of 2000 and 2004 (personal observation).

Alternative Modes of Access: The Free Wireless Movement and Warchalking

In this section our interest is focused on how Wi-Fi technologies bring alternative modes of access to the Internet. Wi-Fi was initially conceived as a wireless alternative for short-range connections between computers within homes and offices. However, it soon became clear that Wi-Fi could also be used to extend the reach of computer networks into public spaces. Moreover, both equipment vendors and wireless enthusiasts realized that, with the appropriate hardware and clever tinkering, point-to-point connections could be made over several kilometers (Bar and Galperin 2005). In this sense, the key fact is that Wi-Fi follows a different business model from traditional telecom and broadband services. Because the network grows incrementally with every new access point and every device capable of receiving Wi-Fi signals, there is no need for incentives to convince a monopoly provider to build expensive infrastructures (Werbach 2002).

Bar and Galperin (2005) highlight the three main factors that explain the successful popularization of Wi-Fi technologies. First of all, Wi-Fi can deliver high bandwidth without the wiring costs. Second, widespread support for standardization, coordinated through the "Wi-Fi Alliance," has contributed not only to lower prices but also to compatibility among Wi-Fi clients. Third, the lack of regulatory restrictions means Wi-Fi networks can operate almost everywhere.[20] In this context, Wi-Fi finds its development through noncommercial actions by individuals and collectives as well as small-scale commercial activities (Sandvig 2003). These are decentralized models of wireless broadband deployment (Bar and Galperin 2005). Our interest, here, is centered on noncommercial initiatives, even if they may be promoted by the public sector or respond to a grassroots movement.

In some instances, governments, and particularly local administrations, are promoting new modes of access to enhance citizens' communication and information capabilities in order to help their citizens overcome the digital divide. Thus, city councils support wireless connection in response to social demands, sometimes for free but usually for a fee. This helps to cover the needs of the institution itself by bringing mobile communication services to municipal employees, public utilities, public transport, and so on. The number of cities deploying wireless broadband networks has been growing very quickly in recent years. Their scale, architecture, and business model vary widely (Bar and Galperin 2005). But by July 2005, there were 88 city and regional wireless broadband networks (38 of them in the United States), while there were 32 citywide networks used for municipal purposes (28 of them in the United States) (Vos 2005). There is also a total of 37 city "hotzones" around the world (22 of them in the United States).[21]

These networks can be found in both big cities and small ones. In the particular case of the United States, it has become increasingly controversial for larger cities to build metropolitan area networks (MANs) which would cover entire municipalities. Incumbent operators have sought legislation to block municipal Wi-Fi projects, and the debate over the proper role of local government in providing wireless broadband services is continuing (Bar and Galperin 2005; Vos 2005). It is important to emphasize that there are communities underserved by existing broadband operators where the role of local governments is of the utmost importance both in developed and in developing countries.[22] In these cases, telecom companies do not appear to have any problem with the public initiative as long as they are not forced to provide services without making a profit.

There are also bottom-up social initiatives, such as the free wireless movements that are particularly strong in Europe, but also present in the United States. This is when different communities voluntarily create areas of free Wi-Fi connection and produce directories of them that can be consulted, for instance, on the Internet. These kinds of initiative can be found in big cities and smaller towns, and aim to create independent networks that are free for public access. Nevertheless, according to research conducted by Sandvig (2003), members of the free wireless movement tend to have high socioeconomic status and live in rich countries. They are not just ordinary people because you need to have sufficient technical know-how and usually over a thousand dollars of personal equipment to participate in a free open wireless community (laptop, wireless card, and so on). In this sense, these networks are more like user groups rather than community networks in the more inclusive sense. One may also see them as a way of providing free access or access to an inexpensive service for the rich.

In technical terms, a free wireless "node" is a hotspot that brings wireless access to the Internet. According to information published by NodeDB, there are 7,408 nodes in Europe. The leading countries in the development of this infrastructure are Greece (5,446), followed by Spain (1,020), Germany (353), and France (172).[23] This social movement may have the support of some local authorities, despite the legal and regulatory problems in providing free Internet access. As a result, a number of Spanish municipal governments have been brought to court because of their sponsoring of free Wi-Fi access.[24]

An interesting aspect of the development of Wi-Fi, as previously stated, is that, while the industry is still trying to work out a viable business model for providing Wi-Fi services, the technology is taking on a community-oriented character. In the United States, for example, there are 38 Wi-Fi clouds and 16 Wi-Fi zones throughout the country, most of which have been set up to enhance the value of communities rather than to generate revenue (New Media Institute 2004). While not all public Wi-Fi systems are free, some communities have made it their objective to make it a free service. This is best exemplified by warchalking activists, who use chalk to draw signs on the wall in order to indicate, publicly, where an open Wi-Fi connection is available.

Services for People with Disabilities

A recurring theme in discussions about mobile communication is that this new technology can revolutionize the lives of disabled people by facilitating greater mobility, independence, and autonomy (see Abascal and Civit 2000; Baker and Bellordre 2003; *Technology Review* 2003/4; Peifer 2005). This expectation is borne out in accounts of various projects under way, involving both public and private entities, to develop new mobile devices and adapt existing ones for people with various physical and mental challenges. People with disabilities, for example, have been found to be more interested than the general public in using Internet-enabled mobile phones (mainly for directory service) and in mobile commerce (Coutts 2002). A study of disabled women in Australia found that the two telecommunication tools they desired the most were Internet access and mobile phones (Women with Disabilities Australia 1999); a study of disabled adults and elderly people in the United States reported that 87 percent of elderly respondents used their mobile phones for emergencies (Mann et al. 2004).[25] Mobile communication technologies have also facilitated improvements in dyslexic children (Skog 2002).

At the same time, some commentators refer to a "disability divide," that is, the gap between those who can effectively use new information and communication technologies and those who cannot (Baker and Bellordre 2003). As mobile communication increasingly becomes an integral part of day-to-day life for most individuals, it ironically becomes more difficult for certain types of individuals to use it (Watanabe 2001). Disabled users want the same thing as the general population – reliable and usable personal communication (Abascal and Civit 2000). Their unique functional needs are, however, often not taken into account when devices are being designed for the general market. Accessibility to communication technologies, including mobile phones, has thus become a rights issue for the disabled community, with some individuals or groups even suing mobile-phone operators for failure to make their devices accessible to disabled people (see, for example, Silva 2003, 2004).

Challenges faced by disabled people in using mobile phones, for instance, include displays that are difficult to read, buttons too small to press, audio difficult to hear, and features that are too complicated to understand (Women with Disabilities Australia 1999; Goggin and Newell 2000; Prometeus 2002; Baker and Bellordre 2003, Howard 2004; Mann et al. 2004; Peifer 2005). One of the more critical problems for people who use hearing aids is that mobile phones emit a high level of magnetic interference that obstructs the performance of hearing aids. In addition, communication tends to be more expensive for disabled people, either because it takes longer to translate from speech to text or vice versa, or because disabled users have to pay for services they do not need (for example, a deaf person has limited use for a voice service but has to sign up for this in order to have the text-messaging capability), or because devices specially designed for disabled people are more costly than regular devices (Abascal and Civit 2000; Harper and Clark 2002). Remedies being implemented to address this situation include cheaper SMS pricing for disabled communicators, and SMS-only packages for those who do not require voice telephony.

Even in their current form, however, it has been possible for some features of mobile technologies to be adapted to the needs of people with particular disabilities. For people who are deaf or hard of hearing, the primary voice functions of mobile phones are largely irrelevant, but text messaging has become a valuable remedy to their communication challenges (Prometeus 2002; Kasesniemi et al. 2003). Studies in Australia showed that 50 percent of deaf people sent at least one text message a day, and 30 percent sent up to ten text messages a day, a usage level about ten times higher than the general population (Harper and Clark 2002). A toy manufacturer in Cape Town uses text messaging to communicate with deaf employees (Samuel et al. 2005), and 50 deaf students in public schools in Toronto are using two-way pagers to stay in contact with teachers, friends, and family (Media Awareness 2005). Video-MMS (multimedia messaging service) can also be used by deaf people to communicate in sign language (Kasesniemi et al. 2003). There are also examples of improvements in handsets which, although not originally designed for these purposes, have proved to be very useful for the deaf; for instance, vibra-call.

Visually impaired people, on the other hand, cannot as easily find adaptable features in mobile phones. For instance, blind people have more problems with mobile phones than they have with traditional wired telephones because interfaces are more complicated. Visually impaired people also cannot read e-mail using i-mode because this system has no screen reader (Watanabe 2001). There are, however, other ongoing efforts to design special devices with buttons inscribed in Braille for blind people.[26]

Some trials have shown that existing off-the-shelf devices are capable of providing access to the telecommunications network for people with even severe disabilities (Garrett 2004). However, it is necessary to match the right technology to each disabled individual. Mobile communication technologies may open up options for people with limited mobility as a result of disability. Although there is anecdotal evidence of the utility of mobile communication to people with disabilities, at present the extent to which these technologies are accessible to the wider disabled community appears to be limited as a result of technical and design issues.

Consumption, Entertainment, and Fashion in Wireless Communication

Just as accounts in the popular media, academic research and marketing reports often associate mobile communication with consumption and entertainment. This is not surprising because early adopters of the technology tend to have higher socioeconomic status, and they include large numbers of young people who are in search of their own lifestyle, often under the influence of the commercial media. In this sense, the mobility and portability of the new technology make it qualitatively different from most other communication technologies, including the PC-based Internet, because the device is lightweight, "personal, portable, pedestrian" (Ito et al. 2005), and has become part of everyday clothing.

"Wearability" makes the mobile phone an item of fashion, ready to be personalized to reflect the identity of the owner. The built-in capacity for customization is a major breakthrough which allows users to play a more active role in the shaping of this particular consumer culture. We would therefore argue that, although corporations around the world are using wireless devices to extend their existing channels of promotion, service delivery, and payment, individual users and the choices they make are, in fact, the ultimate determinant in the appropriation process of the new technology. In so doing, mobile technologies are woven into the everyday behavior of consumers, sometimes by formerly neglected social groups, such as women, in ways not envisaged by manufacturers or service providers. This is because users are no longer just users; they are also producers or "co-creators" (Katz and Sugiyama 2005: 79).

Consumerism as Usual?
User participation influences mobile-enabled consumerism. Entertainment and personal identity can be shaped as a result. But before undertaking an analysis of the interaction between consumption, identity, and mobile-phone use, two points must be emphasized. First, although mobile technologies can be personalized, not all users customize the device to suit their needs and tastes, which means that there are still a significant number of "passive" users who take the wireless device as an "off-the-shelf" item (Katz and Sugiyama 2005: 79). Secondly, corporations have been quick to realize the potential of the new technology and they are among the first to actively engage in the promotion of the mobile device as a consumption item. Overall, the adoption of wireless technologies as a consumer good is a dominant trend in most parts of the world. However, in the United States, the corporate sector accounts for a larger share of its mobile market (ISP Planet 2001; *Revolution* 2003), especially with regard to mobile data services (Sharma and Nakamura 2003). This is because wireless devices and applications were initially envisioned in the United States as business tools (Katz 1998; Standard and Poor's 2003), and the industry has focused mainly on the corporate and upscale market (Collins 2000; Dano 2004). But in most other societies, the consumer market has quickly discovered wireless technology and adapted it to its purposes, leading to the involvement of corporations in the consumer market of mobile technologies.

A prominent example of the consumerist development of the mobile is Japan, where the social uses of the mobile phone center on a set of consumption practices, such as the purchase of latest-model handsets, playing online games, and using m-commerce services offered through the i-mode network. As the latest culmination of the country's "Gross National Cool" (McGray 2002), this consumer culture "has become a staple of the faddish, mobile, mediated, gadget-centered, youth-oriented, licentious lifestyle of contemporary urbanized Japan" (Holden and Tsuruki 2003: 34–35).

To a great extent, the predominance of consumerism has been exaggerated to signal the rise of Asia's "new rich" (Robison and Goodman 1996), not only in developed economies such as South Korea, where users can conduct banking, e-signature, and the purchase of small items based on the mobile phone (Lipp 2003), but also in less developed ones such as the Philippines, where the residents of Manila were reported to have experienced a "mobile mania," especially using texting (Rafael 2003: 404–405; see also Arnold 2000; Strom 2002). Like the Filipinos, Chinese youngsters are most active in using such services as SMS owing to its faddish appeal and much lower price than voice telephony (Sohu-Horizon Survey 2003).

One result of the diffusion of mobile technology is therefore the blurring of the boundary between commerce and everyday life. Viral marketing, for example, is to spread product and service information via interpersonal networks using material and/or reputational incentives. The technique was practiced on the Internet (for example, through e-mail) but gained more momentum with the rise of mobile telephony and messaging services. On the other hand, it is also reported that a high proportion of mobile users in Japan click through banner ads and e-mail ads – 3.6 percent and 24 percent respectively – compared to less than 0.5 percent for PC-based online banner ads (Enos n.d.). In Korea, large corporations are having promotional music sung by popular stars which is then circulated on the Internet and through mobile phones. The Samsung-sponsored music video "Anymotion" was downloaded 3.1 million times onto mobile phones at the price of US $2 a copy, leading one author to conclude that "[i]n Asia, it's nearly impossible to tell a song from an ad" (Fowler 2005).

In the context of developing countries, where credit cards are not fully diffused, multinational corporations are using mobile payment instead. In Zambia, Coca Cola distributors avoid the problems of cash dealings by making payments via text messaging (*The Economist* 2005a). Coca Cola also launched the Coke Cool Summer contest in China during July and August 2002, which generated 4 million SMS messages in 34 days (Cellular-news

2002). Meanwhile, the expansion of the mobile market provides China's three main Internet portal sites (Sina, Sohu, and Netease) with a new business opportunity by allowing them to profit from subscription-based text messages, ring-tones, and image downloads (Clark 2003). Using Internet websites to attract subscribers, then delivering content via SMS, and finally collecting fees as part of the phone bill, the three dot-coms have developed a new business model that has brought them new prosperity. To increase SMS circulation, these content providers also hire a team of *duanxin xieshou* (SMS authors) who put all their creativity into writing jokes, hoaxes, erotica, and congratulatory greetings that are crisp, concise, and fleeting in order to actively target this consumer market (Long Chen 2002: 39).

Yet it is still too early to conclude that business has mastered mobile technology as a means of shaping consumer behavior. The popularization of the new technology has also led to collective action by subscribers; for example, in complaining about the cost of mobile telephony. Different protests have emerged in Italy and Spain, where some consumers decided not to use their mobiles for a whole day in order to transmit their message to the telecom operators. In France, similar protests led to a reduction in SMS prices in 2004.[27] These consumer rights campaigns, mostly organized through the Internet, show that mobile telephony is perceived as a basic service in people's lives, at least in large parts of Europe.

Mobile Entertainment

Entertainment is a fundamental dimension of the media world. However, it becomes a new reality when applied to telephony. The notion of telephone entertainment was very rare until recent times, and it was restricted to activities like phone sex or fortune-telling services. Thus, the emergence of mobile entertainment signals a substantial difference between mobile telephony and traditional telephony. Indeed, with the incorporation of Internet access, and the fast development of audiovisual capabilities in mobile communication devices, mobile entertainment is a key new area of business, technology, and social practice, though it is an area about which we still have very scant, reliable information, beyond the usual hype.[28]

When talking about mobile entertainment, we are referring to:

entertainment products that run on wirelessly networked, portable, personal devices. "Mobile Entertainment" is a general term that encapsulates products like downloadable mobile phone games, images and ring tones, as well as MP3 players and radio receivers built into mobile handsets. The term excludes mobile communications like person-to-person SMS and voicemail, as well as mobile commerce applications like auctions or ticket purchasing. (Mobile Entertainment Forum 2003: 2)

The major components of mobile entertainment include mobile gaming; media content consumption (icons, ring-tones, music, images, movie clips, adult services, gambling, and so on); chat; information services (events, weather, news); and location-based services such as "where is my nearest." While gathering "value-added services" that bring extra profit to companies (content creators and/or telecom), the concept of mobile entertainment may also reduce or exclude peer-to-peer communication, which is in fact the main function of any phone, as best exemplified by the now classic Snake game on Nokia handsets. We will explore the mobile phone and gaming in more detail in chapter 4 because the overwhelming majority of mobile gamers belong to the young generation.

Indeed, entertainment is quickly becoming an important function of mobile communication.[29] This trend is largely technology-driven, as manufacturers learn to pack more capacity into the device, and as providers are eager to offer new services and products to expand the market.[30] It is also conditioned by the demographics of the mobile marketplace, dominated in most places by young people, who are more likely than older generations to be attracted to mobile entertainment.

However, the entertainment function does not preclude other uses for mobile communication devices (Moore 2003). Work-related activities and personal interaction continue to be paramount in the use of the mobile phone. Thus, rather than moving toward domination by the entertainment function, what we observe is the growing multipurpose of mobile communication devices.[31] The new communication system is characterized by its ability to switch from work to sociability and to entertainment in the same time and space. The user-centered structure of the communication network means that all these dimensions of life are constantly installed in the practice of the individual, and that it is his/her choice or availability that determines the exact mix of the various practices integrated into the mobile communication device. Therefore, the most successful devices will be those whose technology and feeder services system allow the user the maximum range of choice and mixing of the various functions. This explains the importance of having enhanced image and audio processing and transmission capacity. Mobile communication devices are the multipurpose, multi-channel connecting points in the network of communication of which everybody becomes a personal node. Entertainment is therefore not a specialized function, but an optional practice integrated in the time and space of the overall range of social practice. The spatial and temporal separation between work and leisure is superseded by their coexistence in the mobile communication networks.

The use of mobile devices for entertainment purposes is also affecting the entertainment industry, as products are newly packaged for their consumption in the new format. This repackaging is both cultural and technological, and is highly related to the adoption process (Skeldon 2003; Wiener 2003) of mobile communications. And, of course, major names such as Disney and the Sony Corporation are directly entering the mobile space, as well as licensing brands, titles, and artists for use in the mobile space (Wiener 2003).

The development of mobile entertainment, in any case, will depend on the regulation of this sector (Skeldon 2003; Wiener 2003), bearing in mind that those regulations affect both the content and the final price of the entertainment service provided through the mobile telephone. With regard to content, significant growth is expected in the mobile adult services sector based on text and images (Skeldon 2003). This trend will probably require new content regulations. Other hurdles this sector may face are device limitations, spam, and negative public perception (Skeldon 2003).

On a related trend, besides developing games, music, and other content specifically for the handset, the motion picture industry is also beginning to draw on people's experience with the new technology as raw material for their big-budget entertainment products. There are, therefore, movies such as New Line Cinema's 2004 thriller, *Cellular*, starring Kim Basinger. The Chinese film *Shouji* (*Cell Phone*), as mentioned above, topped box-office records in China during the Lunar New Year of 2004. And there are a series of horror movies from Southeast Asia to Japan and South Korea that use the mobile phone as a key link in their storylines (see below). The entertainment value of mobile communication is therefore not restricted to the handset and the content and services that can be delivered to it, but is of a more general scope that overlaps with the entertainment industry at large.

Fashion and Identity: Users as Producers of Meaning

The mobile phone is often personalized through a process shaped, to a great extent, by the user. This differs from the traditional mode of customization done by product or service providers because there are unlimited ways in which one can change the handset by putting on new ring-tones or setting one's own digital pictures as the wallpaper of the phone. While some researchers see personalization as an opportunity to promote personal autonomy and identity, others regard it as strengthening the trend toward individualization and making people more self-centered (McVeigh 2003: 24–32). Whatever the evaluation, it is clear that mobile technologies have become closely involved in the processes of personal identity construction, and that the mobile phone is not only a utilitarian tool for communication but also "a miniature aesthetic statement about its owner" (Katz and Sugiyama 2005: 64).

As today's wireless technologies are portable and wearable, like a watch (Ling 2001; Fortunati and Manganelli 2002; Oksman and Rautiainen 2002; Kasesniemi 2003), it is only natural for mobile devices to become fashion items with all kinds of decorative, expressive, and symbolic functions. This is, in fact, a cross-cultural phenomenon: empirical evidence from the United States, Japan, Korea, Namibia, and Norway have all indicated the transformation of the mobile phone from a pure communication instrument to a centerpiece of self-awareness and public display (Cohen and Wakeford 2003; Katz et al. 2003; Katz and Sugiyama 2005). Hence, people "wear" mobile phones as a new addition to their clothing (Fortunati 2002a: 56). From Europe to Japan and China, all kinds of accessories are available to personalize the device (Oksman and Rautiainen 2002; Skog 2002; McVeigh 2003; Yue 2003). This is part of the process of market differentiation because, as the technology becomes omnipresent, "there has been a figurative arms race toward ever more lavish mobile phones," such as the US $26,000 Vertu handset made in London, which has platinum casing and a sapphire crystal screen, and the TCL jeweled phones, some studded with diamonds, which had a sales record of 12 million units between 2001 and mid-2003 (Katz and Sugiyama 2005: 74–75).

Users can also personalize their gadgets with the content and services provided by websites targeting mobile Internet users – not all types of content and services though, but only those that can attract users' immediate attention and therefore the most convenient to be commodified. This is most obvious in the case of ring-tones, which can be downloaded to mobile handsets. This is a good example of how a market niche can be exploited to generate revenue, as shown in the United States when, in 2003, mobile users spent US $80–100 million on ring-tone downloads (Tedeschi 2004). While the bulk of commercial ring-tones are based on existing music, ranging from classical pieces to popular songs, users can also program their own ring-tones. Speeches can also be re-packaged into ring-tones, such as the alleged secret phone-call made by Philippines' President Arroyo, which was circulated in the country as a way of protesting her abuse of power (Robles 2005).

When users are empowered to take the role of producers, at least in part, they can collectively change fashion, which can then lead to changes in well-established companies. It was because of this process that Nokia, the world's largest mobile-phone manufacturer, changed its handset design strategy and introduced its first clamshell phones as a result of changes in consumer preferences (Reinhardt et al. 2004). This signaled a market evolution and handsets began to be available in ways that allowed the end-user to change some aesthetic elements, which, as a result, created a new fashion. This is one of the many examples that can be given of the mutual influence between the creation of individual identity and the formation of fashion when we examine the phone, whose adaptability to personal preferences far exceeds that of other wireless technologies, whether pager or Wi-Fi laptop.

Moreover, a common perception is that the more a provider can cater to the needs of personalization, the more commercially successful it will be, and NTT Docomo is often used as an exemplar. As Ms Matsunaga, a key leader of Docomo's marketing force, articulates:

For me, i-mode is a declaration of independence. It's 'I' mode, not company mode. That's the message I wanted to deliver: this is me in individual mode. Japan's system of lifetime employment, which always meant you had to live your life for the company, is crumbling. The 'i' in i-mode is about the Internet and information, but it's also about identity. (quoted in Stocker 2000)

Indeed, it is a global trend that considerations of identity have been deliberately incorporated into the design and promotion of new mobile phones and wireless services. R&D taskforces worldwide have routinely consulted women and teens to discover their cultural needs in using mobile phones. The findings are integrated into gadget/service design, which then goes through experiments or trial use by members of the target social group for further improvement, as, for example, in Nokia's research on consumer needs in India (Konkka 2003).

Despite corporate efforts to cater for consumer needs, the process of personalization and identity formation remains ultimately a matter of personal decision and choice. Although NTT Docomo, arguably the most successful mobile Internet provider in the world, is often credited for the diversity of its information and service provision, some evidence suggests that a much larger range of content is provided to i-mode users outside the official websites contracted by Docomo[32] even though the official ones are free, while the unofficial ones charge a fee. The question of cost, in Japanese terms, does not explain the difference in usage. This would imply that the range of services wanted by Docomo users is much broader than the uses imagined by Docomo marketing planners, as clever as they are.

Nevertheless, some of the applications that were envisaged as the most promising have turned out to be failures. A case in point is the e-book. A reason for this could very well be that the printed book is already mobile, so that there is no need for a more expensive alternative with very few positive distinguishing features (MGAIN 2003a: 14). This illustrates a fundamental point in our analysis: new technologies are not adopted because they are new, but because they make new uses possible, and new services that are unavailable or more difficult otherwise.

Mobile Communication in a Troubled World: Technology, Risk, and Fears

Mobile communication, owing to its ubiquity, accessibility, and adaptability, permeates all domains of life. Consequently, its diffusion has triggered a whole range of social concerns from anxiety about the fast pace of life to issues of public etiquette and from the blurring of public/private and work/personal boundaries to dangerous driving and the health implications of wireless technology. This section looks at these concerns by examining, first, the worry that mobile-phone use may cause more traffic accidents and more health problems. We will then look at issues of mobile spam, worms, and viruses, and the inappropriate use of the camera phone, in conjunction with the public and private solutions being developed to regulate its usage. This will be followed by discussion of surveillance via mobile devices, and the relationship between wireless technology and the sex industry.

Some of these concerns are more real than others, such as the abuse of the camera phone by "happy slappers" in the UK. Nevertheless, it is important to bear in mind that, more often than not, the perceptions of threat are exaggerated in narrative accounts from both media coverage and interpersonal channels. Yet, as the history of communication technology shows, public perceptions contribute significantly to policy-making processes and to the formation of usage patterns in the long run. We will therefore discuss, at the end of this section, media representations of the mobile phone as a technology of fear and alienation, as exemplified by a series of popular horror movies in recent years.

Dangerous Technologies?

Dangerous driving and emission-caused health risks are among the most widespread concerns publicized by the print and broadcasting media, as well as by e-mail, SMS, and face-to-face interactions. The risk is more clearly established when it comes to driving and using the mobile phone at the same time. For example, about 44 percent of Americans have a cell phone in their car and, although a majority of them recognize the danger of using a cell phone while driving, they say they still do it (Selian 2004). To a large number of them, this practice is justified because time expended at the wheel is perceived as lost time, and thanks to the perpetual contact provided by the mobile phone it can be used for other purposes (Ling 2004). This argument may make sense to those who are caught in a traffic jam everyday. But as so many car accidents are directly attributable to mobile-phone usage, the policy debate in various parts of the world is leaning toward seeing it as a life-threatening activity. Several US states have passed laws banning the use of handheld devices while driving, as has New York City. This is also happening in Europe where the European Union instructs incoming tourists that "using a mobile phone while driving . . . is either explicitly or implicitly forbidden in all EU countries."[33] Alternative solutions are being considered, such as using headsets. Meanwhile, the legal provisions are leading to changes in technology design, such as the development of more convenient hands-free devices, voice-activated dialing, and integrated voice messaging (Beaubrun and Pierre 2001; Hahn and Dudley 2002).

Radio emission is the major health risk in the perception of the public. Radio waves are necessary to support all kinds of wireless communications and they are produced by both handsets and the telecommunication antennae that have proliferated in recent years. Published results from different medical studies are, however, controversial, and provide inconsistent findings. Following Sánchez et al. (2001), there is no clear evidence about the epidemiological consequences of radio-frequency emissions in the short term, but the authors highlight the lack of information on long-term consequences. Indeed, the authors also state that in relation to population studies, the only clearly established risk from an epidemiological point of view is that of traffic accidents, which is not related to radio-wave exposure. Attention must also be paid to complaints about the installation of base-station antennae by mobile operators. Despite community protests against installation, authorities tend to decide in favor of telecom companies because the radiation levels used by these antennae are usually under formal legal limits.[34] In the absence of clear scientific evidence, and given the influence of telecommunications operators on local and regional governments, people's fears continue to rise, while companies dismiss their views as a hypochondriac reaction. The jury is still out on both sides of the debate.

Spam, Scam, and Viruses

What is the most serious problem in mobile-phone communication? When we asked this question of ordinary Chinese mobile subscribers in the first half of 2005, most subscribers answered "spam."[35] This is not surprising as e-mail spam has already become a major global problem and it is relatively easy to go from e-mail spam to mobile spam using the same logic and very similar methods. As in the wired world of personal computers, many mobile spam messages are designed to trap people into scams or to distribute worms and viruses, in this case, to wireless handsets. Indeed, many mobile-service providers allow people to send SMS from web-based interfaces. SMS spam is arguably easier than e-mail spam because the target recipients are identified via simple numeric combinations. This was the challenge Docomo faced a few years ago, when their subscribers were given "i-mode" phone e-mail addresses in the format of "phone number"@Docomo.ne.jp. As a result, spammers could disseminate large amounts of junk mail by generating random eight-digit e-mail addresses. In October 2001, Docomo subscribers received some 950 million e-mails every day, of which about 800 million were returned to senders because of unknown addresses, putting a huge strain on Docomo servers. In June 2002, operators received 140 thousand complaints about spam mail through the i-mode network (ITU 2002a: 94).

Mobile spam poses several major problems, one being the waste of bandwidth, a key network resource. It causes significant inconvenience to users, who have to delete the junk messages selling ring-tones, online dating services, Viagra, and so on. Those whose handsets have only small memory may not receive important messages because the memory space is occupied by spam. Moreover, like e-mail spam, many mobile spammers are not just putting out advertisements but scams of all sorts; for example, by sending out congratulations to people and saying they have won a lottery. In order to claim the prize, however, one has to pay a certain amount of tax and handling fees to the spammer, which is in fact a rip-off.

In the context of monopolized or semi-monopolized mobile-service markets such as China, spammers may be the service providers themselves or their business partners. As a result of the lack of market competition and the close relationship between telecom operators and regulatory bodies, there is little chance of protecting subscriber information from being abused. Hence users receive unwanted advertisements, including explicit sex MMS, for example, on 3G phones.[36] Sometimes subscribers even have to pay to receive spam without knowing it; in extreme cases, because they cannot even sign off a spammer's list, desperate users have to change to new mobile-phone numbers (Ma and Zhang 2005).

However, if the voice of disappointed users can be heard, the chances are that even dominant players will want to change the situation because protecting network bandwidth is also in their interests. In July 2001, for example, Docomo urged its subscribers to change to new e-mail addresses containing alphanumeric characters, in order to reduce the opportunity for mass spamming (ITU 2002a: 94). In November 2001, a new system was implemented to block messages sent to unknown addresses. Error messages were no longer returned to senders so that they were not informed of non-existent addresses. In January 2002, Docomo launched another service enabling users to designate a maximum of ten domain names from which they wished to receive e-mails and to block e-mail from others. However, devious spammers have been able to get around this by sending spam mail using fake domains. In April 2002, Docomo upgraded its mail server to block such forged-domain spam mail (ITU 2002a: 94).

The Japanese government was one of the first public institutions to provide countermeasures to mobile spamming. In January 2002, it obliged content providers to display *mi-syoudaku-koukoku* (non-agreed advertisement) in the mail header, so that users could delete these mails without opening them (ITU 2002a: 94). New legislation was also established so that users could "opt-out" and decide not to receive random bulk mail on their mobile phone. The battle against spamming continues. The latest trend is for public and private entities to join forces in order to keep the situation under control.

At the same time, with the diffusion of smart phones equipped with 3G or Bluetooth technology, mobile-data services have begun to be vulnerable to computer worms and viruses. Because smart phones can download programs and run scripts from the mobile Internet in essentially the same way as Wi-Fi laptops, they can be infected with hazardous programs just like an online computer. Reportedly, in Europe, some Bluetooth handsets have been affected by a benign worm called "Cabir" and one article offers advice about the subject:

The possible use of smartphones to provide a way into office networks is of real concern when considering that phones do not enjoy the same kind of virus protection from software as PCs do. And these phones will be connected to PCs regularly for synchronization, a real risk to enterprise networks in terms of both cost and confidentiality. (Visiongain 2004)

There has apparently been much less discussion about how to prevent worms and viruses targeting mobile handsets than about the efforts to stop mobile spamming, not to mention ways to secure wired online computers. Hence, when one of the authors purchased a 3G-phone in Hong Kong, he was told to turn off the Bluetooth function unless he needed to use it for a short period of time. "Otherwise, if you walk with your phone down the crowded streets," the salesperson said, "your phone will be easily infected with viruses or hacked in no time."[37]

Camera Phones

While a built-in digital camera is becoming a standard piece of equipment on mobile phones and gives rise to a new variant of photojournalism (Ito 2003a), unscrupulous use of the camera phone has also started to emerge, which has been met with collective complaints, private-business interventions, and government regulations around the world (Kageyama 2003). Three kinds of "mischief" have raised serious public concern. One is that some people have taken photos secretly up women's skirts or in public bathrooms. Another is so-called "digital shop-lifting" when copyrighted materials, most often fashion magazines in bookstores, have been photographed with camera phones, which has caused a reduction in sales (Kageyama 2003). Japan's private and public sectors are, again, among the first to respond to these problems. The Japanese police have started to pay attention to inappropriate camera-phone use in public spaces by apprehending people who have taken intrusive pictures of unsuspecting women in crowded train stations and stores, one of whom was fined US $4,200 (Kageyama 2003). Public bathhouses in Japan have also prohibited cell phones with built-in cameras (Kageyama 2003). Some Japanese camera-phone makers, such as Yatane, are also selling phones with a loud shutter noise to warn people that they have had their picture taken (BBC 2003b).

The third offensive practice with the camera phone is to use it as a tool of coercion, often accompanied by acts of violence. For example, a group of secondary school students in Hong Kong beat up a classmate and tortured him with a board cutter while taking pictures of the process. The photos finally reached the media and school authorities, which led to a ban on camera phones in many schools.[38] In the UK, because there have been a series of such assaults, violent youngsters equipped with camera phones are now known as "happy slappers." "Happy slapping" is believed to have started in South London but is now more widespread, leading many schools in London to ban mobile phones "because of fears of bullying and robbery. A number of violent attacks on schoolchildren have been filmed on mobile phones equipped with video cameras" (Sulaiman 2005). One of the worst instances took place in April 2005, when two teenage boys at a North London school raped a teenage girl, while a third boy filmed the process with his mobile phone. The footage was circulated among pupils "within minutes" (Sulaiman 2005). Public outrage was so intense that Burberry pulled one of its classic hats from the market in the UK because, somehow, it had become the fashion accoutrement of choice for the "happy slappers."[39]

Surveillance

We can understand surveillance as a means of determining where someone is and what he or she is doing, either in the physical or in the virtual world, at a particular point in time (Lyon 2001; McCahil 2002; quoted in Bennett and Regan 2004). Surveillance serves two main purposes. First, surveillance systems are used to "sort people's activities and characteristics for marketing and profiling purposes . . . in order to more effectively manipulate them" (Bennett and Regan 2004) and, secondly, "to reduce risk of potential harm and/or liability" (Ericsson and Haggerty 1997; Norris and Armstrong 1999; Ball and Webster 2003; quoted in Bennett and Regan 2004). This reduction of risk, which supposedly should lead to higher degrees of security, is related to both public and private spheres.

Even before the 9/11 attacks, there had been an increasing interest in surveillance studies, which, following Lyon (2002: 1), refers to "a cross-disciplinary initiative to understand the rapidly increasing ways in which personal details are collected, stored, transmitted, checked, and used as means of influencing and managing people and populations." This analytical interest relates to the fact that surveillance systems have become so ubiquitous as to include the "monitoring of everyday life" (Bennett and Regan 2004).

New information and communication technologies help in the creation and maintenance of ubiquitous surveillance, although technologies are simply the tools of the policy emphasis on surveillance (Colina 2000; Lyon 2004). Ultimately, society decides how to use technology and "electronic technology does not only amplify surveillance capacities, but also empowers the possibilities of emission, production and communicative participation" (Colina 2000: 35, based on Lyon 1994).

Surveillance is extensively used in market research, and through it "contemporary capitalism aims at including virtually all social life aspects into its valorization process" (Arvidsson 2004: 456). Its use also allows the reinforcement of "lateral surveillance" or peer monitoring to gain information about, and thus control over, friends, couples, or family members (Andrejevic 2005). This kind of control arises in a context of perceived risk or savvy skepticism in which, as stated by Andrejevic (2005: 488), people are invited to become do-it-yourself private investigators:

The proliferation of uncertainty serves as one marketing strategy for the offloading of verification strategies onto members of the general populace. In keeping with the so-called interactive revolution, individuals are invited not just to participate in the forms of entertainment they consume (interactive television) and in the production of the goods and services they consume (mass customization), but in formerly centralized forms of surveillance and verification.

Therefore, Andrejevic continues, more and more surveillance tools exist, for instance, the caller ID, "once a technology paid for by those with security concerns, now a service as ubiquitous as cell phones" (2005: 488). In the same way, deceit can also be developed in new ways. "It is possible, for example, to download background sound for one's cell phone that provides an acoustically verifiable geographic alibi, providing, for example, the background noises of cars honking to bolster the claim that one is stuck in the traffic" (Andrejevic 2005: 488).

Within the family sphere, the specific issue of the surveillance of children's mobility arises, facilitated by the trend toward increased spatial and functional differentiation of urban areas. Fotel and Thomsen (2004) highlight two different means of remote parental control of children's mobility: the first is the imposition of behavioral restrictions, while the second is described as remote control through technology in the shape of the mobile phone (2004: 543). This could include, among others, global positioning satellite (GPS) technologies to allow parents to track their teens (Brier 2004).[40] In a parallel way, the work sphere also undergoes a transformation as a result of the availability of these new channels of surveillance.[41]

A specific field of surveillance concerns movement. Bennett and Regan (2004) state that mobility can be subject to surveillance, first of all, by controlling the thing that moves, that is, the body (the person) or their transactions (the things the person does either as physical actions or as captured in data); and, secondly, by controlling the movement itself. With the surveillance of mobility there is potentially no hiding. All movements and flows are subject to scrutiny, captured, stored, manipulated, and used subsequently for various purposes. The objects we use (cars, phones, computers, electricity) in turn become tools of surveillance. Bennett and Regan (2004: 453) add finally that: "Movement is not a means of evading surveillance but has become the subject of surveillance."

Surveillance, indeed, can have transformative effects on the behavior of the persons surveilled, whatever the sphere affected – whether in the private sphere, at work, or at school.[42] In this sense, "[e]ven if the surveillance is designed not to control but to care and secure, the awareness that one is under scrutiny, or that one might potentially be under scrutiny, can change behaviors in unintended ways" (Bennett and Regan 2004: 453; see also Sweeny 2004).

There is continuing debate about the boundary between privacy and security (meaning surveillance; Lyon 2002; McGinity 2004). Attitudes of resistance have been identified in a number of instances, when individuals have used technologies and particular behavior to return the surveillance toward the surveilling authorities, in a practice labeled "sousveillance" (Mann et al. 2003). Some elements of biometric surveillance, such as fingerprint recognition, have already been embedded in mobile handsets:

In July 2003, Japan's Fujitsu released the world's first biometric-enabled phone for Docomo mobile subscribers, the F505i, which boasts a personal identification system based on a fingerprint sensor. Subscribers can use the sensor to lock and unlock the device, protect data stored on the device, and as a password for access to e-mail or calendar functions and for a number of other actions. (ITU 2004c)

As for data protection of the mobile digital word, Green and Smith (2004) argue, with respect to the UK, that there is a fundamental contradiction between the collection, use, and manipulation of digital data created by use of the mobile device and the regulatory and commercial conditions under which this data collection takes place. The authors highlight the contradiction between the *practice* of data processing and its *narrative*. The first concept refers to the fact that data traffic is assumed to be connected to individual consumers as long as each mobile handset is attached to one specific person, giving economic value to the data. In contrast, the narrative of data processing states that data traffic is anonymous, a discourse that emerges from both regulatory and industry sources.

Nowadays, information-gathering is potentially ubiquitous and any kind of media or market activity can become "raw material" for the production of an information commodity (Arvidsson 2004: 456–457). In this context, different private actors use the narrative of non-personal data traffic to maximize economic opportunities and contain consumer protection (Green and Smith 2004). Meanwhile, government administration and law enforcement agencies, following a desire to extend "traditional" state surveillance and investigative powers into new digital domains, consider that the surveillance of data traffic is merely the translation of existing powers into a new technological sphere (Green and Smith 2004). Thus, surveillance extends the logic of power into the mobile communication realm, but it also links up with the commercial imperative that characterizes the communication economy.

Sex and Sexuality

As mobile communication can be adapted to all kinds of uses, marginal and even criminal groups can take advantage of the new technology within their networks for such matters as drug dealing or the coordination of terrorist activities. Among these underground and semi-underground networks, one of the largest operations is the commercial sex industry which is now increasingly reliant on information technology. A study of sex workers in Hong Kong, for example, revealed that more and more sex workers in the city are now in so-called *yilouyifeng* (one-woman brothels), an individualized mode of operation that is only possible with a combination of Internet advertising and the mobile phone (Liu 2005). The main websites that carry advertisements for "one-woman brothels" also have a special version for mobile handsets, which display photographs along with text descriptions and contact information. Moreover, as the study found out, this new ICT-facilitated sex industry network not only brings convenience and a more discrete mode of communication to customers but also benefits the sex workers by allowing them to get rid of pimps and, in a sense, empowering them to run their own businesses. Of course, basic problems in the sex trade remain. But, as the researcher discovered through her interviews, the utilization of the mobile phone, including such functions as caller ID display, is helping "to a certain extent distinguish their personal life from business life and create some degree of control over who and at what time should get access to them" (Liu 2005).

In neighboring South China, where sex work is illegal (unlike in Hong Kong), almost all sex workers are equipped with mobile phones because the flow of information is of utmost importance for them to survive police raids and keep their customers. Some have multiple phones which they use separately for their personal life and for business. Moreover, a lot of them are part-timers who reportedly only get into the sex trade at times of financial difficulty, sometimes at the end of the month when their phone bills are due.[43] This is not too surprising as mobile expenditure accounts for a large proportion of migrant worker income in this region (Qiu 2004). There are, however, insufficient data to assess the financial strain caused by mobile-phone usage and its pressure on young females to enter the commercial sex industry. But based on sporadic evidence from several countries, we can see a trend that the flexibility of mobile communication is now dissolving the traditionally fixed boundaries surrounding sex work. This trend includes some female migrant workers in South China as well as certain members of the *kogyaru* school girls in Japan, who allegedly date older men for money through the practice of *enjo kousai* (Ito and Okabe 2003: 14).

This is not to say that mobile phones favor the expansion of the sex industry. Rather, it is the growing importance of this industry, for deep social reasons, that finds a convenient platform in the networks of perpetual contact. This does not relate only to prostitution but also to dating services, although sometimes the line between the two is blurred, as has been observed in the *deai* services in Japan (Holden and Tsuruki 2003: 34). However, there are ample ways for wireless technology to be used for purposes of personal contact that have nothing to do with the sex industry, and relate to the practice of family, partnership, and love.

Overall, we wish to emphasize that the role of mobile technology in relation to sex and sexuality has often been distorted and fantasized. This is, indeed, another avenue by which mobile technologies enter popular culture. It has to do with the widespread use of mobile telephony and SMS for intimate communication, when one can talk about things that one would not normally discuss with others, as has been reported in the Philippines (Pertierra et al. 2002) and China (Lin 2005). In other cases, the association between mobile devices and sex work is fantasized, as in an Indonesian book published in 2004 entitled *Sex on the Phone: Sensai, Fantasi, Rahasia* (sensation, fantasy and secrets), which described party-line services and the life of call girls in the country (Syahreza 2004; cited in Barendregt 2005). From SMS love-letter manuals to *Sex on the Phone*, there is an array of popular literature that focuses on the relationship between mobile communication and sex and sexuality. Most of these popular culture products, again, do not necessarily reflect the real-world situation. But they nonetheless deserve our attention because, by focusing on this particular set of uses of the new technology, they are shaping perceptions in the general public and, at times, policy discourse with important consequences, for both service providers and users. Let us examine the question more closely.

"The Ghost in the Phone"

The most dramatic representation of cultural concerns about mobile technology can be found in a series of horror movies in Asia, which is now becoming a new genre of *telepon hantu* (the ghost in the phone), as Barendregt (2005: 64–67) found in Indonesia and neighboring countries. There are certainly other types of movie stories based on the mobile phone, which do not involve supernatural forces, including the Hollywood thriller *Cellular* (2004), the sarcastic Chinese film *Cell Phone* (2004), and the youth romance *Love Message* (2005) released on the Chinese Valentine's Day in August 2005. However, more ghost movies have been made using the mobile phone as a key connection in the plot, and they have been quite popular in countries such as Japan, South Korea, Indonesia, and Thailand. It is important to note that the movies reflect a growing body of popular literature about "the ghost in the phone" in various media, such as newspapers, magazines, books, and sometimes even SMS. As Barendregt notes in Indonesia, while "some of the reports of a Ghost in the Phone typically first appear in rural areas, others have a modern urban background" (2005: 66).

Early products of this new genre include Hong Kong's *Phantom Call* (2000) and *Samurai Cellular*, which is one of four stories in the Japanese film *Tales of the Unusual* (2000). Both of these have a significant ingredient of humor, and the mobile phone is depicted as containing supernatural power that allows humans to communicate with ghosts or with those who lived hundreds of years ago. There was also the highly popular Thai movie, *999.9999* (2002) about the ancient belief in the lucky number and the scary things that may happen when one dials it. Also in 2002, the bestseller *Phone* was released in South Korea, which turned out to be such a success throughout Asia that it soon had a Japanese remake, *One Missed Call* (2004).

Phone is a typical Asian horror movie which treats technology as fundamentally alienating and estranging. It centers on a female investigative journalist in South Korea, who suddenly begins to receive horrifying e-mails and calls to her mobile. She has to hide from the unknown dangers and, to do so, she chooses to live in a newly renovated mansion which happens to contain the dead body of a high-school girl inside a wall of the room in which the journalist is staying. This high-school girl had had an affair with the mansion's male owner and was accidentally killed by his wife. Her ghost then uses mobile-phone calls to take revenge, first, by occupying the body of a small child. At the end of the movie, the family who owns the mansion is torn apart and the journalist takes the ghost's handset and throws it into a lake. While sinking to the bottom of the lake, the phone, again, mysteriously starts to ring.

The "ghost in the phone" movies deserve our attention because they stand in stark contrast to the entertainment-oriented commercial movies such as Hollywood's *Cellular*. By depicting mobile communication as possessing supernatural power, they are able to blur existing boundaries in time and space between humans and ghosts, between this world and the other. The result of this mixture usually leads to chaos, suffering, death, and, in many cases, the breakup of existing relationships in the family unit and friendship networks. In this sense, these movies are expressing the cultural concern that mobile technology, like new technologies of all kinds, is "anti-social," and that its development threatens traditional ways of life because of the innate alienating nature of technology.

While noting these dramatic modes of expression, we have to bear in mind that mobile phones are not always perceived as a horrifying threat. Under other circumstances, mobile usage may be seen as positively related to traditional forms of spirituality. The best example is probably the "Islamic Saudi mobile phone," available since January 2005, which "displays the Qibla direction and prayer times in over 5,000 cities worldwide" using Arabic, English, French, Urdu, Persian, and Bahasa Indonesia (Hanware 2005). We should therefore see the "ghost in the phone" movies strictly as a media product that reveals more about existing cultural concerns about mobile communication than actual practice. These negative media representations have only just begun to attract academic attention, but their significance is not to be underestimated: they shape user practices and public policy in a broad way that includes the whole of society, which is now consuming stories about mobile communication as well as mobile communication itself.

Summary

On the basis of the evidence analyzed in this chapter, we observe, across cultures and contexts, the emergence of a given pattern of social transformation. We can say that mobile communication is, throughout the whole world, a pervasive means of communication, mediating social practice in all spheres of human life. But it is adopted, adapted, and modified by people to fit their own practices, according to their needs, values, interests, and desires. People shape communication technology, rather than the other way around. Yet, the specificity of the technology reflects into the ways in which people conduct their lives. The fact that we are able to be connected from everywhere at anytime (though not forgetting only where there is coverage by telecommunications systems) makes it possible for us to organize the very same practices that constitute our daily experiences of life on a different pattern, a pattern characterized by ubiquitous networking in the family, in social relationships, at work, in social services, in entertainment, on the basis of selective networking. People build their own networks, and they reconfigure them as their ways of life and work change.

For instance, communication technologies materially allow the post-patriarchal family to survive as a network of bonded individuals, in need of both autonomy and support at the same time. As people rebuild and extend their lives along their networks, they bring with them into these networks, and into their networking devices, their values, perceptions, and fears. The risk society goes into the networking mode as well. And so does the secular need for the state to operate surveillance and control. Because mobile communication devices are multimodal, the new communication channels blur and combine languages, mixing voice, audio, images, text, and self-expression in the shapes, colors, and sounds of the device itself. Mobile communication becomes a multimodal layer of communication that embraces every social practice, extending the beat of life into ubiquitous interactivity, thus relentlessly giving rise to new sources of meaning.

4 The Mobile Youth Culture

By youth "culture," we mean the specific system of values and beliefs that inform behavior in a given age group so that it shows distinctive features vis à vis other age groups in society. This culture has to be placed in the context of a given social structure. In other words, we do not refer here to a transhistorical analysis of the meaning of youth, but to the cultural specificity of youth in the social structure that characterizes our time, the network society. For a definition of the network society, and the justification for our statement concerning its structuring role in our time, we refer the reader to the proper sources (Beck 1992; Castells 2000b, 2004; Himanen 2001; Capra 2002), and concentrate here on the issues that are the object of our current research.

A fundamental hypothesis in our analysis is that *there is a youth culture that finds in mobile communication an adequate form of expression and reinforcement.* Technologies, all technologies, diffuse only to the extent that they resonate with pre-existing social structures and cultural values. However, once a powerful technology is adopted by a given culture because it fits into its pattern, the technology grows and embraces an ever-greater proportion of its group of reference, in this case young people. Our analysis will try to specify this proposition. Namely, we will try to identify the content of this youth culture in Europe, the Americas, Africa, and the Asian Pacific, and see how it fits into the pattern of mobile communication, how it transforms mobile communication, and how it is strengthened and influenced by communication technology. We believe that some basic cultural trends are applicable to differing contexts, albeit with the necessary specification of each culture and institutional setting.

Much existing research on mobile youth culture(s) is concentrated on Europe, where wireless technologies have diffused widely among the younger generations, especially in North and West Europe, attracting scholarly attention from Scandinavian countries to Spain and from the UK to Russia. We shall therefore start with the case of Europe, which includes a more general discussion of youth culture in the network society. This will be followed by observations made in the Americas (mainly in the United States), Africa, and the Asian Pacific, which include more findings from journalistic and trade sources, in order for us better to understand the particularities of these regions on top of such discernible similarities as the tendency toward autonomy, the strengthening and transformation of family ties, and consumerism as a major dimension of mobile youth culture. We will try to show the patterns of mobile youth culture from a cross-cultural perspective, but before doing so we will outline a range of data on the diffusion and use of mobile telephony among young people.

The Diffusion and Use of Mobile Telephony among Young People

The diffusion of mobile communication in the 1990s was nothing short of extraordinary. A key factor in the speed of diffusion was the embrace of the technology by the younger generation as the density of mobile communication users reached its high points in Japan and in Northern and Western Europe. The fast rate of diffusion of mobile communication among the young population may be explained by a combination of factors, including the openness of young people to new technology and their ability to appropriate and use it for their own purposes. Indeed, their greater capacity to use new technologies becomes a factor of superiority compared to their elders. In addition, mobile telephony has become a sign of self-recognition among the peer group. It should be noted that the mobile telephone has a special particularity related to ergonomics and age. Owing to the physical features of the mobile handset, elderly people can find it difficult to manage the device (because of the dimensions of the screen and, especially, of the buttons). These difficulties exceed the generational gap common to new ICT gadgets (Lobet-Maris and Henin 2002: 102; Moore 2003: 68).

It is clear that the notion of "young users" brings together very distinct conditions. We may, at the very least, differentiate between young adults (in their twenties and early thirties), teenagers, and children. However, it is our argument that there is a common culture of communication, with various emphases in its manifestation depending on age. We will clarify this analysis after presenting some research findings, while specifying the analysis for each age group. In detailing the available information, we will follow the geographic order previously stated.

Europe

As discussed in chapter 1, the diffusion of mobile phones in Europe in the past decade has moved from, approximately, 1 subscriber per 100 inhabitants in 1992–1993 to the penetration of 71.55 percent in 2004. If we focus on the 25 countries of the European Union, and given similar levels of penetration ten years ago, in 2004, out of every ten persons, more than nine were mobile-phone subscribers (a rate of 93.4 percent). Among these subscribers, young Europeans in the 15–25-year age bracket and young adults from 25 to 34 years of age have the highest rate of usage (77.2 and 75.8 percent respectively). The penetration rate drops to 70 percent for those aged between 35 and 44 years, and then considerably reduces to lower than 55 percent for older people (figures given in Fortunati and Manganelli 2002: 64, according to Eurescom 2001). Young people in their teens, twenties, and early thirties not only constitute the largest proportion of users, they were also the early adopters, who invented uses that had not been foreseen by the initial designers of the technology.[1] Indeed, in this field of communication, the key to success for companies is to identify and follow the innovations of young users.

In Europe, there is a significantly more detailed body of research results available in the public domain. A first picture of the situation in 2002 for selected European countries shows that, regardless of the country's penetration rate, young people are the ones with higher access to mobile telephony (see figure 4.1). Moreover, the older people are, the less likely they are to have access to this technology, a trend that becomes more pronounced for the elderly over the age of 60. Nevertheless, age differences decrease when mobile telephony further diffuses in society. Thus, the Nordic countries tend to have higher penetration rates among those who are 55 years old and above compared to Spain, where the penetration rate is 50 percent for this age group. Moreover, differences in terms of age are not only in access to a mobile phone but also in everyday usage. We have already seen this in the Asian Pacific and the United States (as discussed in chapter 2), but research in Europe supplies more detailed evidence for this observation regarding both young and older people.

Two studies of Finnish children conducted in 2000 placed children and adolescents into five categories according to the relationship they establish with the mobile phone (Mante and Piris 2002; Oksman and Rautiainen 2002). Children under 7 years, first of all, often have either an indifferent (imaginative) or personifying (animistic) relationship with their device. Games are the most interesting feature of the device, which, in fact, can be interesting on its own, although important toys are more significant.

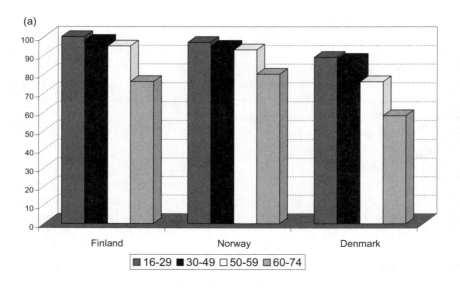

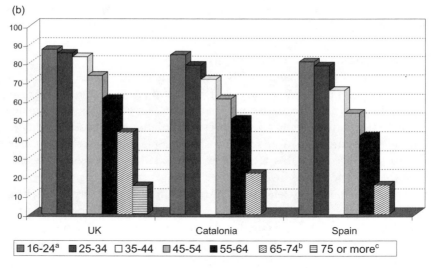

Figure 4.1
Access to mobile telephone by age for selected European countries, 2002 (percentage of population for each age group).[a] UK = 15 – 24;[b] Catalonia and Spain: 65 or more;[c] Catalonia and Spain: data not available. *Sources*: For Finland, Norway, and Denmark: Nordic Council of Ministers, Nordic Information Society Statistics, 2002. © Nordic Council of Ministers, Statistics Denmark, Statistics Finland, Statistics Iceland, Statistics Norway, Statistics Sweden (http://stat.fi/tk/yr/tietoyhteiskunta/nordic_iss_02.pdf); for the UK: Consumers' Use of Mobile Telephony Survey, Office of Telecommunications (www.ofcom.org.gov); for Spain and Catalonia: Household Information Technology Survey, National Institute of Statistics, Spain (www.ine.es); and own elaboration. Source data in online statistical annex (appendix 64).

In the next age group – children from 7 to 10 years old – attitudes begin to differentiate: some children are very interested in the mobile phone as a device, while others remain indifferent to it. Children, for instance, may forget to take the phone with them when they go to a friend's house, while other toys, like Pokemon cards, would not be as easily forgotten. Their relationship with the device is, in fact, quite pragmatic and, given the fact that mobile communication is in itself too abstract for children of this age, they see the mobile phone more as a games machine than a communication tool.

In the pre-teen world (10–12 years old), mobile telephony changes its position, becoming more central and leading to the beginning of the "mobile fever" age that takes place at around the time that the importance of hobbies and friends is increasing, while the significance of toys diminishes. Thus, the mobile phone becomes an important appliance as a communication tool with peers. Pre-teens use it in a creative way; for instance, sending empty (content) text messages as a means of teasing people, or playing various types of "boom call" games.[2] In the fourth category, teenagers, from 13 to 15 years of age, have distinct attitudes to mobile telephony, which can be practical and instrumental or, alternatively, expressive and affective. Moreover, it is at this age that personalization of the handset becomes significant and aesthetics gains importance. Finally, Finnish pre-adults (16–18 years old) tend to decrease their offline use of the device at the same time as the practical and instrumental side becomes more appreciated. Nevertheless, this does not mean that they avoid texting. Indeed, the opposite was reported in the UK, where the same age group would regularly have text-message "conversations" over a number of hours in the evening (Smith et al. 2003).

Taking a wider age range, for European teenagers in general: "the most important thing in mobile communication remains building up and maintaining their social networks" (Oksman and Rautiainen 2002: 28). This observation made in Finland also applies to other countries, such as Norway (Ling 2002) and Spain (Valor and Sieber 2004). It is particularly true "when [young people are] forming their first relationships with the opposite sex" (Oksman and Rautiainen 2002: 28, with reference to Finland) because the technological adoption in this situation combines the coordination and expressive uses of mobile communication to a similar extent (see Ling and Yttri 2002, with reference to Norway; Kasesniemi 2003, with reference to Finland; and Valor and Sieber 2004, with reference to Spain).

Compared to two other age groups (those of 16–18 and 50–60 years), in Britain, the group of young adults (25–35 years old) is the most multimodal (using face-to-face, phone, e-mail, and instant messaging communication), and the only cohort not to report using face-to-face as their most frequent means of communication (Smith et al. 2003). This suggests that they have already developed a set of selective and individualistic networking behaviors. Moreover, young adult males in Norway have an extremely high use of voice telephony accompanying other modes of communication. They spend less time texting and have a lesser degree of device personalization than teenagers (Ling 2002).

Working adults were the first adopters of mobile telephony (Fortunati and Manganelli 2002; Agar 2003; Lacohée et al. 2003). Available information with regard to this age group states that voice use is more common (Ling 2004: 146, with reference to Norway) and coordination uses are the most popular (Frissen 2000, with reference to the Netherlands). It should be added that texting and other creative uses, such as the "boom call," are generally introduced to adults by their daughters and sons. These adults then use SMS to communicate with other adults, although often at a lower intensity than their children (Ling 2004: 146, referring to Norway).

As they get older, people focus their relationships around family and close friends and have smaller social networks (Smith et al. 2003, referring to the UK). This was demonstrated by analyzing the number of names in people's mobile-phone directories, which significantly decreases for people over the age of 60, compared to middle-aged adults and, of course, young people and teenagers (Ling 2004: 109, referring to Norway). In general, old people are not used to communicating via SMS (Ling 2004: 146 for Norway; Smith et al. 2003 for the UK). While we could consider that they are uncomfortable with this new channel of communication, there have also been ergonomic problems (for instance, in buttons and screen dimensions) that prevent the elderly from making extended use of the different features of a mobile phone (Lobet-Maris and Henin 2002; Moore 2003).

Affordability is very important. The European pricing system, together with the availability of a common telephone network within each country and around the continent,[3] greatly facilitates the popularization of the technology even among low-budget groups (as young people are in developed countries). Three characteristics of the European pricing system need to be highlighted. First of all, the call is always paid by the person who initiates communication. Only when "roaming," does a recipient pay part of the cost of an incoming call. Roaming – or the ability to use your cell phone in different systems (Agar 2003: 40) – only applies in Europe outside national borders.

Secondly, the prepaid system has allowed individuals and families with lower purchasing power to have access to a mobile phone. This billing system has low fixed costs, although variable costs are usually higher compared to a monthly contract. The prepaid system, when first introduced, could have been perceived as a system that encourages overconsumption since the commitment (i.e., the payment) is made in advance of the actual consumption act (i.e., the call). Contrary to this idea, the system allows a measure of control over expenditure because the monetary commitment also acts as a limitation on final consumption. Nevertheless, prepaid packages do lead to higher levels of telephone expense because more teenagers will receive a handset, and thus will use it in their everyday life, thanks to the fact that strict parental control over the budget can be applied.

Thirdly, SMS is compatible among different mobile-phone companies. There is no need for the user to know anything about the companies involved in the communication. Thus, all users, regardless of their service provider, can communicate through SMS at the same cost and with the same technical facilities. SMS was built into the European Global System for Mobile (GSM) standard as an insignificant, additional capability (Goggin 2004). Its low cost[4] contributed to its generalized adoption among teenagers, overcoming the inconveniences of the interface. Despite its great success, SMS is the only telecommunication service that has maintained its price (Lacohée et al. 2003: 206) or has even become more expensive. Indeed, nowadays it is almost impossible to send a free SMS through the Web as an end-user.

The United States

In the United States, young people are active adopters of mobile telephony, giving rise to a plethora of different youth lifestyle segments as discovered by research companies (IDC 2003; ScenarioDNA 2004). Given that their young age precludes them from owning a landline, more American youth than the general population have adopted the cell phone as their main telephone line (Greenspan 2003a). However, while a significant body of research on young people's uses of wireless technology is developing in some European and Asian countries, there has been little systematic and academic study of how young people in the United States are using the emerging technology. This is probably because *young people have not been a major element of the US industry until recently.*

After years of focusing on the corporate market, the US wireless communication industry has only just started to target the youth market (Collins 2000; Tan 2001; Fischer 2002; Meyer 2002; Paustian 2002; Sewell

2002; Mack 2003; Smith 2004). Industry analysts have observed the extra-ordinary adoption of wireless technology among young people in other countries, especially Europe and Asia, and hope to replicate those results in the United States (see, for example, TNS 2001b, 2002b; Wireless World Forum 2002; Fattah 2003; IDC 2003; Motsay 2003; In-Stat/MDR 2004a, b; Yankee Group 2004). Since there is some evidence that the general market is reaching saturation point (Henry Fund Research 2003, 2004; Wilson 2003b), the industry is looking for alternative and more segmented markets, of which the youth demographic has been identified as most promising.

However, so far, the largest group of users in the United States has tended to consist of young professionals in the 30-plus age group (Fattah 2003; Horrigan 2003). Other research documents also show that cell-phone users tend to be affluent 25–54-year-olds (Anfuso 2002). In addition, with respect to wireless data applications, the more successful applications, such as the Blackberry, are primarily used by the business community (Fitchard 2002). TNS (2002b) has concluded that the youth market is as yet not critical to the US wireless market. Thus, while in the early 2000s, young people in the UK, for example, were found to have given the mobile phone an inti-mate place in their lives, young people in the United States still tended to see their cell phones mainly as pragmatic communication tools (Wakeford and Kotamraju 2002).

An accurate picture of the youth market in the United States is thus yet to be established since, even where studies exist, there is little compatibil-ity in the data collected by different researchers. However, statistics and commentary in business journals, newspapers, and academic research provide some insights into how children, teenagers, and young adults are incorporating wireless communication technology into their everyday, increasingly mobile, lives. Although there is a perception that young people *naturally* gravitate toward wireless communication devices, such as cell phones, there is a variety of opinion as to the real driving force. Some people believe that teens' busy and increasingly mobile lifestyles are behind the trend; others think that parents are the main drivers; while others suggest it is the mobile-phone industry that is creating demand for the product through clever marketing (Dunlap 2002; Selingo 2004).

Nevertheless, there is some evidence to suggest that the youth market is gradually expanding, not least because of a concerted effort by the indus-try to target products at young people, despite misgivings about the relia-bility of this demographic. The increasing number of companies turning out prepaid phone-card plans is an indication of this trend: in 2001,

Table 4.1
Wireless service usage by age in the United States, 2001 (%)

Age (years)	2001	
	Cell phone	SMS
12–17	51	43
18–24	61	38
25–29	60	32
30–34	69	25
35–54	62	n/a
55+	50	n/a

Source: Genwireless (2001).

prepaid subscribers constituted about 12 percent of wireless subscribers, up from 6 percent in 2000, with one company – Leap Wireless – offering only prepaid services, targeted mainly at youngsters (Rockhold 2001; Luna 2002; Mader 2003; OECD 2003; Standard and Poor's 2004). The prepaid market is designed to cater to the credit-challenged user, of which young people are a significant part.

Available data suggest that somewhere between 29 and 87 percent of young people in the United States own a cell phone, depending on which segment of the youth demographic is taken (table 4.1 shows the data for 2001 from one source). The vast range of this statistic is illustrative of the lack of comprehensive and uniform data on the youth cell-phone market. One report states that 32 percent of 5–24-year-olds owned a cell phone in 2001 (Wireless World Forum 2002); another states that 79 percent of young people owned a cell phone in 2002 (Allardyce 2002, regarding the NTCA/FRS survey). Jorgensen (2002) reports that 40 percent of children in the United States own a wireless device, mainly cell phones but also Palms, PocketPCs, pagers, and lightweight laptops. TNS (2004) reports that, in 2003, 29 percent of children aged 6–14 years had a cell phone, and a study by Selian (2004) finds that 87 percent of youth in college have cell phones.

Young people are found to have a greater interest than older people in non-voice uses of wireless communication technology, such as SMS and the wireless Internet (TNS 2002a). For example, about 14 percent of young adults have used text messaging to vote in a contest or participate in a poll, compared to about 4 percent of adults (according to the Yankee Group's 2003 Mobile User and Mobile User Young Adults surveys, as reported in

TWICE 2004). Young people were the first group to appropriate SMS following the promotion of texting via television entertainment shows that encourage their (mostly young) audiences to send voting messages via their cell phones, a phenomenon now known as the "American Idol effect" (In-Stat/MDR official, quoted in 3G Americas, n.d.). On the other hand, subscribers above the age of 25, or even 35, years were found to be the most active texters in American text-messaging sites such as Verison TXT Messaging, Yahoo! Mobile SMS, and SMS.ac, although the 18–24-year age group was the most active among those who subscribe to Spring PCS Messaging.

While asserting that the mobile phone is "an icon of the youth generation," the Mobile Youth 2002 report stressed that the youth culture "is complex and not easy to decode at first glance. It changes radically between genders and the different age strata ... the cultural norms determine the acceptance of new services, their associated levels of 'cool,' how the phone is used, why it is used and, importantly what meaning the device has in the future" (Wireless World Forum 2002: 10). Sefton-Green (1998: 9) declares that: "there are now different kinds of childhoods or youths."

Notwithstanding the various subcultures within the youth demographic, one can identify certain characteristics that arguably define youth in general, and are evident in the use of wireless communication technology by young people. Among these are the desire for independence, community and connectivity, entertainment, personal identity, and "coolness." These values can be seen in the personal, social, and political uses of wireless technology, especially cell phones, by young people in the United States. According to other research, young people use the cell phone mainly to play games, send text and e-mail messages, download ring-tones, and send pictures – in that order (Selian 2004). In addition, the most popular reason for owning a cell phone is convenience. The most popular location of usage is at the store or on public transport. Other areas/times are at mealtimes, in the bathroom, on a date, at a concert, at a library, at work, in a hospital, in meetings/class, and at places of worship.

Latin America

Data for Latin America are very scarce, though we have some evidence of the takeoff of the SMS market, which foreseeably means the popularization of that service among young people. Table 4.2 shows the growth of SMS revenues in Latin America as a whole and in some selected countries. Aggregated data show that revenues were twelve times higher in 2003 than

Table 4.2
SMS revenues in Latin America (US$ millions), 2000–2003

	2000	2003
Mexico	0	171
Brazil	27	148
Venezuela	6	121
Colombia	0	28
Argentina	9	23
Peru	1	20
Guatemala	0	15
Puerto Rico	4	13
Chile	0	9
Ecuador	0	9
Panama	0	8
El Salvador	0	5
Uruguay	0	3
Bolivia	0	2
Paraguay	0	2
Nicaragua	0	1
Costa Rica	0	1
Honduras	0	0
Latin America total	47	578

Source: Pyramid Research, 2005: Mobile Forecasts for the following Latin American countries: Argentina, Bolivia, Brazil, Chile, Colombia, Costa Rica, Ecuador, El Salvador, Guatemala, Honduras, Mexico, Nicaragua, Panama, Paraguay, Peru, Puerto Rico, Uruguay, Venezuela (http://www.pyramidresearch.com/browse_forecasts.htm).

they were in 2000. Moreover, in all the countries except one – Honduras – the SMS market is already developed, although figures are not always high.

The countries that contributed most to the revenues for 2003 were Mexico (29.5 percent of the total amount of US$ 578 million), Brazil (25.6 percent), and Venezuela (20.9 percent). As seen in other regions, compatibility among the systems of different mobile companies is crucial for the popularization of the SMS service. This is especially valid for Venezuela, the first country in Latin America to require the interconnection of the SMS platforms running in the country. In November 2002, interconnection was possible between two mobile operators, while the third joined them in 2003, leading to a "significant upsurge in the volume of traffic in Venezuela" (Tamayo 2003: 4). Mexico followed the example of Venezuela

(Tamayo 2003), and the measure is achieving similar results. Also in Brazil, compatibility among operators is now fully functional.

In the specific case of Bolivia, available data show the same collective behavior as in other countries with respect to mobile telephony. That is, young urban users are above the general average in terms of possession of mobile-phone handsets. A survey conducted in September 2004, in four major cities in Bolivia, showed that up to 44 percent of the young population, from 18 to 30 years old, had a mobile phone (Apoyo 2004).[5] It was also observed that the higher the income level, the more likely the possession of a mobile phone: 72 percent of richer youngsters (socioeconomic levels A and B) had a handset, but this figure slumped to 34 percent for the poorest (socioeconomic level E; Apoyo 2004). This contrasts with the penetration rate for the whole country, which in 2004 consisted of 20.07 handsets per 100 inhabitants (ITU 2005).

Among the youth surveyed, the prepaid system is the most popular, with a presence of 93 percent of total subscriptions. As in other countries, preference for the prepaid system is not the same for each socioeconomic level. At the lowest socioeconomic level (level E), mobile expenses are covered through pre-payment, while only 86 percent of youngsters in socioeconomic levels A and B choose this payment system (Apoyo 2004). Moreover, the prepaid system in Bolivia, considered as a whole, amounts to 74.9 percent of total subscriptions (ITU 2005), which is a common observation internationally.

Africa

The specific nature and distribution of mobile-phone use among young people in Africa are largely undocumented. So far, the only available survey that has measured use by age shows that, in general, the use of mobile phones in Africa tends to be higher among young adults, especially those aged 20–35 years. Between 50 and 60 percent of mobile-phone users in South Africa and Tanzania fall within the 24–45-year age group (Samuel et al. 2005).

Despite the absence of strong evidence, however, some deductions can be made about the possible character of trends based on what is observed in the general population. Clearly, even with the lowering costs and greater accessibility of mobile phones in the region, mobile communication is still out of the reach of most young people, who are either in school or unemployed. Wealthier countries, and wealthier families within countries, are, however, likely to have a greater diffusion of mobile communication devices among their young people. For example, young people in South

Africa are a significant population of mobile-phone users (Mutula 2002). Parents with adequate resources tend to purchase prepaid phones for their children as a way of limiting their use of the residential fixed line and making children responsible for their own usage levels.[6] Furthermore, as tends to be the case everywhere, young people in Africa are highly appreciative of the lifestyle aspects of mobile phones, such as entertainment and camera features (Lowman 2005). In addition, there are anecdotes about young people, and even children, who, compelled to find some orm of employment in order to survive, have found mobile phones useful in conducting their business activities, such as market trading and street-hawking.

The Asian Pacific

Although mobile telephony began as a business-oriented technology in Japan, it is the younger generations, especially students, who constitute the most prominent mobile-phone user group (Kenichi Fujimoto, cited in Ito and Okabe 2003: 10). Cell-phone penetration is much higher among high-school (76.8 percent) and college (97.8 percent) students than the general population (64.6 percent; Yoshii et al. 2002, cited in Ito and Okabe 2003: 5). Students also pay higher monthly cell-phone bills (on average JPY 7,186 or US$ 67.5) than the general population (JPY 5,613 or US$ 52.7; IPSe 2003, cited in Ito and Okabe 2003: 5–6.). In Tokyo, young urbanites spend an average of US$ 150 on their cell phones each month to exercise their consumer power, resulting from the fact that "a generation of declining birthrates has filled Tokyo with one-child families" (McGray 2002: 52).

Besides financial dependence on adults, other fundamental structural constraints include: (a) the tiny size of the average Japanese household, which means that urban youth usually have to socialize in public spaces such as the street; (b) the prohibitively high cost of setting up a landline (from US$ 600 and up, about twice the cost of a cell phone); and (c) the tradition that parents use the home phone to monitor and regulate children's relationships with their peers (McGray 2002: 9–10). Moreover, there is evidence to show that Japanese young people in their thirties and below are more likely to subscribe to i-mode, Docomo's mobile Internet service, including the latest 3G applications (NTT Docomo 2003).

Indeed, the centrality of the youth market is also observed in South Korea (S-D. Kim 2002: 63–64). Young users are particularly keen to adopt SMS. According to South Korea's Cheil Communications (2003), in May 2003, 93 percent of young Koreans between the ages of 17 and 19 sent or received SMS at least once a day. The percentage decreases with age: 92

percent for ages 20–24, 79 percent for ages 25–29, 58 percent for ages 30–34, and 47 percent for ages 35–39. In order to target the youth market more precisely, all three mobile operators in the country offer specialized rate plans for college students (ages 18–23) and high-school students (ages 13–18).[7] And, what is more, operators are also adding new services, such as mobile-based entertainment, to meet the demands of the mobile youth culture.

The Philippines differs from Japan and South Korea in being a developing country with lower income levels, more religious diversity, and a multifaceted colonial history, which impinges upon the local language to create particular manifestations of its mobile youth culture, such as the shorthand texting language. In this particular case, the income factor is behind the popularity of prepaid phone-card usage among Filipino youth, as well as SMS and texting (Arnold 2000; Salterio 2001; Hachigian and Wu 2003). For instance, "In the Philippines, texting has been the preferred mode of cell phone use since 1999, when the two major networks, Globe and Smart, introduced free and, later on, low-cost messaging as part of their regular service" (Rafael 2003: 404). Media accounts in the country have often claimed that the Philippines is "the text messaging capital of the world" (Kaihla 2001; Hilado 2003). In 2001, about 100 million text messages were circulated every day in the Philippines, which "puts the country well ahead of previous world leader Germany by 1 billion messages a month" (Bociurkiw 2001: 28).

Inexpensive texting, together with prepaid mobile-phone cards, "give cell phone providers a way to attract a broad spectrum of users from different income levels" (Rafael 2003: 404). Urban youth groups in Metro Manila are also known for their texting "mania" (Rafael 2003: 404–405; see also Arnold 2000), for which they are often called the "Generation Txt" (Rafael 2003: 407). This has a lot to do with the cheap price of SMS, which was initially free in the Philippines. However, nationwide, the age of active Filipino texters is significantly higher than in more developed markets. The contrast is that, nationwide, it is the professionals in their mid-30s (ages 33–36) who turn out to be the most active texters, according to a survey of five hundred users in nine regions (Toral 2003: 174). This is quite different from the case of more wealthy countries where teenagers and college students are the most active users, but it is understandable due to income discrepancies among nations.

Reports show that the penetration of SMS is also higher for Chinese mobile users under the age of 35 years (Xinhuanet 2003), many of whom have become highly proficient texters who would call themselves the

"Thumb Tribe" (*New Weekly* 2002), a phrase originally coined for texters in Japan, the *oyayubisoku* (Rheingold 2002: 4–8). Despite the size of its market, so far there has been little research into the country's mobile youth culture and the influence of adoption upon young people's existing social relations at school, at home, and within peer groups. Apart from our preliminary study involving young migrant workers, the majority of reports are journalistic, which give a sketch of the patterns of young people's mobile-phone usage that seem to be not too different from what we have learned in other Asian Pacific countries. The various elements in combination are giving rise to a youth culture that draws much attention from journalists and ordinary urban residents. Just as in many other countries, there is growing concern among the older generations about the emergence of this new culture, which nonetheless is at the very core of the processes of identity formation among young people. But what is the significance of this deep connection between youth culture and mobile telephony?

Youth Culture in the Network Society

Wireless communication technology is only the latest in a long line of technological changes that have successively driven hopes and fears about the impact of technology on the young. Books, radio, television, and the Internet have all been associated with perceived transformations in the relationships between young people and the rest of society, especially as agents of socialization (for example, Arnett 1995; Buckingham et al. 1999; Demars 2000; Srivastava 2004b). Contemporary information and communication technologies, however, are seen as particularly significant in the changes they have made possible. Holmes and Russell (1999: 69), for example, state that new "interactive and wearable technologies" have brought about "a tectonic shift in the contemporary formation of adolescent identity." Essentially, these new technologies move young people away from the sphere of influence of traditional socialization structures, such as the home, educational system, and broadcast media, while providing an ever-widening range of socializing and identification options, thus contributing to the "crisis of boundaries" (Holmes and Russell 1999: 75). At an even broader level, according to Holmes and Russell, the new digital and mobile technology engenders a qualitative change in users' experience of everyday life, as the technology becomes incorporated into routine activities. In the case of adolescents who are highly immersed in the digital lifestyle, the result is a "technosocial sensibility," that is, "the state in which technology and nature are brought together" (Holmes and Russell 1999: 73). The concept of technosociality emphasizes communication technologies not as tools but as contexts, environmental conditions that make possible "new ways of being, new chains of values and new sensibilities about time, space and the events of culture" (Holmes and Russell 1999: 73).

It has been observed that the mobile phone – as "a personal device and communications that are a constant, lightweight, and mundane presence in everyday life" (Ito 2004: 1, for Japan) – is having a profound influence on the patterns of young people's social networks and their relationships with their elders. This tendency gives rise to moral panics related to fears about the disappearance of childhood as new technologies continually give children access to information and participation options that either blur the boundaries between childhood and adulthood or weaken children's ties with family and other social institutions (Sefton-Green 1998). Yet new communication technologies are also associated with optimistic views about their potential to enhance the lives of children, and through them society at large, especially in the areas of education and civic participation (for example, Wilhelm 2002).

We can observe many similarities among mobile youth cultures in different countries in terms of their consumerism, faddish trends, cultural identity, peer-group formations, relationship with existing social institutions (for example, family and school), and the tendency toward more "flexible" social networking in space and time (or the "softening of time"; Rheingold 2002: 5). However, on a more specific level, the context in which the mobile youth culture emerges is also significantly different, both in the Euro-American experience and among the four countries of the Asian Pacific region. As discussed earlier, wireless diffusion rates and social differentiation patterns are quite diverse, given the economic disparities among the countries and among groups of young subscribers of different socioeconomic status, who either use the latest 3G and MMS services or else prepaid and other low-end services, such as China's Little Smart.

Given the above reasons, if we could see a consistent pattern of mobile culture among youth populations across the different countries we have studied, this would add greatly to our argument that a youth culture is emerging globally, which "finds in mobile communication an adequate form of expression and reinforcement." However, at this point, we can only give a very tentative answer – which is, yes, there is a global mobile youth culture – because academic research is insufficient regarding young people and mobile phones in the Americas and the Asian Pacific, and especially in the less-developed markets. Thus, with reservations about our ability to generalize, in the context of structural networking, cultural individualization, and autonomy-building around self-generated projects that characterize the network society, a mobile youth culture is emerging around the management of several processes along a number of social dimensions.

A particularly important process in the development of new family relationships is *the management of autonomy vis à vis security*. The crisis of the patriarchal family leads to the weakening of traditional forms of parental authority and to the early psychological and social emancipation of young people. Nevertheless, the family continues to be an essential source of security and support for young people, both in functional terms (safety, income) and in psychological terms (guidance, emotional support). However, the crisis of the patriarchal family leads to increasing tensions between parents, with increasing rates of dissolution, and high levels of emotional instability. Children become the primary source of emotional reward. But this comes at the price of relinquishing authority in exchange for companionship. In addition, an aging and unequal labor market, in the particular case of Europe, places young people at a clear disadvantage vis à vis the older population, in spite of the higher educational levels of young people and their much greater acquaintance with the new technological environment.

The net result of all these trends is that the older population needs the emotional support of the young, while keeping them economically dependent, and the older generation is not able to exercise its authority because of the crisis of patriarchalism and the early process of individualization. For the younger population, they feel their autonomy as individuals very early, but they need security from their family until very late in their development. Thus, managing the relationship between autonomy and security in their family relationships is an essential condition of their existence. Although this is particularly important in the teenage years, it is also present among children of overworked parents who often return from school to an empty home, and it is also present among young adults who make all the decisions in their everyday lives, but need their parents to solve life's problems.

There is a transformation of sociability in the network society, as has been shown in studies of the social uses of the Internet (Wellman and Haythornthwaite 2002; Castells et al. 2003; Katz and Rice 2003), which leads to *the construction of a peer group through networked sociability*. The culture of individualism does not lead to isolation, but it changes the patterns of sociability in terms of increasingly selective and self-directed contacts. Thus, the new trend is the emergence of networked sociability. The medium of this sociability may vary. It includes, naturally, the Internet and mobile phones, but it can also be face to face. The critical matter is not the technology, but the development of networks of sociability based on choice and affinity, breaking the organizational and spatial boundaries of relationships. The social outcome of these networks is a dual one. On the one hand, from the point of view of each individual, his or her social world is formed around his or her networks, and evolves with the composition of the network. On the other hand, from the point of view of the network, its configuration operates as the point of reference for each one of the participants in the network. When a network is common to a number of its members, it becomes a peer group. In other words, networked sociability leads both to an individual-centered network, specific to the individual, and to peer-group formation, when the network becomes the context of behavior for its participants.

The consolidation of peer groups around shared values and codes of meaning for the members of the group leads to *the emergence of collective identity*. Youth culture(s) are signaled by the presence of these codes: for instance, a shared language, as in the practice of texting in wireless communication and in the adoption of new forms of expression in the written language. It is an open question, probably varying in each society, whether there is a shared youth culture or a series of specific subcultures. Probably, there is a common, global, youth culture, built around various distinctive attributes diffused by the global media, and then a number of specific national cultures and subcultures. The key question, though, is that each one of these cultures will need a set of specific codes of self-recognition, including its own language, as well as protocols of communication with other subsets of the youth culture. Communication is crucial in the formation and maintenance of young people's collective identity.

There is also an observed trend to personalize behavior within the youth culture. In other words, in parallel with the affirmation of a collective identity, there is also *a strengthening of individual identity* as a distinctive attribute of this collective youth identity. What is distinctive to contemporary youth culture is the affirmation of each individual who shares the culture: it is a community of individuals. Thus, there will be signs of individualism in each process of communication. Each person in communication will personalize his or her message and sender or receiver position.

Consumption is an essential value in our society. Thus, seeking status through symbols, and the stratification of individuals in relation to the

hierarchy of valued symbols, are important dimensions of youth consumerism. *The youth culture is not only a consumption culture, but consumerism
(that is, the high value attributed to consumption) is an important dimension of
the culture.* However, the patterns of consumption, and the valuation of
what ought to be consumed, change. These changes, which derive from a
combination of symbolic innovation, crowd psychology, imitation, and
status-seeking, are *modeled in patterns of signs that constitute a fashion.* While
the commercial value of fashion is essential for its diffusion, fashion is not
generated by commercial enterprises. Rather, the market is shaped by those
companies that identify early signs of a fashion and diffuse them under
their label. So fashion is related to collective identity, but it is not the same:
both identity and fashion are embodied in codes that are defined collectively. But identity comes first, and expresses itself in fashion. Fashions
change, collective identity is more stable.

Related to the above, but not the same, is the construction of collective
identity. To some extent, collective identity in the youth culture is a break
with the dominant culture. Fashion is the personalization of this rupture
to achieve an individualized identity. The sum of these individual identities, which express the collective identity in different ways, produces
fashion. To give an example: piercing is a break with the dominant aesthetics, and also an affirmation of the value of self-inflicted pain (it actually started in the sadomasochist movement of the early 1980s). So, it is
part of collective identity. But there is a personalization of piercing, using
different parts of the body, different devices, different colors and shapes.
The variety of piercing forms evolves into a fashion of piercing. We will
see below that the personalization of mobile phones can be linked to a
similar phenomenon.

*The culture of the network society is fundamentally characterized by the impor-
tance of projects of autonomy as a principle of orientation for people.* This has
individual and collective manifestations, including politics, for instance,
in the mobilization characteristic of the anti-globalization movement.
Social actors aim at building their autonomy in all dimensions of
their lives, thus defining their culture in their practice, and starting
not from who they are but from who they want to be. This search for
autonomy is particularly strong among the younger segments of the population, because they are the ones looking to influence and change society
rather than adapt to it. Therefore, here, the practice tends to be instrumental, but we need to understand instrumentality in a broad range
of meanings. It can be professional, it can be communicative, it can be
geared toward control of your own body, it can be political as well, but

what matters is that it is the actor who decides the purpose of the instrumentality.

Finally, *entertainment, games, and the media* are also important elements of youth culture, and, following the same pattern as other activities or services, they are becoming part of teenagers' wireless surroundings. Mobile entertainment is a leisure activity that rivals other activities. The fact that young people constitute the fastest growing segment of mobile communication users favors the entertainment function of mobile phones, as this is precisely the social group most predisposed toward amusement. Again, what can be observed is that technology fits a preexisting, latent need, distinctive of the networked society. And hence: our hypothesis is *that wireless communication provides an exceptional technology of communication to support the dimensions that characterize youth culture in the network society*, as presented above. Let us examine the extent to which this hypothesis is supported by the evidence.

The Management of Autonomy vis à vis Security

Traditionally considered in most cultures as both vulnerable to societal ills and highly risk-prone, young people (especially teenagers) tend to live under the close supervision of parents and guardians. Yet this is also the time when young people's sense of, and desire for, independence and privacy are growing. Wireless technology provides a means for parents and children to resolve this tension in ways that were not possible before. In this respect, we can find empirical evidence from different countries that seems to support our hypothesis.

First, studies in Europe show that small children use the mobile phone within the boundaries of the family. For those up to ten years old, there are regulations and rules to be followed, which means that parental permission is required for making a call, while children commonly read SMS messages (Oksman and Rautiainen 2002: 29, for Finland). Then, among pre-teens (10–12 years old) and teenage users, privacy and autonomy are respected by parents (Oksman and Rautiainen 2002: 30). These findings fit in with the cross-cultural fact that teenagers primarily use the mobile phone to organize their everyday lives and to maintain social relationships (Oksman and Rautiainen 2002: 27; see also Yoon 2003b for South Korea), and they do that autonomously and beyond parental control (Fortunati and Manganelli 2002: 60, for Italy).[8] Wireless phones enhance young people's sense of independence from the family by allowing them to distance themselves from parents and move closer to friends. To them, the cell phone is a source of empowerment (Elkin 2003, for the United States).

It is interesting to note, though, that this ability to exercise independence does not mean that young people neglect their ties with their family. The autonomy they enjoy as a result of owning wireless communication devices still operates within a framework of parental rules which set boundaries for how these devices are used (for example, in terms of cost and keeping in touch), especially when parents are paying the bill. The World Youth Report (United Nations 2003: 322, for the United States) notes that: "mobile communication creates what one might call an extended umbilical cord between youth and their parents." Thus the mobile phone is, in a paradoxical way, the keeper *and* the breaker of family ties (Lorente 2002: 6–8, for Europe). Indeed, there is a kind of "mime" or sham within the family, as can be seen from Fortunati and Manganelli (2002: 62), referring to Italy:

In reality, the mobile in children's hands may solve problems of organization and logistics, calm down parents' anxieties, such as knowing where their children are, but, however, it cannot solve the problem of quality and the flow of communication between parents and children.

The development of a sense of responsibility also passes in a limited fashion through the use of this instrument, inasmuch as most of the time, adolescents "mime" with the mobile in a public area and simulate autonomy and responsibility without actually enjoying them. Having often received the mobile as a gift from parents and being financially supported by them for their own use, adolescents are obliged to show continuous gratitude and acknowledgement toward too generous, permissive parents . . .

In the same way, their parents "mime" with a respect towards freedom as for the children which, in fact, they are very far from expressing, since they actually often would seem to feel the obligation of closely monitoring them.

In contrast, there is also some anecdotal evidence from the United States that, while cell phones may reduce the amount of control parents have over their children's physical movements, they may also be fostering better parent/child relationships (for example, increased trust, quick feedback, and ability to deal with issues on the spot) through the instantaneous communication links they enable (Dunlap 2002).

What is clear, from all the countries for which there is evidence, is that the availability of wireless communication technology *modifies* but does not eliminate the power relations between parents and children. In fact, as children become even more susceptible to being tracked down via their wireless devices, such as the cell phone, parents can get real-time information on their children's whereabouts.[9] Surveys in the United States indicate that young people maintain a high degree of communication with their families via wireless technology. For example, Fattah (2003) notes that US teenagers communicate wirelessly with their parents as often as they do with friends. This finding is supported by data from TNS (2004), which show that young people use the mobile phone mostly to call friends (60 percent) and family (59 percent; see also Bautsch et al. 2001). It appears that young people in the United States accept parental surveillance (the need to check-in often) as a part of their existence.

When focusing on children, the picture varies from one context to another, since what parents appreciate most is the *possibility* of constant communication (Oksman and Rautiainen 2002: 28, for Finland), not the constant contact itself. In Finland, for instance, communication through the mobile phone is strengthened when the child is home alone after school. Calls made by children to parents typically deal with everyday matters: "Can I have some ice cream, mom? Can I go out to play?" (Oksman and Rautiainen 2002: 29). On the other side, communications from parents to children are usually instructions, advice, and schedules (Oksman and Rautiainen 2002: 28). In the case of the United States, a practical illustration of this is that schoolchildren no longer need to go straight home after school so that their parents can call the landline from the office to check on them (Elkin 2003). Thus, while giving parents the security of a lifeline to their children, wireless phones also give children greater levels of the privacy and independence they crave.

However, the evidence for South Korea differs from what has been seen until now. Contrary to the idea that mobile phones would dramatically increase the autonomy of teenagers, Yoon discovered that the adoption of the mobile phone plays a major role in reinforcing traditional structures of family, school, and youth peer group under the *cheong* networks (Elkin 2003: 329). "*Cheong* is an expression of affective and attached relationships between people closely related to one another" (Yoon 2003b: 327). It is "an extended form of familism," maintained by deep commitment and prolonged communication that lasts for many years or decades (Yoon 2003b: 328). Hence, Yoon's main argument is that social adaptations surrounding the mobile phone in fact "immobilizes" the new technology. The authority of parents is maintained regarding the use of the mobile phone because a large number of teenagers receive their handsets as gifts from the older generation. Parents can exert significant influence over mobile-phone use through their control of phone bills. Mobile phones also help parents to track down their children constantly and give orders anytime, anywhere

(Yoon 2003b: 333–336). It is interesting that Yoon chooses the phrase "immobilize the mobile" or "retraditionalizing the mobile," which essentially describes the same phenomenon of the mobile culture reinforcing existing power relations. However, he uses a much stronger term to imply that the mobile phone has been appropriated to such an extent that it is losing some of its innate qualities, such as mobility or the potentiality to subvert tradition. This assessment, although possibly overstated in the context of other countries, may be accurate for South Korea, given the strong hierarchical social structure, especially the father-centered family order, which stands out in comparison not only with Europe and America but also its Asian Pacific neighbors.

Finally, it must been added that Yoon's ethnography of high-school students may not readily apply to other age groups (including college students), given the especially dependent social position of secondary-school teenagers. Since he chose to look at "mainstream" students, it is unclear to what extent the theme of "immobilizing the mobile phone" would fit with more marginal youth subcultures. But Yoon's analyses have systematically shown the ways in which mobile-phone adoption may not challenge the existing social order of families, schools, and peer-group networks. Instead, the new technology enables the strengthening of tradition.

Indeed, a mobile phone is often a present for young users, among others, in a number of European countries (see Fortunati and Manganelli 2002: 61, for Italy; Vershinskaya 2002: 145, for Russia; Höflich and Rössler 2002: 90, for Germany; UMTS Forum 2004: 25, for France; and Ling 2001: 8, for Norway), in South Korea (Yoon 2003b), and even in Brazil where "Children's Day" (or Dia da Criança) appears to be one of the days of the year that stimulates sales of mobile phones (ANATEL 2004). The general evidence, in fact, is that the growing ownership of cell phones by children and young people is driven largely by parents. For teens, and pre-teens especially, cell-phone expenses are borne by parents, and parents are heavily involved in the purchase and use of cell phones (see, for example, Ryan 2000, for the United States; Yoon 2003b, for South Korea; Fortunati and Manganelli 2002, for Italy). In the particular case of the United States, this may partly explain the high levels of wireless communication between young people and their families that we have previously seen.

The two main reasons stated by parents to justify the purchase of a handset for their children are, first and most important, safety (Oksman and Rautiainen 2002: 29, for Finland; Haddon 2002: 118, for the UK;

Noticiaswire n.d. and Teenage Research Unlimited 2003, for the United States), and, secondly, what we can call the "technological dowry" that parents consider is important to give to their children in order for them to avoid a possible "technological divide" (Fortunati and Manganelli 2002: 61, for Italy; Oksman and Rautiainen 2002: 29, Finland). Parents in the United States prefer to acquire cell phones for older children – the average age of a child whose parents provided the phone is 19 years (In-Stat/MDR 2004b) – and Mobile Village (2003) reports that people are more likely to buy a cell phone for a college child than for a teen or pre-teen. To some extent, owning a cell phone has become a rite of passage, like getting a driver's license, and is indeed often associated with beginning to drive (Wakeford and Kotamraju 2002; Fattah 2003). A similar situation had previously been reported for Europe, but the fast adoption of mobile telephony and the high penetration rates, especially among young people, gives a different current situation (Castells et al. 2004: 78–82). From our point of view, the different stages of penetration of mobile telephony could explain the different ways in which society treats this means of communication in the United States and in wealthy Europe. We can state, thus, that once the mobile-phone penetration rate of a specific country reaches a given level and the technology becomes a commodity, then the way in which parents introduce the device into their children lives changes.

It is noteworthy that, as seen for Europe (Castells et al. 2004: 78–82), American children are, however, beginning to acquire phones a lot earlier, even as young as eight years old (Selingo 2004). This is due to several reasons, mostly related to safety and security, including the attacks of 9/11, incidences of school shootings, and the generally increasing pressure from children as more of their peers acquire cell phones (Lewis n.d.; Wetzstein 2003; Noguchi 2004). For instance, a father interviewed by *The Washington Post* stated that his children had cell phones because, after 9/11, he wanted always to be able to find out where they were (Noguchi 2004, as seen above for Europe).

However, once the handset becomes a device of everyday life, daily use differs from the initial purchase motivations following a dynamic process of domestication (Frissen 2000: 72, for the Netherlands; Fox 2001, for the UK). Affective and sociability uses emerge among children besides the safety/security reasons that prompted the initial purchase of the device, thanks to the permanent contact with peers made possible by mobile telephony (Fortunati and Manganelli 2002: 72, for Italy). This communication from home to home can also be clearly seen in the UK, where a "bedroom culture" has developed as a response to parents' desire to keep children in safe places where they can be monitored (Haddon 2002: 118).

Finally, it must be highlighted that, in general, as with the Internet or computers in general, the trend has been observed that "young users" teach parents how to use mobile phones (Fortunati and Manganelli 2002, for Italy; Vershinskaya 2002, for Russia). With information technology, the whole culture of learning is undergoing change (Vershinskaya 2002: 146, for Russia), leading to the transformation of the patriarchal model. Thus, although the education process remains vertical, roles have been reversed and the flow of knowledge has changed direction to rise from the younger generation up to the older (Fortunati and Manganelli 2002: 71–72, for Italy). Thus, the traditional top-down communication dominated by the older generation is starting to be eroded as more youngsters are adopting the mobile phone (S-D. Kim 2002).

The Construction of a Peer Group through Networked Sociability

Young people may seek to be independent, but not completely. An aspect of their lives is also to build their own communities and to be connected with those they choose. Hence, there appears to be a unique differentiation in their uses of different communication technologies for different purposes. For example, some studies have found that young people use e-mail mainly for making contact with non-family adults, such as teachers and employers, while other methods, such as cell-phone conversations, instant messaging (IM), and SMS, are used to communicate with family and friends (Schiano et al. 2002).

Indeed, mobile telephony is an appropriate tool, and a highly important one, in relation to the creation and maintenance of peer groups (Lorente 2002). For teenagers, the mobile phone appears to be an everyday object that may even possess certain human characteristics: it is, after all, the gadget that enables the owner's social network to be continually present online (Oksman and Rautiainen 2002: 25, for Finland). Mobile telephony sets up one of the channels through which peer groups of young users are maintained. It features the creation of new spheres of intimacy, new ways and moments of communication which are at the core of the mobile youth culture. In fact, having the opportunity to be connected in multiple ways appears to have had a significant impact on the way in which young people communicate (Smith et al. 2003: 554). This "perpetual contact" (Katz and Aakhus 2002: 2) has been favored by the fact that handsets are personal, individual devices which are always carried by the owner and tied up inextricably with the everyday life of young people (Andersson and Heinonen 2002: 7; Katz and Aakhus 2002; Lobet-Maris and Henin 2002: 111; Lorente 2002; Cohen and Wakeford 2003), allowing new flexible and creative ways of use, as young people have already discovered, which help break the organizational and spatial boundaries (Ito 2004) of relationships.

What is more, there are a variety of ways for mobile phone-facilitated youth networks to materialize, as recorded by several studies. While McVeigh argues that the use of mobile phones strengthens the trend toward individualization, in which young people are becoming increasingly self-centered (McVeigh 2003: 24–32), Matsuda sees the emergence of a "full-time intimate community" (cited in Ito 2004: 11). In addition, while Ichiyo Habuchi is concerned about "tele-cocooning," i.e., the production, through mobile communications, of social identities in small, insular social groups (cited in Ito 2004: 11), Ito (2003b) describes how mobile-phone use creates an "augmented co-presence" that functions at the center of youth groups before, during, and after their social events.

The fixed telephone is a community communication tool within the family structure (Lobet-Maris and Henin 2002: 111, for Belgium), and has traditionally allowed parental surveillance (S-D. Kim 2002, for South Korea; McGray 2002: 52, for Japan). By contrast, mobile phones are characterized by their personal, private, and, in certain cases, intimate use (Lobet-Maris and Henin 2002: 111, for Belgium). Thus, the handset itself can be understood as a communicational node – always attached to a person – of the social network. Within this view, we can distinguish two different ways in which the mobile phone is used to create, maintain, and reinforce relationships among young users: *through* the mobile phone and *with* the mobile phone. Let us explain.

Creating, Maintaining, and Reinforcing Links through the Mobile Phone

Despite other features the device may have, the mobile phone is above all perceived as a communication tool (Moore 2003: 65). In this sense, and following the taxonomy posed by Ling (2000), two roles that mobile telephony has for many adolescents must be highlighted here. The first one is *accessibility* (or availability), and the second is *micro-coordination*. Both functions are of major importance for the creation, maintenance, and reinforcement of peer links through the mobile phone.

On one hand, accessibility is attainable because the mobile phone allows the remarkable combination of social availability with intimacy. Nowadays, teens can decide when, and where, they can be reached. Micro-coordination, on the other hand, is described as the "nuanced management of social interaction" (Ling 2004: 70). Indeed, lovers and friends are adopting a more flexible manner of making appointments (S-D. Kim 2002: 73). Thus, the use of mobile telephony to glean information on when and where to meet friends is key to a lifestyle wherein "mobility and flexible scheduling and, perhaps, the desire for greater privacy in telephone communications are central" (Gillard et al. 1996; quoted in Ling 2000).

Meanwhile, evidence from South Korea shows similar patterns, as Yoon's (2003b) research shows. The author interviewed teenagers aged 16–17 years from four schools in Seoul on a one-to-one basis or in small groups, with e-mail follow-ups. A number of diary-type self-reports were also collected. Yoon found that the teenagers used mobile phones for three main purposes: first, the mobile phone plays a supplementary role to school relations based on current, face-to-face communication; secondly, it helps sustain relationships with old friends and peers in other schools, whom the teenagers cannot physically meet on a frequent basis; thirdly, the mobile phone is also used to maintain newly acquired relationships (Yoon 2003b: 330–331).

One particular means of communication that has emerged as the most important among young people is SMS. SMS, like voice calls, accomplishes the two roles stated above. Communication can be both instrumental and expressive. This, of course, is also valid for SMS as a teen can both coordinate a meeting and/or say goodnight to a boyfriend/girlfriend after turning off the bedroom light (Fortunati and Manganelli (2002: 75), or even put into circulation certain messages, for instance, those of moral support during term exams (Yoon 2003b: 336–337). However, in the UK, a study observed that texting is mainly expressive (Continental Research 2001).[10] In Norway, there has also been a remarkable move from instrumental to expressive use of the mobile phone among teenagers (Ling and Yttri 2002: 166). These all exemplify what we understand as the fundamental features of expressive texting: it can be local; it will be used as long as it is affordable; and it is perceived as entertainment. Thus, texting acts as a catalyst for the construction and reinforcement of peer groups.

Besides this, there are also examples to illustrate how mobile telephony can both reinforce peer groups and create exclusion at the same time. In our ethnographic study of Chinese migrant workers, we found a female factory worker, aged 19, who did not have a mobile phone at the time and complained that texting was taking up most of the free time of her dorm-mate:

Every night and every weekend, she just sits on her bunk bed spending hours and hours staring at the handset, typing, smiling, and making all kinds of facial expressions. It was annoying because, while doing this, she completely ignores the rest of us, whatever happens in the dormitory. We [including all other dorm-mates] all find it annoying. It's disrespectful.

In this case, several cultural processes were taking place. The texter was using her handset to bypass the immediate built environment surrounding her – the factory, dormitory, and bunk bed – to become connected with her intimate social networks. But in so doing, she was also silently showing off to her peers her supposedly higher social status, signified by the gadget, her capacity to type with her thumb, and the fact that so many people in her network were exchanging messages with her. Meanwhile, this was triggering tension within the dormitory, making the non-adopters feel neglected, annoyed, and possibly pressured to buy their own mobiles.

Other interesting processes performed through the mobile phone are those particular practices that break the reciprocity of communication in peer networks. One example, arising from South Korea, is so-called "chewing out" (*Ssibgi*) which refers to the practice of ignoring calls or messages from others. As one respondent from Seoul said: "The most upsetting thing in using the mobile is to receive an insincere reply and to be chewed out by the person to whom I have sent a message" (Yoon 2003b: 338). The importance of m-etiquette to avoid chewing out was expressed on another occasion when a student explained why she removed the battery of her mobile for a while: "If someone asks me 'Why did you discharge the battery? Why don't you just ignore it when you receive bothering texts?' I will say, I am a quite honest person and I'd rather die than chew out messages from others. Ha ha" (Yoon 2003b: 338). By contrast, there is substantiation of the fact that young people are also using the mobile phone to transmit messages that otherwise might cause embarrassment if they were delivered face to face. This is evident in the following quotation from a female student:

When I bought a mobile for the first time, I made peace with my friends with whom I had quarreled, you know, she also had a mobile. Because . . . I was so excited . . . so, I sent a message to the friend . . . I messaged, "I did something wrong to you last time . . . sorry about that" . . . and then, we began to get along together again. (Yoon 2003b: 337–338)

Nevertheless, not everything is texting. Indeed, we can see a different use for mobile telephony among young European users of a slightly older age. Young adults give up intensive use of SMS (Ling 2002, for Norway;

Smith et al. 2003: 554, for the UK; Valor and Sieber 2004, for Spanish Internet users) and adopt voice calls as an important part of their communication, which would be mainly conducted during the afternoon or evening (Valor and Sieber 2004). This, indeed, seems to be valid for richer countries, while, in the case of less wealthy countries, young adults are the main users of this means of communication, as previously seen in the Philippines. Our hypothesis is that young users quit texting when they can afford other mobile services relative to their income. In any case, mobile telephony is retained as an important tool of communication among those who give up texting.

Creating, Maintaining, and Reinforcing Links with the Mobile Phone

Peer groups are formed in the physical, everyday world, and relations are partially maintained through the mobile phone, as we have already seen. Young people also use the device locally, sharing it mainly within the peer group and, sometimes, using it to establish new contacts (Weilenmann and Larsson 2002, for Sweden). In this sense, a collaborative use of the mobile phone has been observed within peer groups in Sweden, while Yoon (2003b: 337) found a range of sharing activities, such as the borrowing and lending of handsets, and collectively receiving and sending messages, in South Korea.

Taking into account whether the device is physically lent or not, Weilenmann and Larsson (2002) describe two main types of mobile-phone sharing: minimal sharing and hands-on sharing. The *minimal sharing* category refers to a strict sharing of content; that is, information. In this case, the device remains in the owner's hands and the collaborative use of the mobile phone can have different forms. An SMS can be read aloud, or be written together; in the same way, an MMS can be shared with surrounding peers. Finally, a conversation through the mobile phone can also be shared with present peers. While speaking on a mobile phone in public spaces has created a conflicting situation in which the communicator is simultaneously both there and elsewhere (Kasesniemi 2003: 26, for Finland), the behavior described above works in the opposite sense by linking peers present around the mobile-phone speaker with the other end of the line, by means of transmitting the conversation or by using the hands-free feature of the device. Moreover, in oral conversations, nonverbal communication could also be relevant as a way of communication among teens who are physically together during the telephone interaction.

Secondly, the *hands-on sharing* category includes situations in which a person borrows the mobile phone from another, a behavior that only happens if there is enough confidence between the lender and the borrower. We may also add that a new set of etiquette rules has been formed regarding the sharing of handsets in that, if one member of the group refuses to share his or her mobile phone, it is often considered "really not cool" and "irritating" (Kasesniemi 2003: 337), and predictably the group will punish free-rider behaviors.[11]

However, it is worth noting that ownership of a specific technology can also exclude individuals from the processes, described above, in which links are managed with mobile phones. As evidenced in the United States, and following the same pattern previously pointed out in China, the "new digital divide" developing between young people with cell phones and those without is seen in concrete form, as inclusion in social networks is frequently tied to ownership of a means of wireless communication (Batista 2003). In addition to maintaining relationships through face-to-face contact, young people are able to strengthen, solidify, and coordinate these relationships through the seamless, always-on link afforded by wireless communication devices. Without ownership of one of these devices, a young person cannot expect to be a meaningful part of the social networks of cell-phone owners. It is clear that the wireless industry has understood and is fueling this trend with its in-network pricing packages. An advertisement which ran on US television in 2004 illustrates this mindset: a group of young girls state that they do not talk to people who do not have the same phone plan as they do, not because they are mean, but because they are better shoppers. Thus, it is now not only a matter of having a means of wireless communication, but there is also differentiation around which wireless plan you have. Of course, this is purely an industry construction; whether young people will allow the wireless network system to dictate their social networks remains to be seen.

All the types of group behavior seen allow young people to "engage in reinforcing their feeling of family-like friendship" (Yoon 2003b: 336–337). Thus, they increase collective identification and a feeling of bonding, thanks to the creation of a greater sense of intimacy between peers. In this sense, and because of its importance, the next section is devoted to the issue of collective identity.

The Emergence of Collective Identity

The way in which young people adopt and appropriate mobile-phone technology contributes to the construction of their own culture by means of differentiation from adults and, particularly, from parents. Texting is probably the most evident way in which the collective identity of mobile youth is created and maintained. This is the reason why, in the following, we will discuss SMS as well as the different ways in which texting and other mobile-phone features are used.

Thanks to SMS, young people have created their own private, exclusive language to communicate through short text messages of 160 characters. Despite the existence of so many examples of the use of SMS language (see chapter 6 and online appendix 65) and so many books and web pages available on this topic, each youth group has its own language which generates differences to distinguish them not only from adults but also from other youth groups.

Texting is a creative way to use the mobile phone in a context in which the technology is available but young people have very limited budgets to use this communication tool. Indeed, SMS was dismissed by the telecommunication industry for private users because it is hard to use: considerable time is needed to write a message which cannot contain much information (Lobet-Maris and Henin 2002: 103). The costs of texting were seen as high when compared to the possibility of oral communication (Agar 2003). And, of course, business users take less account of the monetary cost than teenagers, who, on the other hand, have plenty of time to "waste" on this kind of activity: "Boys describe how they will spend 15 to 30 minutes composing a single message if they consider it significant" (Kasesniemi and Rautiainen 2002: 184).

As in many other cases, technological innovation was followed by innovation in use and then by a cultural innovation. In this case, a cheap and cost-controlled way of communication grew, allowing a means of reinforcing collective identity that is similar to the "brief, frequent, spontaneous" social networks of pre-industrial communities (Fox 2001). Mobile telephones can thus be considered the "new garden fences" (Fox 2001). Writing, sending, and receiving SMS, composing and receiving MMS, or even "boom calls" belong to social activity. While texting can have an instrumental purpose, it has also been developed as a means of expressive communication, as seen previously in this chapter. This communication, of course, is performed in a different way, and has different connotations, from face-to-face or voice interactions (see chapter 6).

Actually, teenagers send and receive a large amount of SMS (Kasesniemi 2003: 82, for Finland). In the Philippines, for instance, texting has given rise to "an apparently novel social category: *Generation Txt* . . . An obvious pun on *Generation X*, Generation Txt was first used as an advertising gimmick by cell phone providers to attract young users" (Rafael 2003: 407).

It was then picked up and popularized by journalists to refer to young Filipino texters. With their help, Generation Txt has become a central signifier in symbolizing the collective identity of SMS-equipped urban Filipino youth. In China, the diffusion of SMS among youngsters is also most impressive as a result of the faddish appeal of the service as well as its low price (Sohu-Horizon Survey 2003). Picking up the Japanese phrase, urban magazines are also celebrating China's own youth "Thumb Tribes," for example, in the southern city of Guangzhou (*New Weekly* 2002).

Young people may conduct a text-message conversation over a number of hours, maintaining contact with friends when they are away[12] and writing what they consider to be significant messages (Kasesniemi and Rautiainen 2002: 184). However, they can also expend hours texting locally, that is, with other teens who are in the same room, as described by a 14-year-old Finnish girl:

When they had the campaign that allowed you to send SMS for two cents a piece,[13] we pretty much sat there all day with the mobile and probably sent a few hundred messages in all. We could be seated on a bed next to each other typing messages to one another. For three or four hours we just sat on the bed sending messages to one another. (Kasesniemi 2003: 21)

What is more, texting can also be seen as a way of killing time as a teenager may expend hours sending and receiving messages to while away a boring evening or during a trip (Haddon 2002, for the UK; Oksman and Rautiainen 2002, for Finland). In some situations, texting is better than calling, not only because of the connotations of the communication channel, but because more time is expended in the activity itself. If you are using SMS in its expressive function, why kill the moment with a call? A call, indeed, is more specific, and once ended there is perhaps no justification for sending a new SMS. Meanwhile, there is also evidence to show that SMS is usually used in the local surroundings of young people and mostly at home (Grinter and Eldridge 2001, for the UK). This service, therefore, will continue to be used as long as it maintains its low cost and is perceived as an entertaining activity.

Apart from the activity of texting itself, there are other practices that also create, maintain, and reinforce collective identity among teenagers. Some of these practices are related to texting, but others are not. We are talking, for instance, about "boom calls," teasing SMS, chained messages, collective writing or reading of SMS, SMS collection, MMS, and even games. Some of these practices are analyzed below.

A "boom call" is a mobile-phone call that is designed not to be answered and, thus, entails no cost for the caller. As long as handsets can identify the incoming number, a ring on the device becomes significant for the receiver. Boom calls can be used with an expressive intention (meaning, "I am thinking of you") or for coordination purposes (for instance, meaning "Hello, I've just arrived. Are you coming?"). A previous accordance of content should have been arranged, as when defining all kinds of private languages. This is reportedly one of the more popular practices among university students in some African countries, who use it for such purposes as reminding classmates that it is time for a class or an exam (Donner 2005a). In addition, because of the financial constraints faced by consumers in Africa (which are likely to be even more pronounced for young people), there is a tendency for young mobile-phone users to use this form of contact, especially when there is a need to communicate with their parents. Boom calls, or even empty SMS (Oksman and Rautiainen 2002), can be used in a teasing way, and some "boom call games" have been developed within teenagers' groups (Kasesniemi 2003). In addition, as shown in chapter 2, the adult population also uses boom calls.

Other practices among teenagers include teasing SMS and chained messages. They can be of different types, and can include both text and simple images created with the characters allowed by the handset. In Nordic countries, besides ordinary jokes, there have been SMS that made pejorative remarks about users of others mobile-phone companies in a humorous way (Puro 2002). Sometimes, because mobile devices do not have much memory, messages are kept and shared among peers (Kasesniemi and Rautiainen 2002; Puro 2002; Weilenmann and Larsson 2002). There are various kinds of SMS that a teenager can collect. One category is jokes. Then there are also private messages that most times are kept to recreate texting conversations with, for instance, a boyfriend or girlfriend (Kasesniemi 2003). This helps to reinforce the relationship, or revive it when there is a crisis. Finally, there is collective writing and/or reading of SMS, which has been discussed in the previous section on the collaborative use of the mobile phone.

Strengthening of Individual Identity and the Formation of Fashion

Personal identity is important to young people, especially teenagers (Wilson 2003). This can be seen in their attitudes to mobile technology and in their preferences for its products, such as cell phones, ring-tones, wallpaper, and icons. Ownership of a wireless communication device affords autonomy. This is followed by personalization of the device, as a way of attaching individual character to such an ubiquitous device. Music, for example, is an important form of self-expression for young people in the United States, and wireless communication technologies, such as music phones, allow them to express their identity more visibly, virtually on a constant basis (Petroff 2002), and in tune with current fads and fashions.

Nowadays, the mobile phone has become a symbol of youth identity in many countries (Ling 2004: 103). For children and teenagers, and also for young adults, although with less intensity, ownership of the "correct" type of mobile phone is a subject of relevance (Ling 2001, 2004; Fortunati 2002a). Nevertheless, having the appropriate device is not enough; it must also be personalized (Oksman and Rautiainen 2002; Kasesniemi 2003; Ling 2004) because, ultimately, it is an expression of personal style and way of life (Oksman and Rautiainen 2002). Indeed, as stated by Fortunati (2002a: 56): "With respect to other mobile technologies, it is the one item that specifically presents us with the problem of wearability and, thus, of its relation with clothing."

A mobile phone, then, can be compared with, and treated as, a piece of clothing, linked to temporary collections because it is a product of limited life that is always attached to the body. Within this strict point of view, we can agree with other authors who identify the mobile device with a watch (Ling 2001; Fortunati and Manganelli 2002; Oksman and Rautiainen 2002; Kasesniemi 2003). And, most important, mobile technologies become closely involved in the processes of self-conscious display, self-assessment, and self-improvement (Cohen and Wakeford 2003).

The personalization of handsets in Europe was initially achieved by painting them with different kinds and colors of paints or, perhaps, by making assorted pouches in which the device could be kept and carried (Oksman and Rautiainen 2002; Skog 2002). This, in some way, led to a market evolution, and handsets began to be made available in a way that allowed the end-user to change some aesthetic elements, which, in the end, created a new fashion. This is one of the many examples that can be given of the mutual influence between the creation of individual identity and the formation of fashion. A similar method of personalization has been seen in China, where young people also customize the cell phone into an "artifact" to demonstrate individuality, using all kinds of "hand-phone cosmetics," made from crystal, feathers, and silver, in the shapes of "Hello Kitty," "Garfield," flowers, and animals (Yue 2003).

Following Skog (2002), it can be stated that teen users are not only consumers but producers as well, since they are free to create an individual phone by combining downloaded ring-tones, logos, pictures, and games, as well as different external elements. The flexibility and social contact allowed by the technology means that it has become harnessed to the cause of many teenagers' identity projects.[14]

The whole concept of individuality may also have some limitations, however. For example, some industry analysts have concluded that what works for the teenage market is "prepackaged individuality," having found that cell-phone features that give users publishing capabilities (for example, composing your own ring-tone) are not popular with teens (Lee 2002b). As discussed in chapter 3, Japanese R&D taskforces have conducted research among members of their target social group so that they can deliberately incorporate considerations of identity into the design and promotion of their new handsets and wireless services. This was also the case with Nokia, the world's largest mobile phone manufacturer, who altered their classic design to introduce clamshell phones in order to reflect the changing tastes of final consumers (Reinhardt et al. 2004).

There are other aspects of mobile youth culture in Japan, such as its ultra consumerism and the "cute culture" or "culture of *kawaii*," which we discussed in chapter 2 because these trends relate significantly to general characteristics of the whole society. Of particular relevance to our discussion here is the case of *kogyaru* (high-school girls), which is "a label attached to the newly precocious and street savvy high school students of the nineties who displayed social freedoms previously reserved for college students" (Ito and Okabe 2003: 6). As observed elsewhere, we can see the familiar "emancipation" of young people, aided by the new technology, but it is of specific significance in Japan because it "flies in the face of mainstream norms that insist that young women be modest, quiet, pale, and domestic." Yet, on the other hand, most Japanese schoolgirls, including full-blown *kogyaru*, "tend not to have oppositional relationships with their parents and teachers" (Ito and Okabe 2003: 6). They often maintain a split personality and hide their *kogyaru* identity in front of their elders, an observation that also echoes our analysis of youth culture in the network society in terms of young people's management of autonomy vis à vis security.

The changing power dynamic is most manifest among cell-phone-equipped *kogyaru*:

In certain city centers, *kogyaru* continue to be highly visible, sporting platform sandals, brightly-colored fashions, sun-tanned faces, colored hair, and often a highly decorated mobile phone hanging from their necks. Unlike the male *otaku* (techno-geeks) associated with video games and computers, media savvy girls have been associated with communication technologies such as pagers and mobile phones. *Kogyaru* are commonly thought to be the social group that pioneered and popularized uses of mobile communications, first with their appropriation of pagers in [the] early nineties, and then with mobile phones [in] the latter half of the nineties. Within a space of a few years between 1995 [and] 1998, mobile phones shifted from [an] association with business uses to an association with teen street culture. This shift coincided with the high visibility of *kogyaru* in the media and on the streets. (Ito and Okabe 2003: 6–7)

Projects of Autonomy as Principle of Orientation for People

On this topic, we must first distinguish between two kinds of projects of autonomy: the individual project and the collective project. With regard to the *individual project of autonomy*, mobile devices tend to help the development of the project itself, although, as in the case of computers, some new ways of surveillance may also emerge as the technology develops (Rule 2002). Indeed, an increase in private surveillance has been predicted as an outcome of the popularization of mobile telephony, although such perceptions are not unvarying (Rule 2002; Vershinskaya 2002).

We have discussed above the individual project of autonomy as related to the emancipation of youth from parental control. With regard to other particular points, Skog (2002) found that young users belonging to working families saw the mobile telephone as one of the instruments that could help with the development of their autonomy. Moreover, it is common for some young people to use the mobile handset as their only telephone, usually during those more nomadic periods when a transition toward an adult way of life is taking place (Ling 2004), as we also found for Chinese migrant workers.

With regard to what can be called the *collective project of autonomy*, the free Wi-Fi collectives should be mentioned. Spreading across Europe, this project claims a cooperative use for Wi-Fi connections with the argument that, as happened with the Internet, non-profit-oriented cooperative behavior will better benefit society. Indeed, we are talking about civil associations whose objective is to create areas of free wireless access to the Internet. And, finally, also related to the collective project of autonomy but with great importance at present, we should mention the use of mobile telephony for political mobilizations, which will be discussed in chapter 7.

However, while the rise of *oyayubisoku* – a term meaning "Thumb Tribe," which describes youngsters who can type cell-phone messages by moving their thumbs at extraordinary speed, sometimes without even looking at the handset – continues to amaze Western scholars (Rheingold 2002: 4–8), Japanese researchers, such as Tomoyuki Okada, see *keitai* cultures as developing "out of the fertile ground of youth street practices and visual cultures and a history of text messaging that extend[s] back to youth pager use from the early nineties" (Ito 2004: 8). These practices, while being facilitated by new technologies, reflect the structural conditions that constrain the social activities of Japanese youth. According to Ito and Okabe (2003: 1), "Teens use mobile phones because they enable new kinds of social contact, but also because teens are limited in access to adult forms of social organization." In more specific terms, this means that:

While youth do have large amounts of discretionary time, energy, and mobility that is the envy of working professionals and parents, they are limited in their activities by their weak social position and limited access to material resources. Their lives are governed by certain structural absolutes, such as dependence on parents, educational requirements, and regulation in public places. (Ito 2004: 8)

Indeed, the most notable manifestation of the power of mobile-equipped youth is without doubt the Nosamo movement of 2002, which played a crucial role in electing Roh Moo-Hyun to the presidency in South Korea. This was an important event because it effectively mobilized "the 20- and 30-somethings who might have otherwise sat out the election because of cynicism about the political process" (Demick 2003; Rhee 2003: 95; see also *Korea Times* 2002). On the other hand, in the Philippines, when a similar process happened, it was a kind of shorthand Taglish[15] that was used to compose millions of messages transmitted during People Power II, the dramatic showdown in 2001 between Generation Txt and the country's sitting president, Joseph Estrada. In addition, the demonstrations in Spain of March 13, 2004 followed the same course. We will have more to say on the sociopolitical aspect of these three cases in chapter 7.

Entertainment, Games, and Media

We have already discussed, in chapter 2, the significance of mobile entertainment in everyday life within developed societies; here we will focus our attention on its influence on youngsters. The fact that the young population constitutes the fastest growing segment of mobile communication users favors the entertainment function, as this is precisely the social group more predisposed toward entertainment. In Japan, for example, more than half of i-mode's data traffic points to entertainment content, such as games (Lindgren et al. 2002: 61).

The year 1997 was an important one in the history of mobile entertainment when Nokia first released the game of Snake for free, embedded in their mobile phones (MGAIN 2003a: 13). It was a success thanks to the following features: it was affordable (in fact, free), accessible if you had the proper device, and appropriate for young users by allowing them to play and have fun, kill time, compete with peers, and go beyond the pure communication function of the technology.[16]

Preliminary analysis of patterns among young people shows that when they make spending choices, *entertainment is key* (TNS 2004). A study reports, for instance, that US teenagers' primary objective is to have fun while they are young, in anticipation of a more restrictive and responsibility-laden adult life (Teenage Research Unlimited 2000). Technology plays a central role in this pursuit of fun. Thus, the number one answer in a study of two thousand respondents describing their peer group was "we're all about fun," while the number two answer was "high-tech is such a [huge] part of our lives." One aspect of this is high-tech games, which seem to be particularly important to this group. American youngsters aged 6–14 years spend most of their money on games (63 percent), clothes (31 percent), and CDs (27 percent; TNS 2004). As can be seen in table 4.3, entertainment is a major component of young people's wireless communication usage. Games accessible via cell phone provide US teens with immediate gratification, peer influence, and the convenience of not having to carry an additional electronic gaming device (Petroff 2002).

Table 4.3
Uses of cell phones by children 6–14 years in the United States, 2003

Activity	%
Call friends	60
Call family	59
Download games	41
Download ring-tones	38
Use Internet	38
Text friends	36
Take pictures	34
Text parents	30
Text/call TV shows/contests	30
Download pictures	30
Have cell phone	29

Source: TNS (2004).

Thus, despite lagging behind their peers in other countries in their uptake of wireless technology, US youth are reported to download more wireless games than young people in Europe and Asia (TNS 2004).

One finds young people straddling the border between childhood and adulthood; hence the high importance of entertainment in their mobile usage, alongside other more utility-based uses. As Sefton-Green (1998) notes, new digital technologies provide children with realms of "adultification" and "juvenilization," that is, the ability to act as adults while remaining immersed in the world of leisure and games. While some may criticize the predominance of entertainment in young people's wireless usage, one should also acknowledge that play is a valid part of human existence, which just happens to be more apparent in our youth (for example, Sandvig 2003: 179).

There is a type of personalization that belongs to the entertainment sphere that industry has made good use of: ring-tones and other multimedia contents, which can be downloaded to the mobile phone (as discussed on p. 113 in the context of the strengthening of individual identity). Indeed, this is a good example of how a market niche could be exploited to generate good revenue. For instance, after voice communication with family and friends, and downloading games, downloading ringtones is the next most popular activity among US youngsters aged 6–14 years (table 4.3; TNS 2004). Their willingness to pay for this type of wireless data service has been observed in a number of surveys (Petroff 2002; Dano 2004), indicating that young people consider personal expression to be important enough to wish to pay for the ability to do so in unique ways. According to entertainment and cell-phone executives, "the biggest market for ringtones is teenagers, for whom simply owning a cell phone is no longer distinctive" (Tedeschi 2004: C5).

According to research by the Yankee Group, in 2003, 41 percent of young adults and 22 percent of teenagers in the United States downloaded at least one ring-tone per month (Marek 2004). During the same year, cell-phone users spent between US $80 and 100 million on ring-tone downloads (Tedeschi 2004). The importance of this element of personalization may decline with age. Schiano et al. (2002) found, in their study of teens and pre-teens in affluent Palo Alto, that younger teens were more concerned with personalization, entertainment, and the "coolness" factor, while older teens were more concerned with utility. Although American children are reported to have large amounts of disposable income – teenagers spend about US $174 billion a year (Anfuso 2002; Teenage Research Unlimited 2004) – most of them have to earn their money through formal employment or household chores, rather than get it as pocket money: only 25 percent of US children receive pocket money (Anfuso 2003; TNS 2004). It is not surprising, therefore, that, in addition to style and features, American youth are most sensitive to its (high) cost when it comes to buying wireless communication devices (Garcia 2004).

What is clear is that teenagers must decide where to spend their income, just as adults do. Moreover, there is some empirical evidence that supports the hypothesis of substitution among entertainment goods and services. It seems, specifically, that an observed decline in teenage smoking is correlated with mobile-phone ownership, as reported by Lacohée et al. (2003). There was a sharp decline in smoking among British boys and girls aged 15 in the late 1990s during which time mobile-phone ownership sharply increased. Among other reasons, the authors highlight the fact that mobiles consume teenagers' available cash, particularly through the need to top up pay-as-you-go cards (Lacohée et al. 2003: 208).[17]

Meanwhile, what is peculiar in terms of youth mobile-phone culture concerns the explicit role of commercial promotions, often involving mass-consumption multinational corporations. The phenomenon, which also exists in other countries, surfaced quite prominently in the Chinese case, as reflected in the conference "Marketing to Teenagers in China."[18] One example was the Coke Cool Summer contest (Cellular News n.d.). During July and August 2002, Coca-Cola China hosted this interactive contest that generated 4 million SMS messages in 34 days. To win, users had to correctly guess the highest temperature in Beijing everyday, based on the highest temperature of yesterday which was text messaged to their handsets on a daily basis. The prizes included one year's free supply of Coke and new Siemens mobile phones. Participants could also download the Coke advertisement jingle and mobile coupons that gave away free ice creams at McDonalds in Beijing and Shanghai. By the end of the event, 50,000 downloads of the Coke ring-tone and 19,500 downloads of McDonalds mobile coupons had been recorded. KFC, among others, also launched similar SMS-based advertising campaigns.[19]

These advertising campaigns highlight an aspect of the mobile youth culture that has so far been inadequately addressed. Many existing studies emphasize the fact that cell-phone adoption gives young people independence and autonomy (Katz and Aakhus 2002; McVeigh 2003); others contend that cell-phone usage does not necessarily cause clashes with the older generation, but, rather, it may even reinforce existing power relationships in the household (Ito and Okabe 2003; Yoon 2003). While both arguments may be true, there is a third dimension of this consumption culture which clearly shows that the cell phone can further empower large corporations in shaping the consumerist identity of young cell-phone users. Given the susceptibility of youth, and peer-group pressure now materialized in the hyper-fast networks of the mobile phone, individual youngsters now have very little autonomy or independence not to get involved in the commercial vogue created by the likes of McDonalds and KFC. Besides this, the expansion of the SMS market among Chinese youth is traceable to the promotion of content by all three of China's Internet portal sites (Sina, Sohu, and Netease), which are the main daily content providers of subscription-based text messages, ring-tones, and images (Clark 2003).

Finally, we should mention the "American Idol effect" (In-Stat/MDR official, quoted in 3G Americas, n.d.). This phenomenon refers to the fact that, in the United States, young people were the first group to appropriate SMS following the promotion of texting via television entertainment shows that encouraged their (mostly young) audience to send voting messages via their cell phones. Nowadays, indeed, it is very common, in both television and radio, to use SMS as a communication tool to complement other communication means. Participation may be through a closed-structured message sent for voting purposes (as in the Eurovision Song Contest) or in order to win a prize; it may also be an open-structured text sent to state an opinion, to ask for something, or to contribute to a debate with audience members in a live program.

Summary

This chapter has shown that, overall, a youth culture is emerging across Europe, the Americas, and the Asian Pacific when it comes to the ways in which young people use the mobile phone. Ample evidence has been collected and analyzed with regard to the following characteristics of this mobile youth culture:

1. Young people across the world are quick to adopt and appropriate mobile technologies, as long as they can afford them, because in general they use these new services with more intensity for all kinds of purposes in their everyday life. As a result, they become a major social group that is constantly networked through wireless communications, and in so doing they reveal more quickly the potential uses of the technology compared to people of an older age.

2. Mobile-phone usage is transforming youth cultures around the world via two interconnected processes. While the technology enhances the autonomy of young people as independent, communicating selves, it generally does not lead to the weakening of the dependency relationship between young people and traditional social institutions, especially the family in the form of financial support and/or parental surveillance. Mobile-equipped young people in different societies thus face the same central question: how to manage the new opportunity for autonomy under the existing structural conditions imposed on them, most importantly within the family and at school.

3. The mobile youth culture, as a new set of values and attitudes that informs practice among the younger generations, is a typical networked culture. Peer groups formed at school or in residential neighborhoods often serve as its basis. What the spread of the mobile phone does is to reinforce and extend existing youth networks and drive them toward a higher level of networked sociability, where face-to-face interaction is equated with mobile-based communication, and where a "full-time intimate community" takes shape. The process of reinforcement is also a process of selection because, as researchers have found on many occasions, communication in mobile youth networks is usually restricted to a small circle of close friends.

4. Mobile youth culture in many countries, such as those in the Asian Pacific, is characterized by a strong consumerist tendency. This is manifested materially in the appropriation of the mobile handset as a fashion item and in the entertainment dimension of young people's mobile usage. When used in the public space, mobile devices are also displayed as notable objects of consumption.

5. With the diffusion of technology, the mobile phone has become a central device in the construction of young people's individual identity. This is particularly so as a result of the new opportunity to personalize handsets and messages, as well as one's autonomy to decide how to use the technology anytime, anywhere.

6. A new collective identity with global relevance is emerging from the mobile youth culture. It is reflected and reproduced in a shared language, such as SMS codes, which is a prominent indicator of mobile youth culture in all three regions we examined. Such a collective identity does not suppress personal identity, but affirms it. At times, this kind of community, such as the Generation Txt in the Philippines or youth groups in the anti-globalization movement, may be instantly mobilized as a force for social change.

In addition to the above characteristic constants, we can also see a few variables in the formation of the mobile youth culture by comparing Europe, the United States, and the Asian Pacific.

1. The emergence of the mobile youth culture is influenced by the position of young people in the mobile-phone market. Although in most countries young people are among the most prominent users, the market situation varies, especially with regard to the willingness and actual strategies of mobile operators to meet the needs of youth groups. The market setup may influence not only the diffusion rate among young people but also the actual processes of appropriation among teenagers, college students, or young professionals, who may use one type of service or another to attain different goals.

2. The purchasing power of the young is another important variable as we have seen that young people of different social strata are adopting the technology with different patterns of usage and different kinds of networked sociability. This is why American children download so many games with the cell phone, whereas young migrant workers in China stick to SMS. It is, however, not yet clear how youth groups operating within a lower budget differ from those with higher purchasing power with regard to their basic values, attitudes, and norms.

3. The last, and least researched, variable is the extent to which existing youth cultures and subcultures of different societies are shaping the mobile youth culture. This would be an important question given our initial hypothesis regarding the cross-cultural significance of the phenomenon. But it is a question rarely raised in existing research. Although we know, for example, the *kogyaru* subculture among Japanese schoolgirls, in-depth analysis of the social shaping of mobile-phone usage in similar groups is largely lacking at the present stage.

In sum, there is a clear correspondence between the emergence of a global youth culture, the networking of social relationships, and the connectivity potential provided by wireless communication technologies. The three processes reinforce each other.

5 The Space of Flows, Timeless Time, and Mobile Networks

Time and space are the fundamental, material dimensions of human existence. Thus, they are the most direct expression of social structure and structural change. Technological change, particularly change involving communication technology, critically affects spatiotemporal change, but the influence of technology does not act in isolation from broader sources of change. An investigation of the structure and dynamics of the network society has shown the emergence of new forms or processes of space and time: the *space of flows* and *timeless time* (Castells 2000b: chs 6 and 7). Simply put, the *space of flows* is the material organization of simultaneous social interaction at a distance by networking communication, with the technological support of telecommunications, interactive communication systems, and fast transportation technologies. The space of flows is not a placeless space; it does have a territorial configuration related to the nodes of the communication networks. The structure and meaning of the space of flows are not related to any place, but to the relationships constructed in and around the network processing the specific flows of communication. The content of the communication flows defines the network, and thus the space of flows and the territorial basis of each node (Wheeler et al. 2000; Graham 2004). *Timeless time* refers to the desequencing of social action, either by the compression of time or by the random ordering of the moments of the sequence; for instance, in the blurring of the lifecycle under the conditions of flexible working patterns and increased reproductive choice[1] (Gleick 1999; Green and Adam 2001).

The diffusion of mobile communication technology greatly contributes to the spread of the space of flows and timeless time as the structures of our everyday life (Fortunati 2002b, 2005a; Katz 2004; Meyrowitz 2004a, b). Mobile communication devices link social practices in multiple places. Even if the majority of calls are to people living in the same town, and often inhabiting a nearby place (Fortunati 2005b), the space of social interaction becomes redefined by creating a subset of communication between people who use their space to build a network of communication with other spaces. Because mobile communication relentlessly changes the location reference, the space of the interaction is defined entirely within the flows of communication. People are here and there, in multiple heres and theres, in a relentless combination of places. But places do not disappear. Thus, in the practice of rendezvousing, people walk or travel toward their destination, while deciding which destination it is going to be on the basis of the instant communication in which they are engaged. Thus, places do exist, including homes and workplaces, but they exist as points of convergence in communication networks created and recreated by people's purposes.

The concept of rendezvousing we are using here refers to the informal, geographic coordination of small groups of friends, family, and teammates that takes place in the physical world. Indeed, it is precisely in the context of a wireless networked society that this concept achieves its full sense, because the purpose of a rendezvous is for individuals to come together to participate in a subsequent group activity. Thus, it includes "meeting a friend for lunch," "collecting the kids from school," and "pausing at an intermediate way-point to re-stock and plan the next phase of activity," but does not include formal or anonymous attendance at institutions, such as "reporting to the Tax Office for interview" or receipts for service, such as "Pizza delivery" (Colbert 2001: 16).

This phenomenon has also been described by other authors using different terminology. Indeed, what Ling and Haddon (2001) call micro-coordination through mobile communication influences the travel patterns that are modified in real time by the instructions or negotiations related to the micro-coordination process. The "freedom of contact" provided by the mobile phone allows people to free themselves from the place-based context of their interaction, shifting their frame of reference to the communication itself; that is, to a space made of communication flows, based on the availability of the technological infrastructure that makes it possible (Crabtree et al. 2003). Among Korean professionals who go to after-work parties, for example, Shi-Dong Kim (2002: 70–71) calls this a new "nomadic" way of life, because appointments can be made anytime to allow someone to go to several parties during a single evening as the night unfolds. When such a mobile mode of communication and networking becomes prevalent, those who are not equipped or forget to bring their handsets with them will be socially disabled, as Misuko Ito (2003a) observes in the Japanese context: "To not have a *keitai* [cell phone] is to be walking blind, disconnected from just-in-time information on where and when you are in the social networks of time and space."

In Japan, not only do the mobile phone and the wireless Internet play a role in the integration and disintegration of communities, they also provide a generic social space in which collective practices become regularized and formalized, giving rise to social norms that shape future development in the social uses of the technologies (Ito et al. 2005). More than anything else, this line of change starts with the changing notion of time, or the "softening of time" (Rheingold 2002: 190–198), under certain spatial conditions; for example, the "burst of information during 'in-between' time: while waiting for a train, riding in a taxi, sitting alone in a coffee shop" (Larimer 2000: A29). In Fortunati's words, we have become snails: "we carry our relational house in the back" (2005b: 217).

However, the spatial practice of the mobile phone is changing. According to European research analyzed by Fortunati (2005a), in 1996 about one-fifth of mobile-phone calls were from a fixed place, work or home. In 2004, in studies of high schools in Northern Italy, the majority of calls came from a fixed place, namely the home, the workplace, or school. Fortunati goes on to explain:

Why has the mobile phone changed gradually from being a primarily mobile technology to a rather sedentary technology? To answer this question, we have to appeal to the theory of the co-construction of technology and society [Latour and Woolgar 1979], according to which, if it is true that on one hand ICT design brings with it user design, it is equally true that ICT users and their patterns of use are increasingly able to invent functions and services and then to dictate future ICT developments. (Fortunati 2005a: 63)

The mobile phone has become an individualized tool of communication, used in all spatial contexts to build a new space, the space of selective communication, connecting to wherever the other communicators are located at any given time. Users of the mobile phone have privileged connectivity over mobility. In Fortunati's analysis, the place of the mobile phone is the individual him/herself. Kellerman (1999) conceptualizes this trend as "person–place convergence" (quoted in Fortunati 2005a: 64), with the home often becoming the privileged location for mobile communication. According to studies reported by Fortunati (2005a), this is particularly the case for teenage girls, compared to boys, because of the greater intimacy provided by the home environment. In fact, it would seem that with the diffusion of mobile communication and the enhancement of its technology, sedentary uses for the mobile phone take precedence over its mobile uses. This is particularly the case in developing countries, where the mobile phone is often the only available communication device for low-income households (see chapter 8). For instance, Ureta (2004) titles his ethnographic study of the mobile-phone uses of low-income families in Santiago de Chile, "The Immobile Mobility." He shows that the mobile phone is generally used as the phone of the whole family, with the mother taking precedence in carrying it with her when it is taken out of the home.

Yet Fortunati (2005a) summarizes the results of a number of studies, including her own, in Europe and Italy, to conclude that the mobile phone is being increasingly perceived as an instrument of global communication, in spite of the fact that most of its uses are local, even sedentary, and related to close personal interaction. This trend seems to be linked to the evolution of the technology, the advent of GSM, and a growing connection to the Internet, which truly enables users to feel the immediacy of the local/global connection. If we bring the two observations together – that is, the nearness of use and the perception of global reach – it can be hypothesized that the diffusion of mobile communication does not cancel space, but creates a new space that is local and global at the same time. This is what we have conceptualized as the space of flows.

Thus, wireless communication does not eliminate place. It redefines the meaning of place as anywhere from which the individual chooses or needs to communicate, even if these places are often the home or the workplace. Places are individualized and networked along the specific networks of individual practices, a trend pointed out by several scholars, such as William Mitchell (2003). Ubiquitous connectivity rather than mobility is the fundamental process in the redefinition of space. Places are subsumed into the space of flows, thus losing their meaning in the space of places. In concrete terms, the places from which people communicate with their mobile devices become a backdrop of communication, rather than the locality of communication (Meyrowitz 2004b).

Timeless time as the temporality that characterizes the network society is also enhanced by mobile communication. The availability of wireless communication makes it possible to saturate time with social practice by inserting communication into all the moments when other practices cannot be conducted, such as the "in-between" time during transportation, in a waiting line (Larimer 2000: A29), or simply during free time. Thus, teenagers use their time at home, under the surveillance of the family, or in school, under supervision, to transcend the institutional barriers of control and create their own space for interaction, thus filling in time during nonchosen activities. Or else, any waiting time becomes a potential communication time and the general notion of time is "softened" to accommodate all kinds of activities, sometimes in a simultaneous manner (Rheingold 2002: 5, 190–198). Personal networks of communication never leave the individual. The professional worker constantly interacts with his or her office or receives calls and instructions. Everybody becomes only one call away from his or her working environment, family duties, or personal connections. Everybody transports their world with them. Under such conditions, we witness the emergence of what Nicola Green (2002) calls "mobile times." Green (2002) argues that connections between mobile space and time, as articulated in multiple, heterogeneous places and rhythms, are not constant and do not have equal effects for all. Thus, three main rhythms of mobile time can be defined: rhythms of device use; rhythms of everyday life; and rhythms of institutional change.

First of all, *rhythms of device use* refer primarily to the duration and sequencing of interaction between an individual and the device. Elaborating on the concept, Green says that time spent using communicational devices makes relationships durable and continuing, rather than "fragmented." Secondly, taking the duration of the activity as a measure of significance for the usability of the device may not be as salient as previously thought. Nevertheless, and in order to qualify the first statement, we must add that, when discussing usability, a distinction needs to be made between time spent because the user wants to expend it (i.e., in texting, playing, and so on) and time compulsorily spent in order to complete certain activities. Although these categories are subjective, and could also appear in a mixed form with different combined effects for professionals, students, and housewives, for example, only in the second case will usability be eroded by activities of long duration. In the end, usability, as is the case with the Internet (Katz and Rice 2003), is related to the option of doing what you want as quickly as you can.

Rhythms of everyday life, the second category described by Green (2002), refer to the local temporalities associated with social and cultural relationships in which a specific device use is embedded. Indeed, these aspects of device use are integrated into the emerging patterns of organizing mobile communications and relations in everyday life (see the analyses presented in chapter 3). *Rhythms of institutional change*, finally, refer to the historical and infrastructural elements that enable mobile use, including such dimensions as the institutionalization of travel, cycles of technological development, or the time taken to establish and maintain network technologies. In this sense, "mobile devices act as 'Lazarus' devices –

devices that 'resurrect' mobile time that would have previously been considered 'dead'" (Perry et al. 2001, quoted in Green 2002: 290).

This has already been perceived by teenagers, who fill this empty (or dead) time by playing, sending SMS, or listening to music, and call them "killing-time" (Moore 2003: 71) activities, which, whatever the context, will help to prevent boredom. So the "resurrection of mobile time" affects working time, leisure time, domestic time, and so on, contributing to the blurring of time, thanks to the desequencing of activities that is allowed by "perpetual contact" (Katz and Aakhus 2002) or the "space of persistent connectivity" (Ito and Okabe 2003: 1, 4, 19). In addition, and as long as they provide utility to the user, new activities that arise during "dead time" create value both for people and for business. By saturating "dead time" with communication, people compress and, ultimately, deny time. As Meyrowitz (2004a: 101) writes: "Paradoxically, the more our new technologies allow us to accomplish in an instant, the more we seem to run out of time."

As in all spatiotemporal configurations, the transformation of space and time also results in a rearrangement of their relationship. Hence, in the "real-time city" (Townsend 2000), there emerges a reconfiguration of the spaces of urban social life that introduces opportunities for new continuities across space and time, previously disjoined through centralization (Green 2002: 290). Indeed, as stated by Green, individuals organize their activities around flexible compartments of time, rather than compartments of time associated with particular geographic spaces. Thanks to mobile communication, a kind of spatial and temporal "boundary rearrangement" becomes possible (Green 2002: 288). Moreover, it is this time-based (rather than space-based) organization of activities that defines "accessibility," leading to a redefinition of "public time" and "private time" into "on time" and "off time" (Green 2002: 288). Thus, in the United States, most people use text messaging to communicate with friends (73 percent), family (70 percent), and, less frequently, with business contacts (26 percent; eMarketer 2003). They do so often while taking advantage of undetermined time or time defined for a different activity. Wireless technology enables people to make "productive" use of "downtime," as well as to subvert time that is supposed to be used productively. The volume of usage in busy meetings and classrooms illustrates the latter point (see table 5.1). In fact, some organizations have instituted "no-laptop" policies during important company meetings to address this problem (Boyle 2002), and the Pew Internet and American Life Project (2002a) anticipates tensions between students and professors as the wireless Internet expands into the classroom, just as has been the case with cell phones.

Table 5.1
Locations where mobile instant messaging is most frequently used (%)

Crowded public transport	36
Sporting events	25
Busy meetings	23
Campus classrooms	22
Hospitals	12

Source: AOL and Opinion Research reported in Greenspan (2004a).

In South Korea, the ways of organizing social gatherings are also chang-ing. Among colleagues who socialize together after work, the old custom was to call each other using their office phones to make arrangements for the evening at around 5 or 6 o'clock. The spread of mobile phones, however, not only makes such calls easier, and enables appointments to be made at any time during the evening, but it also enables people to go to several parties in the "nomadic" way discussed above (S-D. Kim 2002: 70–71). The same applies to lovers. Previously, people would make arrange-ments for the next date at the end of a romantic get-together. Now, young lovers say "call me later" instead (S-D. Kim 2002: 73). The notion of time is therefore becoming more flexible with these easy calls and easy appointments.

As it becomes more and more integrated into everyday existence, the pervasiveness of wireless communication exposes cultural concerns about the changing pace of life engendered by this technology. In the United States, concerns range from general anxiety about the increased pace of life to specific issues of public etiquette, the blurring of work and personal boundaries, dangerous driving while using a mobile phone, and the health implications of wireless technology. Wireless communication technology has made it possible for people to occupy their every potentially idle moment, whether by checking e-mail at the bus stop or while waiting for a flight, sending text messages when bored, or conducting clandestine con-versations or personal research during meetings. Yet, the saturation of time by wireless technology is culturally specific. According to various surveys, American respondents are less willing to receive calls from employers outside working hours than Chinese respondents, and use screening devices, such as pagers and caller ID, to maintain a separation between work and personal time (Caporael and Xie 2003).

In sum, by making interactive communication possible around the clock and across space, whether local or global, regardless of the location of the nodes in the network, wireless communication homogenizes space: being ubiquitous means redefining space into the space of communication. By compressing and desequencing time, it also creates a new practice of time. But space still exists and so does time, because social practices are material practices that need material support for their existence. However, this material support is embedded in communication systems, and in the social geography and cultural context of these communication systems. There is a new spatiotemporal formation made of communication flows and their infrastructure. Because this infrastructure depends on place-based nodes (the access points) and their networking, the space of flows shapes timeless time. Where you are determines your ability to transcend time and space. The spatial structure of wireless communication determines the capacity of people and functions to access the new spatiotemporal configuration of our age. The more information systems and databases can be accessed and interacted with from mobile devices, the more access to the space of flows becomes the decisive feature of social organization.

6 The Language of Wireless Communication

A subject that has begun to attract research interest is the transformation of language as wireless communication, especially the short message system (SMS), spreads throughout everyday life and is appropriated to transmit all kinds of messages, not only among young people but also in the population of mobile-phone users at large. This is one of the basic ways in which technological evolution acts as a factor of cultural and behavioral change in the mobile society. In this sense, the oral and written languages used in mobile communication, as well as other forms of expression, such as the "smiley," are reflecting this transformation. By language in this context, we do not refer only to text language or verbal communication, but to all codes and forms of expression that are communicated in a multimodal wireless system. It is precisely the multimodal nature of wireless communication, including the ability to recombine modes in the same process of communication, that creates a new language – the language of the mobile hypertext.

Texting

First of all, our attention will focus on the *texting* phenomenon. The short message system (SMS) is most prominently influencing the writing skills of children and teenagers. While the multimedia messaging system (MMS) is opening up new possibilities in communication, two main factors need to be kept in mind in the case of SMS: the limit of 160 characters per message; and the challenge of the interface – mobile-phone keyboards are far from being comfortable for writing. Thus, as a major consequence of intensive texting, young users improve their ability to synthesize. Young users need to summarize their messages in order to optimize each SMS sent, as if they were *"haikus"* (Ling and Yttri 2002: 158). In addition, they develop a new language that can be defined as a new "written orality" based on symbols and abbreviations (Fortunati and Manganelli 2002). More important, the new language is based on phonetics because reproduction of the oral language helps to save characters, a scarce resource when texting. According to Kasesniemi and Rautiainen:

> Messages often bear more resemblance to code than to standard language. A text filled with code language expressions is not necessarily accessible to an outsider. The unique writing style provides opportunities for creativity. A mistake in one letter, a typing error, can produce a new term of endearment, which may remain in the SMS language either for a short time or permanently. (2002: 183–184)

This creative activity is nevertheless characterized by some common patterns, such as the elision of vowels between consonants, the omission of word spacing, new ways of using punctuation, and dispensing with capital letters. This represents, according to Fortunati (2003), "the discovery of the charm of writing" among adolescents because, despite the short length and informality of these messages, they are still a form of writing. Hence, "in many cases adolescents have surrendered to writing, discovering the attraction of the written word, its power (when the word is written it loses its sound, but takes on greater density of meaning) and its permanence in time (transcribing it in a diary can become social memory)" (Fortunati 2003: 9).

A common trend within SMS language around the world[1] is the general use of English expressions combined with local languages, including not only other European languages but also native languages of all kinds; for example, the shorthand *Tanglish* language used by the Filipino Generation Txt that combines Tagalog, English, and Spanish (Rafael 2003: 407). In neighboring Indonesia, the SMS language used in Yogyakarta is called *bahasa gaul*, which is "a mix of Indonesian, English and Javanese." As in other places where SMS communication uses simplified language, the traditional nine-mode Indonesian language is reduced to just one mode, the intimate mode, in *bahasa gaul*. But this simplified language is nonetheless an important signifier: "[f]or those economically less well off, interestingly, language is the cheapest way of marking oneself as trendy and modern . . . Texting in this sense is not just exchanging chitchat, but chitchatting in a very modern way, both through the technology and the language that are used" (Barendregt 2005: 56–57).

SMS allows discreet and asynchronous communication, though this asynchrony is nuanced by the fact that, in many cases, communication is actually more instantaneous than e-mail, and an answer is given promptly to be polite to your interlocutor. This answer could be just a "boom call," that is, a short signal call not intended to be answered. A final feature that should be highlighted about SMS is that direct contact is not necessary. So texting turns out to be a more "relaxed" way of communicating (or explaining) feelings or sensitive subjects (see Mante and Piris 2002: 51, for the Netherlands; Lobet-Maris and Henin 2002: 104, for Belgium). Without SMS, users may be more embarrassed to communicate the same information. Thus: "[t]exting helps teenagers (and some adult males) to overcome awkwardness and inhibitions and to develop social and communication skills – they communicate with more people, and more frequently, than they did before mobiles" (Fox 2001, with reference to the UK).

This function of SMS, of course, is not limited to teenagers, but applies to people of all ages. One popular genre of SMS literature among migrant workers in South China, for example, includes love letters and sex jokes, which are often circulated via inexpensive paper-based "SMS manuals," teaching migrants about romance and love (Lin 2005). Because it is less embarrassing to talk about sex using SMS, even middle-aged housewives in countries like Indonesia have been reported to flirt with strangers via text messages in order to kill time, although this did not necessarily result in any affair in the real world (Barendregt 2005). Hence, the transformation is not only in the format of language, in such things as grammar and spelling, but also in the content and the immediacy with which one can talk about topics of an intimate nature which used to be reserved to a very small circle of family and close friends.

While texting has become very popular in various European and Asian countries, it is still in the early stages of growth in the United States. However, young Americans have also developed shorthand languages for quick messaging, especially based on their familiarity with PC-based instant messaging services. Notwithstanding the genuine innovations that are taking place in text language, it should be noted that the creativity being observed among SMS users is not totally new, and that compression of text dates back to shorthand writing, for example for taking class notes (Lorente 2002) and secret codes among friends. What is new, perhaps, is the extent to which this has become an interactive and formalized medium among a larger population. Though perhaps not as prolific as their European and Asian counterparts, young Americans in particular are becoming part of the community that has text-based communication as a significant part of its culture. Not surprisingly, older generations are uneasy about the effect that this will have on Standard English as messaging language seeps into formal writing tasks, especially in school (Trujillo 2003), which is already happening (Lee 2002a).

Besides shorthand writing, a noteworthy development is the use of "smileys" and a variety of facial expressions, as in the list of non-standard orthographic forms in the Netherlands.[2] Again, many of these non-verbal expressions originate from online chatting, instant messaging, and e-mail. Mobile communication therefore extends existing practices and, by so doing, further shapes the technology for the purpose of intimate relationships. The multiple emotions that can now be expressed through coded language broaden the scope of interpersonal communication beyond the constraints of standard written language.

From SMS to MMS

Although not yet as popular as SMS, another element that will contribute greatly to the process of language transformation is the multimedia message system (MMS). Thanks to MMS, users can now send and receive images accompanied by text. Thus, peer-to-peer communication is extended, and it is foreseeable that, if pricing and system conditions are favorable, the same creativity experienced with texting could also emerge for "imaging" in the near future. MMS is an asynchronous service that, despite its similarity to SMS, is less textual and more visual (Mobile Streams Ltd 2002). It opens a multimedia channel of peer-to-peer communication, which users are just beginning to exploit (Kurvinen 2003).

Users of multimedia messaging are not only going to be young people because taking a photo, or making a short video, requires different skills from writing an SMS. So there may well be different demographics for end-users who will develop different uses for the system. For instance, camera phones have been used by real-estate agents to enable them to forward pictures to prospective buyers, giving them an edge in a competitive market; or, at a Welsh hospital, senior doctors have allowed interns to send them X-ray pictures in order to speed up the process of diagnosis and treatment.[3]

Taking a wider view, the wireless Internet should also be considered because it can be seen as the next step in the development of multimedia communication. This new opportunity for perpetual contact through the Internet is opening up other possibilities for language transformation, as computers become less stationary (Kasesniemi 2003: 37). Therefore, when the computer, with its full range of capabilities, becomes "wearable" and attached to the human body, it will truly become a personal device (Crabtree et al. 2003: 6).

Mobile Orality

Meanwhile, oral language, in addition to written and visual ones, is also in the process of transformation because of the new situations that mobile telephony generates in everyday life. Conversational analysis gives us some interesting clues on this subject (Weilenmann 2003). The identification of individuals involved in a mobile-phone conversation differs from the situation when a call is made to a fixed-line telephone. Almost always, when a mobile-phone conversation starts, both the caller and the receiver know with whom they are talking. This is because mobile telephones have a caller ID display, and, as handsets are personal and each telephone number is usually associated with one person, it is reasonable to expect that the individual who answers the phone is precisely the one whom the caller wants to talk to.

However, there is still an asymmetry of information (Schegloff 2002: 290) between the caller and the receiver, as is the case with the wired telephone, although its components have changed. While the identification process changes and less time is devoted to it, other parts of the conversation increase in importance. In this sense, more conversational time will be devoted to giving orientations about the location, activities, and/or availability of the person called. Sometimes it will be enough to state one of these three points to give very rich information, but this is not always the rule. Indeed, sometimes it is crucial to inform the caller about the context because of the intrusion the call may cause. Once this piece of information is given, the talk can either finish or continue with its real topic.

Despite the actual availability of the individual called, there is also the possibility for the receiver not to answer the incoming call (Schegloff 2002: 296) or, even, to hang up the telephone. Each behavior gives a different message to the caller, but the second one may show a higher degree of intimacy with respect to the caller. There are different cultural formations of m-etiquette, as we saw in chapter 3 on the transformation of sociability, which require more systematic in-depth investigation that goes beyond existing observations of m-etiquette on a country-by-country basis.

A New Language?

As Fortunati (2003: 1) maintains, in her study of mobile usage among adolescents in Italy: "the territories of orality and writing are completely

restructuring, in the sense that in this case, for example, writing has pen-
etrated in[to] the territory of the [mobile phone], the oral instrument par
excellence." Based on evidence from other parts of the world, we concur
that texting is indeed changing language through its widespread use in
wireless communication. From an evolutionary view, we are seeing a new
case in which the adoption of new technology affects language, including
vocabulary and rules of grammar in user practices. And these practices ultimately affect the common language, and language itself.

In some cases, new forms of written expression are the signs of subcultures, and the expression of innovation from users. In fact, creative uses
of language become a form of personal and group expression. However, in
most cases, we are seeing the simple adaptation of language to the format
and limits of the technology, including strategies to reduce the cost of
transmission. Thus, what originally existed as "shorthand," with limited
instrumental and personal uses (for example, taking quick notes during an
interview or lecture), has now evolved into a fully-fledged language system
used widely within the wireless culture for social interaction. We are
already at the point where new texting-oriented vocabularies can be listed
for different languages, on the basis of observed practice.[4]

Furthermore, the multimedia capacity of wireless communication technology (as is the case with the fixed-line Internet) displays a multimodal
form of communication, with text, image, and audio being used from multiple locations. Observation shows that the combination of these different
modes of communication, particularly by young users, creates new forms
of meaning, characterized by the mixture of methods of assigning
meaning, for example, by using texting only for personal commentary or
for emphasis, while sound and images are supposed to be self-explanatory.
The merging of text and audiovisual media is now diffused more widely
in various contexts of communication through the distributed communicative capacity of wireless technology. We are also beginning to see the
texting vocabulary spill over into Standard English, French, Spanish, and
Japanese writing. Teachers are beginning to complain about students using
SMS words in their essays. Since language is closely related to the formation of culture (the systemic production and communication of meaning),
we are clearly in a process of cultural transformation associated with the
spread of wireless communication, although the lack of academic research
on the subject precludes us from knowing, for the time being, the contours and directions of this transformation.

7 The Mobile Civil Society: Social Movements, Political Power, and Communication Networks

Wireless communication provides a powerful platform for political autonomy on the basis of independent channels of autonomous communication from person to person. The communication networks provided by mobile telephony can be formed and reformed instantly, and messages are often received from a known source, enhancing their credibility. The networking logic of the communication process makes it a high-volume communication channel, but with a considerable degree of personalization and interactivity. In this sense, the wide availability of individually controlled wireless communication effectively bypasses the mass-media system as a source of information, and creates a new form of public space.

Without prejudging the merits of this political autonomy (because, of course, it can be used to support very different kinds of political values and interests), we have observed a growing tendency for people, in different contexts, to use wireless communication to voice their discontent with the powers that be, and to mobilize around these protests by inducing "flash mobilizations" ("flash mobs" in Howard Rheingold's 2002 terminology), which in a number of instances have made a considerable impact on formal politics and government decisions. To illustrate this tendency, in this chapter we will examine some examples of political mobilization in which wireless communication played a significant part. These are the ousting of President Estrada in the Philippines in 2001, the electoral defeat of the Spanish Partido Popular in 2004, and the voting into power of Korean President Moo-Hyun in 2002. Additionally, we will discuss the organization of a series of protests during the 2004 US Republican Party's National Convention, and the low level of sociopolitical uses among mobile subscribers in Japan, and in China during the SARS epidemic of 2003. On the basis of these case studies, we will elaborate on the broader implications of the sociopolitical uses of wireless communication, which are being adopted around the world.

Social Networks and Civic Uprisings: People Power II in the Philippines, 2001

In January 2001, thousands of cell-phone-touting Filipinos took part in massive demonstrations, now dubbed "People Power II"[1] This four-day event is generally considered the first time in history that the mobile phone has played an instrumental role in removing the sitting president of a nation-state (Bagalawis 2001; see also Salterio 2001: 25). On June 30, 1998, Joseph Estrada was sworn in as the thirteenth president of the Philippines. Son of an engineer in Manila, he dropped out of college at the age of 21 to become an actor, a profession that so deeply troubled his parents that they forbade him to use his family name, Ejercito. He therefore adopted the screen name "Estrada" (meaning "street" in Spanish), and the nickname "Erap" (or "*pare*," meaning "friend," spelt backwards), which he continued to use in later political life. During his movie career, Estrada played the lead role in more than a hundred films and produced more than 70, most of which were popular action and comedy movies that brought him huge fame. In 1969, he began to serve in the public sector, first as mayor for 16 years, then as senator and vice-president. As a politician, he continued his on-screen Robin Hood-style image as the friend of the poor, especially low-income Filipino farmers (BBC 2003a). With strong support from them, he won the presidential election of 1998 with a landslide victory of 10.7 million votes (Lopez 1998).

But from the beginning of his presidency, Estrada was subject to allegations of corruption, including mishandling of public funds, bribery, and using illegal income to buy houses for his mistresses. The most serious charge that led to his ousting came in October 2000, when he was accused of receiving US$ 80 million from a gambling-payoff scheme and several millions more from tobacco tax kickbacks. On October 12, Vice-President Gloria Macapagal Arroyo, a Harvard-trained economist and the daughter of former president Diosdado Macapagal, resigned from the cabinet and later become the leader of what would soon become People Power II (*Pamantalaang Mindanaw* 2000). On October 18, 2000, opposition groups filed an impeachment complaint against Estrada with the House of Representatives. Protests started to emerge in Manila. In less than a month, dozens of senior officials and lawmakers from Estrada's ruling party withdrew their support, including the president of the Senate and the speaker of the House of Representatives. On December 7, the Senate impeachment trial formally began. Multiple investigations took place, revealing more and more evidence to the disadvantage of Estrada.

Soon, a violent disaster disrupted the political life of the entire country. On December 30, 2000, five bombs exploded in Manila, killing 22 people and injuring more than 120 (*Philippine Daily Inquirer* 2001). The explosions were synchronized to hit the city's crowded public spaces, including the airport, a light-rail train, a bus, a hotel, and a park near the US embassy (BBC 2000b). Police investigations accused the Muslim rebel group, *Jemaah Islamiyah*, which was later linked to *al-Qaeda* (Associated Press 2003b), although many suspected at the time that the explosions were linked to Estrada's impeachment trial.

At a critical meeting for the trial, on January 16, 2001, senators voted 11–10 against opening an envelope that was believed to contain records of Estrada's secret transactions. Within hours, enraged Manila residents – many of them following instructions received on their cell phones – gathered at the historic shrine at Epifanio de los Santos Avenue, also known as EDSA, the site of the People Power revolt of 1986, to protest against perceived injustice and demand the immediate removal of Estrada from the presidency.

The massive demonstrations of People Power II lasted for four days, from January 16 to 20. The group of senator-judges serving at the impeachment trial resigned on January 17 and the case was suspended indefinitely. With increasing pressure from protesters, led by Gloria Arroyo and other former officials, the Defense Secretary and Finance Secretary resigned on January 19 to join the opposition. By then, the Estrada cabinet had basically collapsed with most of its key posts being abandoned, and, most importantly, the military had sided with the demonstrators. On January 20, 2001, Estrada was escorted out of the Malacanang Palace by the Armed Forces Chief of Staff and Vice Chief of Staff. By the end of the day, the Supreme Court would declare the presidency vacant, Gloria Arroyo was sworn in, and People Power II was concluded on a triumphant note.

News coverage of the demonstrations invariably highlights the role of new communication technologies, especially SMS and the Internet, in facilitating and enabling the protests. By one account, anti-Estrada information began to accumulate in online forums as soon as he took office in 1998, which culminated in some two hundred websites and about a hundred e-mail discussion groups by the time People Power II started (Pabico n.d.). A famous online forum was E-Lagda.com, which collected 91,000 e-signatures to support the impeachment through both the Internet and SMS (Bagalawis 2001). Besides information, a large number of Internet and text messages were jokes and satires making fun of Estrada, his (allegedly) corrupt life, and his poor English.

While this kind of semi-serious communication continued for more than two years, allowing for the widespread expression of discontent, it was texting that made the swift gathering of tens of thousands possible immediately after the crucial vote of January 16. According to a member of the Generation Txt who joined the demonstrations, she was out on a date in the evening when the news broke (Uy-Tioco 2003: 1–2). She first received a message from her best friend: "I THNK UD BETR GO HME NW [I think you'd better go home now]." But by the time she got home, already pretty late in the evening, she had received numerous messages from others such as: "NOISE BARRAGE AT 11PM," "GO 2 EDSA, WEAR BLACK 2 MOURN D DEATH F DEMOCRACY." She then quickly followed the instructions:

I barely had time to kick off my high heels and slip on my sneakers when my mom, brother, and I jumped into the car and joined the cars in our neighborhood in honking horns in protest. And then to Edsa we went. At midnight, there were a couple of hundred people. Families clad in pajamas, teenagers in party clothes, men and women in suits fresh from happy hour, college students clutching books obviously coming from a study group, nuns and priests.

During the week of People Power II, Smart Communications Inc. transmitted 70 million text messages, and Globe Telecom, the other main SMS operator, handled 45 million messages each day as opposed to its normal daily average of 24.7 million (Bagalawis 2001). The demonstrators were using text messages so actively that it caused serious strain on the networks covering EDSA. According to Smart's public affairs officer, "The sudden increase in the volume of messages being handled at that time was so tremendous that sometimes the signals were not coming through, especially in the Edsa area." High-level representatives from Globe admitted similar difficulties, saying that mobile cell sites had to be transferred from the Senate and rural Bicol to ease equipment load, alleviate congestion, and provide back-up contingency (Bagalawis 2001).

Most English-language Filipino media regard the overthrow of Estrada as a positive development in the country's democratic life. Comparing People Power II and the People Power movement of 1986, they argue that there was less violence and military involvement (Andrade-Jimenez 2001), and that the demonstration was more centered on information and IT. "[T]he wired and wireless media became effective messengers of information – be it jokes, rumors, petitions, angry e-mails or factoids – that made People Power II much wider in scope and broader in reach than its predecessor" (Bagalawis 2001). Moreover, the speed of IT-based mobilization was much faster. Whereas Marcos managed to continue his rule for almost two

decades despite serious allegations of corruption and human-rights viola-
tion, Estrada was ousted in only two and a half years, less than half of his
six-year presidency (Pabico n.d.; Andrade-Jimenez 2001). For these reasons,
Helen Andrade-Jimenez (2001) has claimed that: "People Power II showed
the power of the Internet and mobile communications technology – not
to mention broadcast media – not only to shape public opinion but also
to mobilize civil society when push came to a shove." According to these
descriptions, the victory of People Power II was the victory of new technologies, especially the mobile phone and the Internet.

These media accounts, however, need to be treated with caution. After
all, "[n]early all the accounts of People Power II available to us come from
middle-class writers or by way of a middle-class controlled media with
strong nationalist sentiments" (Rafael 2003: 401). Written in the immediate aftermath of the protests, most of these accounts are excessively cele-
bratory, glossing over many issues important to our understanding of the
role of the mobile phone in this political movement.

First, characterizing People Power II as non-violent and information-
centered oversimplifies the case. The military was not a negligible factor
in the process. It was only after the armed forces sided with the protesters
that Estrada retreated and was "escorted" out of his presidential palace by
military commanders. Moreover, the deadly synchronized explosions that
killed 22 Manila residents and injured more than 120 took place only 17
days before People Power II. Given the sensitive timing in the middle of
the impeachment trial, such a violent incident clearly threatened everyone – especially the senator-judges – with an all-out civil war on top of the
ongoing clashes with the Muslim rebels accused of the December 30
bombing. This was quite possible because, despite the corruption charges,
Estrada had overwhelming support in the countryside and among the
poor, as shown in his landslide victory in the 1998 election. In fact, a
seldom-reported story was that, three months after People Power II, on
April 25, 2001, Estrada was formally arrested on charges of graft and corruption. This incident quickly spurred a reaction: "a crowd of perhaps one
hundred thousand formed at Edsa and demanded Estrada's release and
reinstatement" (Rafael 2003: 422). According to Vicente Rafael:

Unlike those who had gathered there during People Power II, the crowd in what
came to be billed as the "Poor People Power" was trucked in by Estrada's political
operatives from the slums and nearby provinces and provided with money, food,
and, on at least certain occasions, alcohol. In place of cell phones, many reportedly
were armed with slingshots, homemade guns, knives, and steel pipes. English-
language news reports described this crowd as unruly and uncivilized and castigated

protestors for strewing garbage on the Edsa Shrine, harassing reporters, and publicly urinating near the giant statue of the Virgin Mary of Edsa. (2003: 422)

Besides showing the potentiality of large-scale violence during the impeachment trial, the Poor People Power calls into question the pro-claimed importance of the new media because, although most poor demonstrators did not have cell phones (let alone Internet access), this par-ticular crowd could also appear in no time.[2] They had to be "trucked in" since, unlike the middle-class protesters, they had no other means of trans-portation (see the above quote from Uy-Tioco 2003 for the use of private cars in People Power II). Meanwhile, as Rafael (2003) points out, the neg-ative descriptions of the Poor People Power was in part due to the class positioning of Filipino English-language newspapers:

Other accounts qualified these depictions by pointing out that many in the crowd [of Poor People Power] were not merely hired thugs or demented loyalists [of Estrada] but poor people who had legitimate complaints. They had been largely ignored by the elite politicians, the Catholic Church hierarchy, the middle-class-dominated left-wing groups, and the NGOs. Even though Estrada manipulated them, the protestors saw their ex-president as a patron who had given them hope by way of occasional handouts and who addressed them in their vernacular . . . Generation Txt spoke of democratization, accountability, and civil society; the "tsingelas crowd," so called because of the cheap rubber slippers many protestors wore, was fixated on its "idol," Estrada. (Rafael 2003: 422–423)

The Poor People Power was finally dispersed by the military after five days of gathering (Rafael 2003: 425). This incident, seldom incorporated into the narrative of People Power II, shows the oversimplifying nature of the "People Power" label with respect to the deep-seated class problems of the Philippines which offer more fundamental explanations for the social unrest above and beyond the over-celebrated power of the new media in and of themselves. Almost 40 percent of Filipinos live on a daily income of US$1 (Bociurkiw 2001). Given the country's total population of 80 million,[3] only about 13.8 percent of Filipinos had access to mobile phones in 2001. The scope of the cell phone's political influence was therefore still quite limited. Although some members of the lower class also took part in People Power II, they were, like the "tsingelas crowd," presumed to be "voiceless" in the "telecommunicative fantasies" about the cell phone (Rafael 2003: 400).

The contradiction of class interests was most acutely presented in a book called *Power Grab* (Arillo 2003), a summary of which was prominently featured on Estrada's official website.[4] It maintains that:

[Estrada] lost his job when white-collar mobsters and plunderers, backed by sedi-
tious communists, do-gooder prelates, traditional politicians, and misguided police
and military generals, banded together and toppled his regime, first, by using
massive disinformation and black propaganda carefully crafted to provide half-true,
misleading, or wholly false information to deceive and anger the public. (Arillo
2003)

Putting aside the highly partisan language, this pro-Estrada writer obviously agrees that communication technologies played a pivotal role, not
to inform and mobilize in a positive sense, however, but to disseminate
"disinformation," "to deceive and anger the public," and to "misguide"
police and military generals.

The question that emerges is, given that Estrada was the sitting president, why did he not prevent the "disinformation" and vicious mobilization against himself? How could this be possible? Is it simply due to the
invincibility of the new technology since "one could imagine each user
becoming his or her own broadcasting station: a node in a wider network
of communication that the state could not possibly monitor, much less
control" (Rafael 2003: 403)? More likely, as Rafael argues, the power of new
technologies, especially of the cell phone, arose because there was a need
for "the power to overcome the crowded conditions and congested surroundings brought about by state's inability to order everyday life" (2003:
403). In other words, the existence of a relatively weak state was a prior
condition for the key role of the mobile phone and the Internet in this
case. The final result might have been very different had there been
stronger state control. Although there were indications that Estrada
attempted to acquire technologies for monitoring cell-phone use, "[i]t is
doubtful, however, that cell phone surveillance technology was available
to the Estrada administration" (Rafael 2003: 403).[5] Besides problems with
technology, this probably also has to do with Estrada's career, first as a successful movie star (therefore over-confident about his image built by film,
television, and radio), then as a long-time, small-town politician (therefore
less prepared for the communication power of the new media in Manila;
Pabico n.d.).

It should also be pointed out that there were other social forces playing
critical roles, especially the Catholic Church, and the radio and other
media resources under their influence. A Catholic nun was among the first
to openly accuse Estrada's family of mishandling public funds (Uy-Tioco
2003: 9). Cardinal Sin, the head of the Roman Catholic Church in the
Philippines, was among the most prominent anti-Estrada leaders from the

beginning of the events of October 2000 (BBC 2000a; see also Gaspar 2001). Moreover, while many were suspicious of the credibility of SMS messages because so many of them were ungrounded rumors, religious organizations were deliberately involved to add legitimacy to anti-Estrada text messages. As one activist reveals in a listserv post:

I was certain [texting] would not be taken seriously unless it was backed up by some kind of authority figure to give it some sort of legitimacy. A priest who was with us suggested that Radio Veritas [the church-owned broadcasting station] should get involved in disseminating the particulars . . . We [then] formulated a test message . . . and sent it out that night and I turned off my phone . . . By the time I turned it on in the morning, the message had come back to me three times . . . I am now a firm believer in the power of the text! (quoted in Rafael 2003: 408)

As mentioned earlier, mobile phones were also used closely with hundreds of anti-Estrada websites and listservs during the movement. In addition to famous online forums such as E-Lagda.com, blogging sites were also involved, such as "The Secret Diary of Erap Estrada (erap.blogspot.com)" (Andrade-Jimenez 2001). It is thus erroneous to give all the credit to texting since mobile phones had to function in a particular media environment, which reflected the middle-class-dominated power structure at the time. It is within this larger framework that we should acknowledge that the mobile phone – as a medium that is portable, personal, and prepared to receive and deliver messages anytime, anywhere – can perform the mobilization function much more efficiently than other communication channels at the tipping point of a political movement.

On the other hand, as a tool of political communication, texting has a serious limitation: it allows short messages to be copied and distributed quickly and widely, but it permits very little editing or elaboration based on the original message. It is suited to simple coordinating messages, such as specifying the time and location of a gathering and what to wear (for example, black clothes in this case). However, it is highly inadequate for civic deliberation. With SMS, the messages were "mechanically augmented but semantically unaltered . . . producing a 'technological revolution' that sets the question of social revolution aside" (Rafael 2003: 409–410). "Texting is thus 'revolutionary' in a reformist sense" (Rafael 2003: 410). If a real revolution were to take place that fundamentally alters a social structure, it would most likely involve other media, including not only the Internet, which has been used in conjunction with the cell phone in political mobilizations in most cases, but also traditional mass media and interpersonal communication.

Finally, there was a global dimension to People Power II. New media technologies, especially the Internet, enabled the global Filipino diaspora to participate more easily (Andrade-Jimenez 2001). Since overseas Filipinos are more sympathetic toward middle-class appeals, they added significantly to the oppositional force. Moreover, Estrada had been an outspoken nationalist for most of his political life. He was named the Most Outstanding Mayor and Foremost Nationalist in 1972 (Alfredson and Vigilar 2001). In 1991, he was the first senator to propose the termination of the American military base in the Philippines. He therefore had little support from global capital or the US government, which would rather see him being replaced by Gloria Arroyo who was more Westernized and represented middle-class interests.

To sum up, during People Power II, the mobile phone, and especially text messaging, did play a major role in message dissemination, political mobilization, and the coordination of campaign logistics. Because it allows instant communication at any time, anywhere, it is most suited to assemble large-scale demonstrations as soon as political events emerge, such as the senators' decision on the impeachment trial on January 16, 2001 (similar to events during the Korean presidential election of 2002; see below). However, the mobile phone was limited in the social scope of its influence due to the "digital divide." It is often a tool serving the interests of the middle class, traditional stakeholders (for example, the Catholic Church), and global capital. It does not always have high credibility or sufficient capacity to spur two-way civic deliberation. For these reasons, the new media of mobile phones and texting have to work closely with other media, such as the Internet and radio, as shown in this case, in order to deliver actual political consequences.

Getting Out the Youth Vote: Nosamo and the Power of Information Technology in South Korea, 2002

On December 19, 2002, South Korea elected a new president, Roh Moo-Hyun, whose victory has been widely attributed in major part to Nosamo, an online supporter group known by the Korean acronym for "People who Love Roh." The success of Roh, and of Nosamo, is now "a textbook example for the power of IT" (Hachigian and Wu 2003: 68), with a combination of the Internet and mobile phone-based communication being systematically utilized. While the Internet-based campaign had lasted for years, providing the core political networks, it was the mobile phone that mobilized large numbers of young voters on election day and probably reversed the voting result (Fulford 2003: 92; see also S-D. Kim n.d.).

Nosamo represented a strategic coalition between liberal pro-reform political forces and new communication technologies in response to pressing issues such as economic growth and the problem of regionalism. Based on the nation's high Internet and mobile-phone penetration rates, it also draws on the historical roots of pro-democracy student demonstrations of the 1980s (Fairclough 2004; see also J-M. Kim 2001: 49). This was a very sensible strategy given that the traditional media, especially newspapers, were predominantly conservative (S-D. Kim n.d.). These "old" media had little appeal to young people in their twenties and thirties, yet this age group is a baby-boomer generation which makes up slightly more than half the total number of voters (J-M. Kim 2001; Rhee 2003).

A self-educated labor lawyer, Roh Moo-Hyun assumed the presidency at the rather young age of 56. He differed from most other politicians in having a more radical reformist agenda that, on the one hand, favored a fundamental overhaul of the *chaebols*, the family-dominated conglomerates that "have long funded the country's political machinery" (Fairclough 2004), while, on the other hand, it attempted to transcend the boundaries of regionalism, a deeply ingrained structural problem in Korean politics (Rhee 2003: 95). In addition to these singular political stances, Roh was also known for his highly idealistic personality[6] because, despite repeated losses in elections (for the mayor of Pusan and then the National Assembly), he refused to compromise or switch parties as many other opposition figures had done. This iconoclastic image won him "an almost cult-like following among young Koreans" (Demick 2003).

Roh's age, policy, and personality gave him strong popularity among young voters "just as President Bill Clinton appealed to many American baby boomers" (Fairclough 2004). At the core of his support was the generation of the so-called "386ers," that is, those who were in their 30s during the presidential election, who grew up in the 1980s amid Korea's pro-democracy movement, and were born in the 1960s at the dawn of South Korea's era of industrialization (Fairclough 2004). Unlike the older generations, the 386ers are "more skeptical of the United States in part because Washington backed the same military rulers they fought against as college students" (Fairclough 2004). In addition, there was also a large number of younger supporters in their twenties, such as Hwang Myong-Pil, a stock trader who quit his well-paid job to become a full-time volunteer at Nosamo (Demick 2003). Together, the 20- and 30-somethings were Korea's baby-boomer generation, accounting for slightly more than half of the voter population (J-M. Kim 2001; Rhee 2003). Most of these young activists regarded themselves as having inherited the revolutionary spirit of student demonstrations from more than a decade ago. At large political gatherings, they would chant songs dating back to the pro-democracy movement of the 1980s, such as *Morning Dew* (*Korea Times* 2002).

To reach this critical block of voters, Roh experimented with online campaigns back in 1995 when he was running for mayor of Pusan. It "fits in with his political philosophy of openness and direct communication with the people."[7] Many of his closest aides in the presidential election were former student activists (Fairclough 2004). This was a highly innovative approach, not only because it used new technology, but also because it appealed to the younger generations in a more substantial way because traditional media, as part of the Korean political machine, were predominantly conservative. Consequently, young people had been feeling cynical and disenfranchised from the political process: "Nearly a third of the nation's twenty-somethings didn't bother to vote in the 1997 presidential election. Less than 40 percent of the 8 million people in their twenties voted in parliamentary elections in April last year [2000], far below the 57 percent national average" (J-M. Kim 2001: 49).

It was at this historic moment of low turnout among young people, when Roh Moo-Hyun lost his second race in the parliamentary election, that Nosamo (www.nosamo.org) came into being. On June 6, 2000, Nosamo was formed when around a hundred founding members convened in Taejon (*Korea Times* 2002). While Roh's campaign team had been actively utilizing the new media, Nosamo was a voluntary organization self-funded by membership fees and only informally affiliated with Roh (*Korea Times* 2002; see also Rhee 2003: 95). Within five months, the membership mushroomed from around 100 to nearly 5,000 in November 2001 (J-M. Kim 2001: 50), and then, within a year, to 70–80,000 by the end of 2002, forming a most formidable political force.[8]

During the presidential election of 2002, Nosamo members raised more than US$ 7 million over the Internet (Demick 2003). They used electronic bulletins, online polls, and text messages to formulate collective decisions and coordinate campaign activities. "All the decisions about their activities are made through an electronic voting system and the final decision-making online committee has its monthly meeting in chat rooms" (J-M. Kim 2001: 50). Among a variety of logistics, one coordination task was to make sure that people wore yellow outfits to attend political rallies, yellow being the color symbolizing Roh's campaign (*Korea Times* 2002).

At times, members of Nosamo could act quite aggressively. For instance, a professor made a comment perceived to be critical of Roh's supporters on a television talk show. He was subject to hundreds of angry e-mails and was widely lambasted in the Nosamo forum (Demick 2003). For this and similar activities, Nosamo was criticized for behaving like "Internet Red Guards" with "violent words in cyberspace and an appeal to populism" (Demick 2003). About a month before the presidential election, South Korea's electoral commission barred the group from raising money for their candidate (Demick 2003), and the organization's website was forced to close until election day (*Korea Times* 2002).

Meanwhile, the rather unconventional approaches of Roh Moo-Hyun continued to work to his deficit. Mainstream media, most of which belonged to the conservative camp, kept putting him in a negative frame (S-D. Kim n.d.; Rhee 2003). A few months before the election, Roh was so far down in the opinion polls that members of his own Millennium Democratic Party (or MDP) tried to force him out of the race (Demick 2003). On the eve of the election, Roh's key campaign partner, the multi-millionaire Chung Mong-Joon suddenly withdrew his support, dealing a heavy blow to the whole campaign at the last minute (*Korea Times* 2002).

As election day dawned on December 19, 2002, Nosamo members were caught in a deep sense of crisis. With their main website having been closed for a month before the election, young activists started the day by posting online messages such as "Let's go vote!" (Rhee 2003: 96).[9] By 11 a.m., exit polls showed that Roh was losing by a margin of 1–2 percent (Fulford 2003; Rhee 2003: 96). At midday, "[h]is supporters hit the chat rooms to drum up support. Within minutes more than 800,000 e-mails were sent to mobile phones to urge supporters to go out and vote. Traditionally apathetic young voters surged to the polls, and by 2 p.m., Roh took the lead and went on to win the election" (Fulford 2003).

Several elements contributed to this historical event when the mobile phone for the first time played a significant part in changing the outcome of a presidential election. First, there was already a large-scale grassroots political network centered on Nosamo, whose members not only had frequent online exchanges but also met offline. Secondly, Roh Moo-Hyun's center-left policies and iconoclastic image energized young liberals, many of whom were highly dedicated and ready to act promptly at a time of crisis. Thirdly, Chung Mong-Joon's sudden withdrawal on the eve of the election and the temporary trailing of Roh created an urgent momentum to rally public support. And the mobile phone – the quintessential grassroots communication gadget that is always on, anywhere, anytime –

turned out to be the best medium for these rallying calls. Given the strength of youth networks (Yoon 2003a, b) and the demographic fact that people in their twenties and thirties make up slightly more than half the total number of voters (J-M. Kim 2001: 49), young people, mobilized through mobile messages, became a most decisive voting bloc. At the end of the day, "sixty percent of voters in their 20s and 30s cast ballots for Roh" (Rhee 2003: 95).

After President Roh took office, Nosamo decided to remain active following an internal poll in January 2003 (*Korea Herald* 2003). Nosamo members continued to "solicit suggestions for appointees to Cabinet positions and engage in debates over topics ranging from North Korea's nuclear program to whether it would be more appropriate for Roh to take up golf or jogging as president" (Demick 2003). In fact, as any long-term civic group, they played a relatively independent watchdog role in observing, and sometimes criticizing, Roh's decisions as president. Back in 2001, a founding member of Nosamo was quoted as saying that "We're using the Net to support him. But we want to say 'no' when he makes any decision which we think is wrong" (J-M. Kim 2001: 50). On March 24, 2003, Nosamo adopted a statement opposing the US-led war in Iraq and Seoul's decision to dispatch engineering and medical troops there (*Korea Times* 2003). Yet the Roh administration proceeded with the plan, causing some Nosamo members to withdraw from the group, one of them saying: "I withdrew from Nosamo because President Roh Moo-Hyun has shown us drastically different aspects since becoming president. I do not love Roh Moo-Hyun anymore. I hate the sight of the president supporting the barbaric war of the United States killing innocent civilians of Iraq" (*Korea Times* 2003).

In spring 2004, Nosamo again played a major role in staging support for Roh during an impeachment investigation against him on charges of violating Korean laws barring partisan remarks within 17 days of a parliamentary election (Len 2004). During this election, the liberal Uri Party, which had Roh's support, utilized the mobile phone for campaigning purposes. Along with the usual policy statements, candidate profiles, and appearance schedules, their website also encouraged supporters to copy "Get out and vote" messages and send them out by mobile phone to ten friends who were then, in turn, asked to forward the message to ten of their friends (Salmon 2004).

Sociopolitical uses for the mobile phone are still on the rise as Korean society further transforms and the technology further diffuses and becomes more mature. Again, in this case, the role of the mobile has to be understood as closely related to other media, especially online bulletin boards. These new media function most importantly as a catalyst for the mobilization of existing youth networks, giving rise to groups such as Nosamo, which are, in one sense, newly formed political forces, whose historical origins, however, can be traced back for at least two decades. This said, it would be to exaggerate to give too much credit to the mobile phone as the sole, or even the most important, device, with some kind of magical, innate political power. Yet it would be equally erroneous to ignore the unique capacity of the cellular phone – as a means of "perpetual contact" – to effect the swift mobilization of certain social groups at critical political moments such as the Korean presidential election of 2002.

Terrorism, Political Manipulation, Political Protest, and Political Change: Spain, 2004

On March 11, 2004, a Madrid-based, mainly Moroccan, radical Islamic group associated with *al-Qaeda* carried out in Madrid the largest terrorist attack in Europe, bombing three suburban trains, killing 192 people (Spanish Minister of Internal Affaires 2005),[10] and injuring over a thousand. The bombing was conducted by remote-control-activated cell phones. Indeed, it was the discovery of a prepaid cell-phone card in an unexploded bag that led to an arrest and subsequent elimination of the phone. *Al-Qaeda* claimed responsibility for the bombing later that evening (Pacheco 2004; Partal and Otamendi 2004; Rodríguez 2004; Sanchez 2005). The attack took place in a very special political context, four days before the Spanish parliamentary elections, which were dominated by debate about the participation of Spain in the Iraq War, a policy opposed by the vast majority of the citizens. Yet the Conservative Party, the Partido Popular (PP), was expected to be returned to power, based on its record on economic policy and its stand on Basque terrorism.

As soon as the Madrid terror attack occurred, and before any evidence had surfaced, the PP government stated with total conviction that the Basque terrorist group, ETA, was behind the bombing (Pacheco 2004; Partal and Otamendi 2004; Rodríguez 2004; Sanchez 2005). As the hours went by, it became increasingly likely that *al-Qaeda* was to blame. Yet the Minister of the Interior, and the government's spokesman, continued to insist on ETA's responsibility until the evening of March 13 (Pacheco 2004; Partal and Otamendi 2004; Rodríguez 2004; Sanchez 2005). In political terms, making the Basque terrorists responsible would favor the PP in the elections, while acknowledging the action to be that of Islamic terrorists would indicate to Spaniards the high price they were paying for their government's policy in Iraq, thus potentially inciting them to vote against the government. In the minds of millions of Spaniards (actually 67 percent of them) the government was manipulating information about the attack to gain political advantage. This widespread feeling was an important factor in the unexpected political defeat of the PP on March 14, leading to the election of a Socialist government and to the immediate withdrawal of Spanish troops from Iraq.

Concerning the actual events of March 11–14, a commission of inquiry of the Spanish Congress produced evidence according to which, at the very least, the PP government had deliberately delayed publication of some critical information, and stated as facts elements that were still under scrutiny. There was clearly an inclination to favor the hypothesis of Basque terrorism and not to follow as a priority the Islamic trail, in spite of early leads by the police in this direction (Spanish Parliament 2004). But regardless of the extent of the manipulation that actually took place, what counted was that thousands of citizens were convinced on March 12–13 of the existence of manipulation, and that they decided to disseminate their views to the entire population through wireless communication and the Internet (Rodríguez 2004; de Ugarte 2004; VV.AA 2004; Francescutty et al. 2005). Because the main television networks were under the direct or indirect control of the government, they supported the hypothesis of Basque terrorism, as did most of the radio networks (but not the largest one) and most of the print media, after the Prime Minister personally called the editors of the main newspapers, giving his personal opinion that the attack was carried out by ETA.

Thus, oppositional views on the actual source of terror had to find alternative communication channels to be heard. The use of these alternative communication channels led to mobilizations against the PP on Saturday, March 13, a "day of reflection," when, according to Spanish law, political demonstrations and public statements are forbidden. Yet, the action of thousands of protesters, most of them young (Rodríguez 2004; Francescutty et al. 2005), made an impact on public opinion, and particularly on the two million new voters, young people who usually have a higher abstention rate or vote for minority parties, rather than for the Socialists or Conservatives. In this election there were 2.5 million voters more than in the parliamentary election of 2000, and about 1 million voters switched to the Socialists in an attempt to punish the government both for its policy on Iraq and for its perceived manipulation of information. The Socialist Party won a clear majority over the PP in an election that saw a 77 percent turnout. This discussion, on the basis of published reports, explores the process through which alternative communications channels were created and used effectively.

The actual process of alternative communication began with the outpouring of emotion that surrounded the Friday March 12 street demonstrations, called for by the government, with the support of all political forces. This is important: it was first in the physical gathering that people started to react and to oppose the official version of the facts, independently of the political parties which remained silent on the occasion (Francescutty et al. 2005). While the demonstration was called against terrorism and in support of the Constitution (an oblique reference to Basque separatism), many of the participants displayed banners opposing the war. The demonstration was intended to mark the end of political statements, leading to the day of reflection on Saturday and to the election poll on Sunday.

Yet, on Saturday morning, a number of activists, mostly individuals (Cadenaser.com 2004) without current political affiliation, and independent of the mainstream parties, began to circulate text messages to the addresses programmed into their cell phones. In the messages, they denounced the manipulation of information and called for a demonstration in the afternoon (at 6 p.m.) in front of the headquarters of the PP in Madrid, and then in other Spanish cities. This was in fact outlawed, and naturally did not receive any support, explicit or implicit, from any party, although some of the participants in these gatherings were members of the left-wing parties, particularly of the United Left (a small party in parliament which includes the remnants of the Communist Party in Spain). But most of the activists were participants in the anti-war movement, and most of the people who gathered in front of the PP headquarters were simply those reached by the network of SMS. The first and most famous of these messages, all fitting within the 160 characters of the SMS format, was the following: *"Aznar off the hook? They call it day of reflection and Urdazi works? Today, 13M, 18h. PP headquarters, Genova Street, 13. No parties. Silent for truth. Forward it! ("Pasalo!")*.[11] The reference to Urdazi needs to be explained: he was the notorious anchorman of Spanish national television, well known for his manipulation of political news (in fact, he was sentenced for such by the court). In the meantime, Spanish television continued to defend the story of Basque terrorism, and, in the evening before the election, changed its regular programming to broadcast a documentary film on the assassination of a Socialist politician by Basque terrorists.

On Saturday, SMS traffic increased by 20 percent over a regular Saturday, and by 40 percent on Sunday,[12] an all-time record for these messages. The critical matter is that, while most messages were very similar, the sender for each receiver was someone known, someone who had the receiver's address in his or her cell-phone address book. Thus, the network of diffusion was at the same time increasing at an exponential rate but without losing the proximity of the source, according to the well-known "small worlds" phenomenon.

The Internet had begun earlier to function as an alternative channel of communication: on March 11 particularly, but also on March 12 (de Ugarte 2004; VV.AA 2004). People used the Internet to look for other sources of information, particularly from abroad (Francescutty et al. 2005). But there were also a number of initiatives, including some by journalists acting on their own, to set up websites with information and debates from various sources. Interestingly enough, the PP started an SMS network with a different message: "ETA are the authors of the massacre. *Pasalo!*." But it diffused mainly through party channels, did not reach a critical mass of known-person-to-known-person, and, more importantly, was not credible to the thousands of people who were already doubting the government's word.

The context provided by the mainstream media was also meaningful. Major television networks were ignored as reliable sources very soon. Newspapers, because of their hesitancy, became unreliable, although *La Vanguardia* in Barcelona emerged on Saturday as legitimizing the version of events that placed *al-Qaeda* at the origin of the attack (Rodriguez 2004; de Ugarte 2004). On the other hand, the major private radio network (SER), under the initiative of its journalists, immediately looked for evidence away from the Basque trail (sometimes too eagerly, as they circulated some inaccurate information). Yet, most of the SER's reports proved to be accurate. As a result, many people referred to the radio (on their portable radios) as their source of information, and then interacted with SMSs and cell-phone calls. Voice communication provided direct discussion with close friends, while SMS was used to distribute personally crafted messages or to forward messages people had received that they agreed with (Cué 2004; Rodríguez 2004; de Ugarte 2004; VV.AA 2004; Francescutty et al. 2005).

Thus, the context for communication was provided by the physical gathering in the streets, at the start of the formation of public opinion, and as a result of the process of political communication: being together in front of the PP buildings was the verification of the usefulness of the message. Then the actions in the street attracted the attention of some radio and television networks (regional television networks and CNN-Spain), and ultimately forced the Minister of the Interior to appear on national television acknowledging *al-Qaeda's* possible role, on Saturday at 8:20 p.m. (Juan 2004; Partal and Otamendi 2004; Rodriguez 2004). Yet later on, the leading candidate of the PP also appeared on national television denouncing the demonstrators – thus unwittingly fueling the crisis of trust that the government had induced. Thus, a mistake in political communication (Rodríguez 2004) amplified the effect of the demonstrations.

The Internet was important in providing a source of information and a forum for debate in the days preceding the demonstrations. But the critical event was the Saturday March 13 demonstration, a typical "flash mob" phenomenon prompted by a massive network of SMS that increased the effect of communication exponentially through interpersonal channels. This happened first in Madrid, but spread to Barcelona, and, ultimately, to all Spanish cities (Sampedro Blanco and Martínez Nicolàs 2005) because, of course, address books in cell phones include friends and acquaintances in other cities.

This experience in Spain, coming three years after the flash mob mobilization that forced the resignation of Estrada in the Philippines, will remain a turning point in the history of political communication. Armed with their cell phones, and able to connect to the world wide web, individuals and grassroots activists are able to set up powerful, broad, personalized, instant networks of communication. Without prejudging the merits of this phenomenon (as it is subject, of course, to the diffusion of harmful, misleading information), this form of autonomous communication rings a warning bell for the control of information by governments and mainstream media.

The three case studies discussed above illustrate the potential of mobile communication technologies to facilitate political upheaval. However, this is not an autonomous characteristic of the technology. There are cases where mobile communication has so far had less than dramatic sociopolitical impacts. Examples can be found in the United States, Japan, and China. They demonstrate that the particular usage of wireless technologies is shaped within the social context and political structures of a given society.

Cat and Mouse Games: Political Activism, Surveillance, and Wireless Technologies at the Republican National Convention, New York, 2004

The Republican Party held its 2004 National Convention (RNC) from August 30 to September 2 amidst heightened expectations of disturbances caused by anti-Bush activists. The run-up to the New York Convention was characterized by reports and rumors of planned and potentially spontaneous protests and how the police and security agencies were preparing to deal with these incidents (Carpenter 2004; Gibbs 2004; Shachtman 2004; Terdiman 2004). Reference was made to the "battle" of Seattle in 1999 when over 40,000 protesters descended on the city from all over the world to protest WTO policies, leading to scenes of violence and contributing to the breakdown of the WTO talks.

What was particularly interesting about these reports was the way in which they took for granted the central role of wireless communication, not just in the protests but in all aspects of the Convention. In the event, there were indeed several (mostly nonviolent) protests coordinated primarily via wireless communication and the Internet, leading to over 17,000 arrests. The Convention itself was hardly affected by the protests apart from a few minor disruptions. In fact, President Bush experienced a bounce of two percentage points in the polls (among likely voters) after the Convention (*The Economist* 2004; Jones 2004). While one cannot yet judge the long-term impact of these events, preliminary examination indicates that this was a case where the use of wireless communication technologies served to enhance efficiency but not to effect change.

Protests began as early as August 27 with the largest on August 29, a march organized by an anti-Iraq War group, United for Peace and Justice. Although the police did not give an estimate of numbers, organizers of the march said that there were about 500,000 people, the largest ever Convention protest (Hauser 2004). Protesters marched past Madison Square Garden, the site of the Convention, chanting anti-Bush slogans, led by prominent personalities such as Jesse Jackson and filmmaker Michael Moore. Other protests followed throughout the four days of the Convention, all helped along by the use of cell phones and text messaging. As one newspaper reported:

All week long, thousands of protesters in New York have deployed an amazing array of Internet and wireless gadgetry to try to outsmart police, share plans, and publish "uncensored" news. Amateur journalists blurt out live reports by cell to Internet radio listeners. Activists zap text messages to armies of colleagues. Cell-phone photographers transmit images of cops videotaping protesters – scenes ignored by what they call "corporate media" – to the web for millions to see. (Becker and Port 2004)

Wireless communication, especially text messages, featured prominently as a means of coordinating the activities of protesters and sending out alerts about ongoing activities, such as spontaneous gatherings or police arrests, at least from the perspective of news coverage of the protests. For example, text messages were used to call a spontaneous rally on September 1 at the pier where arrested protesters were being held by the police (Simon 2004). Other people used text messages to decide which protests they would attend, or to avoid "hot spots" where police brutality was taking place. Especially prominent were messages warning about where police were located and whether they were arresting protesters, as shown in this sample of text alerts from the text-messaging service, Ruckus:

from: TextAlerts (8.31.04 5:59 p.m.)
mounting police presence at public library action: 3 arrested so far, potentially more. protesters are clearing.

from: TextAlerts (8.31.04 6:41 p.m.)
Everyone in streets at 34th st and 6th ave being arrested

from: TextAlerts (8.31.04 7:02 p.m.)
large police presence entering herald square from south

from: TextAlerts (8.31.04 7:21 p.m.)
police planning on penning in all protesters at herald square and arresting: only exits through the south and subway

from: TextAlerts (8.31.04 7:26 p.m.)
16th betw irving and union sq people being beaten by police

from: TextAlerts (8.31.04 7:31 p.m.)
28th and bwy herald sq union sq madison sq pk all confrontation and police activity at all locations

from: TextAlerts (8.31.04 7:32 p.m.)
16th and 34th people being beaten while in handcuffs

from: TextAlerts (8.31.04 8:48 p.m.)
area hot from 27th and park to 28th and bwy arrests et al

from: TextAlerts (9.1.04 3:39 p.m.)
Anyone with eyewitness accounts of arrest/police misconduct please reply. Reporters interested. (Rubin 2004)

The pre-Convention hype about protest activities was to some extent accurate, but also exaggerated the potential for wireless communication to cause major upsets at the Convention. For the most part, the protests were widespread but not decisive. This happened for a number of reasons. First, the use of wireless communication as a protest tool had been so highly anticipated that it was incorporated into the strategies of the security forces. For one thing, security detail used wireless monitoring techniques themselves, such as head-mounted miniature video cameras that transmitted footage from the security personnel's location to a mobile command center (Reardon 2004). Security personnel also allegedly infiltrated protesters' planning meetings and monitored text messaging and other communication services used by activists (Gibbs 2004; Gibson 2004). For example, during the Convention, protesters using indymedia's website to transmit messages soon realized that the "police were on to them." Thereafter, "calls for 'direct action' stayed posted only for a couple of minutes and used code words for location" (Becker and Port 2004).

Secondly, unlike some radical protests that were generated spontaneously, such as those discussed in the previous case studies, there was a high level of central management associated with wireless uses in the RNC context. Most of the protests and protest strategies were carefully planned, some as much as a year in advance (Archibold 2003). In addition, protest groups had to obtain a permit to demonstrate, of which eventually 29 were granted (Archibold 2004). The location and route of the protests were mapped out in detail (Slackman 2004), and each protest was closely monitored by the police. Generally, those who tried to implement protests without a permit ended up being arrested for unlawful assembly and their numbers were never large enough to change the tone of the protest environment. Although thousands of demonstrators gathered at Central Park after the August 29 march, in defiance of a court decision not to allow protests in that area, there is no indication that this gathering was of any critical significance to the progress of the Convention.

Another example of central management was the use of specially tailored text-messaging systems, such as Ruckus, TxtMob (probably the most popular service used at the RNC), which was specifically designed by the Institute for Applied Autonomy for use by activists to broadcast messages during the Democratic and Republican conventions, and MoPort, which allowed individuals to "mobblog" by sending pictures of the protests from their mobile devices to be downloaded onto the Internet. The objective of MoPort was "to join the disparate streams into a collective reporting effort" (Dayal 2004). It is possible that there was a need for such centrally organized services because of the lack of a common standard in the United States to allow people to send text messages to other people on different phone networks. While these types of service effectively brought together communities of like-minded people for the purpose of activism, they lacked the character of direct person-to-person texting, based on interpersonal relationships, because users had to sign up to send or receive messages through the service provider's server. Incidentally, for a period during the convention, users of TxtMob reported problems receiving messages, for which the service provider gave no explanation, leading to conspiracy theories that some cell-phone companies (T-Mobile and Sprint) had deliberately blocked messages. The ultimate theory was that this may have been the work of a spam filter, which tagged messages going out from the same server to more than a hundred people as spam (Di Justo 2004; Lebkowsky 2004). The blackout effectively shut down a flash mob organized by A31 Action Coalition, partly because potential participants did not know where the starting-point was, although it is not clear why other forms of communication, such as mobile-phone calls, could not supplement effectively. This illustrates the limitations of communication technology, especially centralized systems.

The energy of the protests was also affected by the fact that they involved several groups with different agendas, from anti-war through animal rights to abortion rights. Admittedly, the convergence of all these groups in one place against a central political institution would be a formidable force. At the same time, the single-mindedness associated with other protests that have effected immediate change was absent from these demonstrations. This can also be linked to the apparent absence of measurable goals. With the election too far away for them to galvanize action to vote against President Bush, and with no chance of overturning the Republican Party's nomination of Bush as their candidate for 2004, protesters marched on such goals as "to regain the integrity of our country . . . to regain our moral authority . . . to extend the ban on assault weapons . . . for more police on our streets . . . for more port security . . . for a plan to get out of Iraq" (Jackson 2004) or "we want to take charge and reach the right people and influence them to go on and spread the message that this is a corrupt government (protester quoted by CNN 2004).

It seems then that, so far, the use of wireless communication has not had any significant effect on political events in the United States, at least on the surface. Yet, social undercurrents may develop into changing people's minds and influencing their political behavior. Indeed, insofar as the protesters' objective was to peacefully make their voice heard during a central political event, while avoiding clashes with the police, one can say that the protests were successful. However, we do not have evidence to claim any direct impact on the political process itself.

Mobile Messaging vs Mass Media: Managing Information Flow during the SARS Crisis, China, 2003

China is an extreme case given its authoritarian political system which is fundamentally at odds with spontaneous grassroots mobilization. Hence, despite rapid growth in the mobile-phone market, the new technologies have so far seldom been put to sociopolitical use. The few observed sociopolitical uses are instances of state-sponsored experiments. One such instance occurred during the National People's Congress in March 2002, when the Xinhua News Agency teamed up with China Mobile to offer the public a chance to text-message their concerns and proposals to the country's lawmakers (News.com 2002). There was little indication that this trial was successful, particularly due to the limited content capacity of SMS. It may be unrealistic to expect text messages to convey anything more than quick requests or short complaints, much less any meaningful deliberation.

The SARS outbreak of 2003 in China illustrates the limited nature of the sociopolitical uses of the mobile phone in general and SMS messages in particular. At the very beginning, no news media or Internet outlets reported on the epidemic. But victims and their friends and families, especially those who worked in the local hospitals of Guangdong, began to send text messages regarding this strange, deadly disease. The SMS alerts spread quickly among urban residents in Guangdong and then outside the province. But at this time, public hygiene and propaganda authorities in Beijing decided to dispel this "rumor" by launching a mass-media campaign claiming that the infections were no more than a variant of pneumonia, that it was already under control, and that the public panic partially induced by text messages was groundless. This official campaign via traditional media effectively undermined the earlier information disseminated via mobile phone because SMS was perceived to be a medium of lower credibility and there was no other alternative source of information. As a result, most people, including experienced foreign analysts living in South China, chose to believe the official version, only to witness the SARS epidemic in full swing within weeks.[13] Given that the mobile phone was so inadequate to sustain a non-state information system regarding even such an immediate and crucial life-and-death issue, it would be much more difficult for the new technologies to be applied to other autonomous sociopolitical uses with any significant consequences, at least in the short run.

In any case, the Chinese authorities have been seeking to limit the use of new communication technologies, including wireless, among political dissidents. The Telecom Ordinance of 2000 outlawed the transmission of harmful information via any telecom facility (Fries 2000: 43–44).[14] Later known widely for its influence on the establishment of China's Internet censorship regime, this measure was initially designed, in large part, to counter the subversive potential of pagers in the mid-1990s. The ordinance provided the legal basis for further, more specific, controls over uses of the mobile phone.

Although the Chinese authorities are stepping up their regulatory efforts, some elements of Chinese society have nonetheless started to use pagers and cell phones for alternative or even oppositional political organization. Despite the lack of systematic research, it is likely that three social groups may have used wireless technologies for their political ends. First is the Falungong group that Beijing denounces as an "evil cult." Secondly, there have been constant demonstrations by laid-off urbanites or pensioners, such as the massive protests of workers in the petroleum and machinery industries in northeast China in 2002 (Eckholm 2002). Thirdly, in the countryside, there have also been protests against the misconduct and corruption of local officials (Duffy and Zhao 2004). Some members of these movements, especially the organizers, may have used wireless technologies (especially the low-end applications, such as prepaid phone cards and Little Smart) for small-scale coordination. However, this technical adoption is yet to have any significant impact upon the existing power balance because so far all these perceived challenges to the state have been kept under control at the national level despite sporadic outbursts in certain localities.

Whereas there are some small-scale ICT-facilitated urban movements in China, it is unlikely that they will be connected with the country's 800 million peasants (Zhao forthcoming: 18–19). Furthermore, due to the privileged status of the "information-haves," those who have access to the new technologies are "not necessarily the ones most ready to act upon this critical information" (Zhao forthcoming: 20).

A-political Grassroots? When Wireless Communication Is Used for Everything Except for Social Mobilization: Japan

In Japan, despite the very high penetration of mobile-phone and mobile-Internet services, so far (2006) there have been no identifiable instances of grassroots sociopolitical mobilization which utilize wireless communication. The Japanese authorities did make some effort to use mobile technologies as a broadcasting system of some sort, for example, the "Lion Heart" e-newsletter from the Office of Prime Minister Junichiro Koizumi (Reuters 2001a), which had 1.7 million subscribers through PCs and mobile phones by March 2004 (*PR Newswire* 2004).

At the local level, city governments, such as Sagamihara in Kanagawa Prefecture, in the southern part of Tokyo, also launched an m-government experiment in April 2004 which allows users to report any damage or defects they find in the street or on public signs by sending pictures from their camera phones (Suzuki 2004). These are, however, state initiatives that operate from the top down rather than sociopolitical mobilization that begins within the networks of ordinary mobile-equipped citizens and their organizations, as in the other countries we have discussed above. The lack of grassroots-level political usage among Japanese mobile subscribers is an interesting issue that remains to be explored, although at this initial stage we suspect that it has to do with the ultra-consumerist tendency of Japan's mobile culture and the relative inactivity of alternative political forces outside the mainstream in general, which is a result of the larger sociocultural framework of Japanese politics that goes way beyond the mobile culture *per se*.

Conclusion: Wireless Communication and Insurgent Politics

Control of information and communication has been a major source of power throughout history. The advent of the Internet and of wireless communication allows the development of many-to-many and one-to-one horizontal communication channels that bypass political or business control of communication. Therefore, new avenues are open for autonomous processes of social and political mobilization that do not rely on formal politics and do not depend on their framing in the mass media. However, relative autonomy of the content and process of communication does not necessarily lead to social change. The case studies reported in this chapter demonstrate the diverse outcomes that the use of communication technologies can generate. In three of the cases (the Philippines, South Korea, and Spain) the outcome was substantial political change, insofar as it affected the choice of a government. The fourth (the United States) had a mild outcome and did not influence the November 2004 election. In China, the government successfully delegitimized the grassroots efforts to inform about the SARS epidemic. In Japan, the mass diffusion of wireless communication does not seem to extend to political uses given the depoliticization of Japanese youth.

In the Philippines, wireless communication was employed to oust a sitting president before his term of office ended; in South Korea, the same technologies were used to change the fortunes of a presidential contender who was trailing in the polls. In Spain, not only was text messaging used to galvanize people to vote a government out of power, but it was also used extensively to supplant, supplement, and debunk government propaganda and mainstream media. In the United States, text messaging and other wireless technologies were employed as efficient tools to coordinate and monitor protest activities (by protesters and police) during a political convention. Finally, in Japan and China, sociopolitical usage of mobile phones is minimal despite the rapid diffusion of the technology in these two countries.

Although we have not been able to analyze in depth other cases of political mobilization in which wireless communication has played a role, the "Orange Revolution" in Ukraine in 2004, and the revolts that ousted President Gutierrez in Ecuador in 2005, in a social movement that the president labeled *Los Forajidos* ("the bandits"), bear witness to the increasingly important role that youth, armed with their mobile phones, also play in these movements. Thus, this is a phenomenon that is transforming political mobilization, but whose causes and consequences have to be understood by considering their specific contexts. For instance, a critical difference between the United States and the other cases studies here is that while in the Philippines, Korea, and Spain a combination of factors converged to stimulate spontaneous uprisings, in the United States the process was more centrally managed, thus removing, to some extent, the element of interpersonal communication flow based on friendship networks. Significantly, there were no surprises in the US case; everyone had anticipated how wireless communication would be likely to be used. Conversely, in the Philippines, Korea, and Spain, events were less charted, less predictable, and there were no effective countermeasures. On the other hand, the Chinese government had the resources to overwhelm the emergence of an SMS-based information flow in a situation where the demand for reliable information was high.

An episode in Italian politics may illustrate the interaction between the technology and the specific use of the technology in determining its outcome. On June 12–13, 2004, a few weeks after the Spanish election of March 2004, Italy held regional elections. Prime Minister Berlusconi, having learned the lesson of what had happened to his friend Aznar, decided to take the initiative with the new mode of communication. He already owned the three main private television networks, and, through his government, had significant control over the public television networks. Just to make sure that the election would not prompt a wireless communication insurgency, he decided on the eve of the election to send 13 million "personal" messages to cell phones. The strategy backfired (BBC 2004c). People were indignant at seeing their personal and political privacy invaded by the prime minister for electoral gain. Already forced to consume pro-government television, they had, in addition, to absorb unsolicited political advertising in the very private sphere of their phone. Berlusconi lost the regional elections by a larger margin than anticipated.

Although it cannot be proved that the cell-phone incident aggravated the defeat, as many observers claim, we can say at the very least that in this case wireless communication did not help the successful reception of the message sent. Berlusconi did not understand that the key to the success of the Spanish messages in prompting mobilization was that people received them from someone they knew, from the address book of someone who had their name, not from a central register obtained from some company. The viral diffusion of messages in the Spanish case, as in the Philippines, South Korea, Ukraine, Ecuador, and other countries, is related to a network effect on the basis of a few networks containing a few nodes in each case, and then expanding the networking of each one of these nodes exponentially. It is person-to-person, horizontal, mass communication, rather than a new technology for top-down mass communication, that accounts for the difference in the mobilizing impact of a given message. Thus, the context in which the message circulates, its resonance with each person who receives it, and the origin of the message (which provides its credibility) are critical ingredients of the political power embedded in wireless communication technology.

But what is the added value of wireless communication vis à vis the Internet in the process of political mobilization? There are three main differences. The first one is, of course, mobility, and this adds a component of spontaneity to potential mobilizations. It increases the chance of reacting instantly and emotionally to a sudden event that strikes the minds of the people. Secondly, wireless communication is more directly related to person-to-person contact; it is the always-on tool of communication within everybody's network. Thus, it is the appropriate platform to scale up from personal life to social concerns. Thirdly, because of its multimodality, cell phones may transmit voice calls, text, images, sounds, so that they become living eyes and ears, together with minds, to observe events in real time and share them with the network. As a result, numerous cell-phone users become instant media reporters of their experience in order to share it in real time with their friends and, through them, with the world at large. So, while the immobile Internet and mobile communication are complementary rather than alternative means of communication, going wireless adds a significant component to the capacity of new forms of communication to disintermediate the mass media.

However, we must introduce a word of caution against a technologically deterministic argument. As noted in the various case studies, other communication processes and media, both wired and unwired, were also important in these processes. And, of course, revolutionary political mobilizations have occurred in countries where wireless communication was lacking. When wireless communication has the political impetus that we have seen, some or all of these other processes have been in play, including a precipitating event strong enough to arouse anger or other emotions, activist instigators, support from respected institutions such as the church, and supplementary information from mainstream media and/or Internet sources. In addition, people involved feel that they really can bring about change and they tend to have a focused goal, which can sometimes be directly implemented through the voting process. We have also noted that communication is a two-edged sword, and that wireless communication has the ability to speed up the process in both positive and negative ways. Speed of information flow through interpersonal networks that have the ability to move people to act can as easily be used to spread rumors or inaccurate information as to spread truths, to mystify consciousness as well as to denounce abuse. Also, insofar as there is some differentiation in the diffusion and usage patterns of wireless communication technologies between countries, as well as on the basis of age, gender, and socioeconomic status within countries (as we have demonstrated in earlier chapters of this volume), the process of political mobilization using wireless communication could be limited to some groups.

Still, it cannot be denied, based on the observation of recent processes of sociopolitical change, that access to and use of wireless communication technology adds a fundamental tool to the arsenal of those who seek to influence politics and the political process without being constrained by the powers that be. Arguably, other media, such as wired phones, radio, and television, could perform the same rallying function as wireless communication does, but not in as timely a manner, not with the ability to reach people wherever they are, and not free of the production and communication constraints associated with the traditional media.

Wireless communication does not replace but adds to, and even changes, the media ecology, expanding the information networks available to individuals and social groups to emphasize the interpersonal level and to enhance the efficacy of autonomous communication oriented toward political change. When the dominant institutions of society no longer have the monopoly of mass-communication networks, the dialectics between power and counter-power is, for better or for worse, altered forever.

8 Wireless Communication and Global Development: New Issues, New Strategies

The growth in mobile-phone subscriptions in developing countries has attracted considerable interest from the international community, in particular with the expectation that mobile telephony may be the answer to closing the "digital divide" and promoting development (Kelly et al. 2002; Brewer 2005; *The Economist* 2005d). Not only are there new possible scenarios for deploying telecommunication systems, there are already a number of new strategies emerging at institutional and consumer levels to appropriate wireless communication technologies for the purpose of expanding access. In spite of the early success of mobile telephony, however, the next stage of diffusion may be more difficult, as it will concern the greater proportion of the population, which is generally of low income. The continued spread of mobile phones into these groups may depend on the extent to which regulators, governments, and telecommunication companies adopt further initiatives to promote expansion (MacDermot 2005). Social demand is expected to intensify as the few studies conducted on the social uses of mobile telephony in developing countries show its increasing significance for the life of people at large, usually in ways not completely dissimilar to that which has been observed in other areas of the world.

In this chapter, we will first review key issues in mobile communication and development (some of which are not entirely new). We will then examine some of the strategies being deployed in developing countries to meet connectivity needs in low-income communities. This will be followed by more detailed case studies of alternative modes of mobile communication access in Asia, Africa, and Latin America.

Rethinking Development and Communication: Issues at Stake

Leapfrogging Development?
The old Gershenkronian argument about "leapfrogging" arises whenever mobile communication is discussed in the development context. There is no doubt that mobile telephony has provided the means for developing countries to leapfrog fixed-line technology. In most developing countries, state-owned, fixed-line operators have traditionally failed to develop adequate infrastructure, a problem that recent privatization and deregulation have not always alleviated. Thus, in such economies, mobile phones are serving as fixed-line substitutes for large populations, in contrast to developed economies where they tend to be used as a complement to fixed-line telephony. However, as we will describe in the cases of China and India, fixed-line infrastructure has proved useful in the provision of low-cost wireless alternatives to mobile phones. Overall, there is no overwhelming evidence to support the leapfrog hypothesis in terms of eliminating stages of economic development. However, *one of the most important identifiers of the potential developmental impact of mobile telephony could be its contribution to moving developing countries as close as possible to a universal telecommunications service*, which has been shown to be the critical-mass level at which telecommunications began to exhibit significant impacts on economic growth in advanced economies (Coyle 2005).

The Effects of Mobile Telephony on Economic Development
There is an intense debate about the general impact of mobile communication on a country's economic development. An empirical study of mobile telephony in Africa (sponsored by Vodafone) concludes that mobile phones can have significant beneficial impacts on economic growth, in some cases as much as double their impact in developed countries. The results show that in low-income countries, per capita GDP could be 0.59 percent higher if there were on average ten more mobile phones per 100 population. Mobile phones provide significantly higher network effects in developing than in developed countries where fixed lines have already performed this function. According to Waverman et al. (2005: 17), "while in developing countries the benefits of mobile are two-fold – the increase in the network effect of telecoms *plus* the advantage of mobility – in developed economies the first effect is much more muted." In this view, differential economic growth rates among developing countries could be partially explained by the penetration of mobile telephony (Waverman et al. 2005). There also seems to be an important link between telecommunications infrastructure

and foreign direct investment (FDI) flows to Africa, especially in the case of mobile telephony in sub-Saharan Africa (Williams 2005), and Latin America, particularly Brazil and Mexico, which have become sizeable exporters of mobile telephone equipment (MacDermot 2005). To be effective, however, mobile communication technology needs to exist in concert with developments in other areas of economic and social infrastructure (for example, better trunk roads and postal systems).

The Mobile Digital Divide

Although mobile phones have undoubtedly reduced deficiencies in access to telephony in developing countries, there is still a wide gap between developed and developing countries in terms of diffusion levels and the type of technology in use (see chapter 1, p. 10; Kauffman and Techatassanasoontorn 2005). So far, mobile technology has not had much impact on the "Internet divide," although several researchers and scholars have articulated the potential of emerging technologies (for example, Wi-Fi) for the provision of Internet access in developing countries (see, for example, Jensen 1998; Kibati and Krairit 1999; Peha 1999; Dutton et al. 2004; Galperin 2005).

One "divide" that rarely receives attention in the euphoria over the advance of wireless communication is *the rural–urban divide within developing countries*. For example, in Mexico mobile-phone use is significantly higher in urban than in rural areas (see chapter 1; Mariscal and Rivera 2005). A study in Ghana comparing an urban slum (Mamobi) and a rural area (Praso) found high levels of mobile-phone use in Mamobi, as well as a perception of mobile phones as "a means of social connection, coordination and practical information exchange" (Information Society Research Group n.d.). In Praso, on the other hand, there was no telecom coverage (mobile or fixed) and ICTs in general were perceived as widening the rural–urban divide.

As noted in chapter 2, statistics on mobile-phone subscription are generally not disaggregated by location, making it difficult to estimate the extent of this divide in mobile network coverage. Governments have not entirely ignored this disparity, however: although controversial in the free-market system, several countries have extended universal service/access obligations to mobile-phone companies, a requirement that has traditionally been reserved for fixed-line operators. We will observe later in this chapter a number of mechanisms and structures that have developed to address this problem to some extent (for example, shared-access systems and low-cost wireless alternatives).

Mobility vs Connectivity

In developed countries, mobile phones are defined by the term *mobile* and appreciated as a means to communicate on-the-go. However, the immediate benefit for people in developing contexts is that of *connectivity*, associated with having a means of communication whether mobile or not. Thus, considerations linked to mobility, style, and Internet access, for example, are arguably secondary to that of basic connectivity at this stage.[1] As Gough and Grezo (2005: 1) note, "the ways in which mobiles are used, valued and owned in the developing world are very different from the developed countries."[2] It is therefore important to identify the true source of developmental benefit. For people who already own a landline at home and/or at work, mobile phones bring the added benefit of mobility and convenience. To those for whom the mobile phone is the first form of personal communication to be owned, the major prize is to be connected at last; the phone is acquired not in order to be mobile, but in order to be connected, although mobility is an added bonus. This is probably the major difference between mobile-phone owners in areas with high fixed-line teledensity and those with low levels of fixed lines in developing countries. Indeed, there are areas where, because it is the only phone in the household, the mobile phone has a permanent location, and in some instances, because of short battery life, is permanently plugged in (see, for example, Ureta 2004).

Designing Technology for Developing Markets

A testament to the flexibility of wireless communication technology is the extent to which it has been adaptable to conditions in developing economies. It is precisely this flexibility that has enabled the development of a variety of techniques to adjust mobile telephony to developing countries. At the same time, it becomes clear that it is possible to design profitable technology (for example, low-cost handsets) and delivery systems (for example, prepaid) targeted at poor countries, if only equipment manufacturers and service providers are prepared to accept lower average revenues per user.

Even in developing countries, mobile-phone operators have been slow to develop strategies directed at very low-income groups, as they prefer to focus on maximizing revenue from existing business in profitable areas (Shanmugavelan 2004). For example, in South Africa, mobile-phone operators, in anticipation of slimmer profit margins, have slashed their commissions to prepaid card wholesalers, effectively reducing wholesalers' markup from 30–40 percent down to about 10 percent (*Middle East and*

Africa Wireless Analyst 2004). This tendency toward profit maximization in the shortest possible time may ultimately be the most significant obstacle to rollout in African countries. South Africa has Africa's largest GSM network and a competitive industry with two mobile-phone operators, yet it also has some of the highest tariffs on the continent (Minges 1999). The average monthly tariff for 100 minutes of airtime is about three times that for fixed lines; halving the tariff, according to Minges (1999), would still enable operators to recoup their investment within five years.

Cost vs Benefit

Considering the poor state of telecommunication services available to poor people, it is ironic that they tend to spend a higher proportion of their income on telecommunication than do richer people. The ability to exchange information is clearly important whether one is rich or poor. However, some thought should be given to how the use of mobile telephony affects poor people's income in terms of the uses it could be put to. Studies have shown that users do benefit from lower communication costs in comparison with the cost of traveling, for example, to make a fixed-line call or to meet a colleague in person. Mobile phones also have the added benefit of mobility and the ability to monitor and control costs. However, it is also arguable that fixed lines, if extended to poor communities, may provide even cheaper telephony and access to data, thus potentially leaving them with more disposable income for other important needs.

Although it is relatively cheap to establish and get connected to a wireless telecommunication system, the cost of usage (handsets and phone calls) is still high for a majority of the population. It is not clear whether the trade-off between mobile and fixed lines produces greater benefit for poor users. The issue, however, is that in many regions there may not be an alternative to wireless communication because of the high cost of deploying a fixed-line telecommunication infrastructure. Under such circumstances, the regulator should step in to make sure that the monopoly of wireless systems on communication does not lead to abuses in cost or neglect of service.

Social Uses vs Business Uses

In this respect, users in developing countries appear to be no different from those in developed countries. In Africa, as in other parts of the world, it has been shown that people tend to use their mobile phones more for social interaction than for business activities (see, for example, McKemey et al. 2003; Donner 2004, 2005b; Samuel et al. 2005; Slater and Kwami

2005). This includes the use of the telephone at the personal level and in the business context. Research in a number of developing countries, including Ghana, South Africa, and Jamaica, indicates that mobile phones have become a key tool for managing business and extended family as well as diasporic relations and obligations (Information Society Research Group n.d.; Slater and Kwami 2005; Horst and Miller 2005, 2006). It has, in fact, become increasingly difficult to distinguish between social and business uses because interpersonal connections often contribute to social capital formation, which is important in the development of small businesses (Donner 2004).

Alternative Uses and Alternative Modes of Access

As these issues and problems become more resonant in the lives of people in developing economies, some innovative and fairly effective mechanisms and products are emerging to address them. These include the prepaid system, scaled-down products and services, alternative technologies, efforts to develop low-cost handsets, shared access and maintenance models, and strategies to redistribute resources from wealthier to poorer users.

Prepaid Systems

Chapter 2 has detailed the central role played by prepaid systems in making mobile telephony accessible to people on low incomes. This benefit is particularly evident in developing countries where most subscriptions are prepaid. In some countries, however (especially in southern Africa), the prepaid market is already reaching saturation, although not all market segments have been served. This indicates that for the rest of the population, the access, cost, and convenience benefits of even the prepaid system are beyond their reach. Thus, even in the midst of the explosion of prepaid wireless communication, mobile-phone teledensity remains low, and, as we point out below, communities are still seeking additional cost-reduction strategies, such as low-cost handsets and community sharing.

Furthermore, there is the possibility that focusing on prepaid mobile telephony will shift responsibility for service quality and universal access from service providers and government to individual users. Mahan (2005: 69) discusses this probability, noting that "For prepaid, the individual bears all the risk and potential detriments of higher per call charges, up-front investment in future calls, lower quality of service; and frees the service provider from investment in billing and collection." In this regard, proposals in Brazil for regulatory measures that would promote a more equitable situation for prepaid users are noteworthy. Brazil's Congress has been debating a proposal to abolish the 30–90-day limit on the validity of prepaid credit (MacDermot 2005). The resultant proposal would extend the period of validity to six months (Zimmerman 2005a). In addition, this new regulation will allow prepaid subscribers to receive a refund of unused credit during the contract term (Zimmerman 2005b).

Scaled-Down Products and Services

In order to enable at least some access to mobile telephony, some operators in developing countries have begun to offer scaled-down services, such as the Little Smart phone and wireless local loop (WLL) telephony (discussed in the case studies below), which provide wireless telephony with limited mobility. The cost of mobile handsets has also been identified as one of the major obstacles to mobile-phone ownership in developing countries (*The Economist* 2005b, c). In some countries, especially in Latin America, generous handset subsidies are provided for consumers (MacDermot 2005).

To improve the access problem in other regions, in 2004 the GSM Association pioneered an effort to develop inexpensive handsets for distribution in developing countries at less than US$ 40 per unit. The contract to provide up to six million phones was ultimately won by Motorola. Other companies in China are developing low-cost handsets that can provide Internet telephony via Wi-Fi hotspots (Handford 2005). In general, these cheaper handsets come with stripped-down features (for example, data applications may be limited), but show some sensitivity to the needs of users in developing economies. For example, some have extended battery life in recognition of the limited electricity supply in some areas; others have been designed with environmentally friendly packaging and simplified user interfaces to facilitate usage by people with low education (Parker 2005).

Wi-Fi for Internet Access

While products with reduced capabilities make telephony more affordable for people with limited income, the provision of adequate Internet access remains problematic. It is believed in some circles that since mobile phones can carry both voice and data, they may be the best option for Internet access in developing countries. Others suggest that Wi-Fi networks hold huge potential for making PC-based Internet access available in poor communities (Galperin 2005; Galperin and Girard 2005). For example,

Galperin notes: "the new breed of WLAN technologies thus creates opportunities for developing nations to leapfrog the current generation of Internet access technologies, much like cellular telephony allowed leapfrogging the traditional public switched telephone networks" (2005: 50).

In developed markets, Wi-Fi networks are usually provided by telecom operators, government authorities, or other institutional agents. Although such systems also exist (to a limited extent) in some developing countries, some researchers are exploring the potential of bottom-up deployment approaches. The flexible infrastructures arising from local community involvement in the design and use of Wi-Fi networks will arguably be better linked to local conditions. Consequently, there are several pilot projects underway in different parts of the world to harness the potential of this technology. They include a project that has connected El Chaco, a remote village in the Ecuadorian rainforest, to the Internet (ICA 2005), as well as the Xixuau-Xipaina Ecological Reserve Project in Brazil, the Baja Wireless Project in Ensenada, Mexico, the Linking Everest project in the Khumbu Region, Nepal, and the Internet Project Kosovo in Pristina, Kosovo (Jhunjhunwal and Orne 2003).

Shared Access and Maintenance

Unlike most developed societies where telecommunication is acquired and used on an individual basis, in developing countries systems of shared access, such as telecenters and cybercafés, are being experimented with as alternative methods of expanding access to telecommunication. Wireless telephony offers a new avenue to achieve these aims as can be seen in the variety of commercial and noncommercial sharing arrangements emerging around this technology. One of these arrangements, so far observed exclusively in developing countries, is the mobile payphone service, which will be discussed in more detail below. Other shared-access systems, discussed in chapter 2, include family and community sharing of a single mobile phone (Carrasco 2001; Konkka 2003; McKemey et al. 2003; Wachira 2003; Ureta 2004), and the use of separate SIM cards with one handset.[3]

Communities also come together to maintain mobile phones; for example, by having a designated individual take all the phones to a location with access to electricity to charge the phone batteries (Samuel et al. 2005). Yet another innovation emerging in China is the setting up of public mobile-phone battery charging units where mobile-phone owners can charge their batteries in 5–10 minutes for a fee (Siew 2003; Marth 2004; Weigun and Shibiao 2004). There are reportedly thousands of coin-operated charger consoles in major cities. The service has met with a mixed reaction. Generally, it is seen as useful mainly for emergency situations since most people have chargers at home, and tend to keep their batteries fully charged.

Resource Redistribution: Beeps and Remittances

"Beeping," as discussed in chapter 2, is a popular practice in developing counties which enables poorer mobile-phone users to communicate at the expense of wealthier owners. A potentially more mutually beneficial application for mobile phones, however, is the recent introduction of mechanisms by which mobile-phone owners can transfer both airtime and currency to other owners. The facilitation of arrangements for financial remittances appears to be an especially important use of telecommunications in developing countries (see, for example, Bayes et al. 1999; McKemey et al. 2003; Skuse and Cousins 2005), and may be an important factor in the adoption and use of mobile telephony in African countries.

Mobile phones provide a quicker, more convenient, and safer alternative to traveling or the use of public fixed lines for arranging remittances. In a number of developing countries, people wishing to make or receive remittances can now transfer prepaid minutes to other subscribers using text messaging (Wade 2004; Commission for Africa 2005; Mobile Africa 2005c). This use of prepaid airtime as a "cyber currency" (Mobile Africa 2005c) enables people to move funds around within seconds without the need for a bank account, and also avoid bank charges for money transfers. Some analyses estimate that the transaction costs of money transfers could be reduced from 12 to 6 percent through the use of mobile telephones (Commission for Africa 2005).

The View from the Ground: Case Studies in Asia, Africa, and Latin America

Having discussed some of the emerging strategies for the circulation and adoption of mobile communication technologies, we will now present a number of case studies of some of these emerging trends that are occurring in the unique contexts of developing economies. The Asian cases illustrate the development and use of alternative technologies to provide quasi-mobile-phone solutions, as can be seen in China and India. The African cases describe the development of shared-access models, using existing mobile-phone technology and infrastructure. Here, mobile payphones have become a ubiquitous feature of mobile telephony in developing countries. Originating in organized form in the Grameen village payphone system in Bangladesh,[4] this system has been adopted in the same or modified form in some African countries, of which Uganda, South Africa, and Ghana will be discussed here. Finally, Chile provides an illustration of the ways in which low-income families adopt the mobile phone in a Latin American country.

The Success of Little Smart (Xiaolingtong) in China

Little Smart was first launched in China in 1997. Its technical backbone is a variant of the wireless local loop (WLL) technology, "a micro-cellular system that provides connectivity between the end user and the local switching center (local loop or 'last mile') where traditionally, copper wires had been used to connect these locations" (Frost and Sullivan 2003: 3). It is based on the PAS (personal access system) that uses the global PHS (personal handyphone system) standard, first developed and deployed in Japan. PAS offers consumers "the convenience of a mobile phone, with the cost advantages of a fixed-line phone" (Frost and Sullivan 2003: 6).

Little Smart has been a big hit in China since 2002 with a sales record of US$ 2 billion in 2003, when 25 million subscribers were added to the Little Smart business, surpassing the growth of both China Mobile and China Unicom.[5] By the end of June 2004, the number of Little Smart users had reached 50 million (*PR Newswire* 2005). The first trial site of Little Smart was built in Yuhang, a small city in east China. It then expanded to other small cities and towns in mountainous areas. Kunming and Xi'an (in inland, western China) were the first two provincial capitals to adopt Little Smart. Beijing, Guangzhou, and Shanghai were the last to deploy Little Smart, with the technology finally being introduced into the urban areas of Shanghai in May 2004 (see figure 8.1).[6] Not only is the service popular in nearly 400 cities and 31 provinces in China, but it is also expanding to Southeast Asia, South Asia, Latin America, and Africa (Kuo 2003).

To start using Little Smart, an average user would need to pay around RMB 250 (US$ 30.2) for the handset and put a deposit of RMB 200 (US$ 24.2) in the account. This represents about a quarter of the average cost of a regular cell-phone handset among migrant workers in the three southern cities we surveyed in 2002. The operational cost is also much lower, with per-minute rates being 50–75 percent lower than regular mobile services (*China Daily* 2003). Furthermore, there is no fee for receiving a call. With Little Smart, users also have access to Internet telephony for domestic long-distance calls at a price even lower than prepaid Internet telephony for fixed lines.[7]

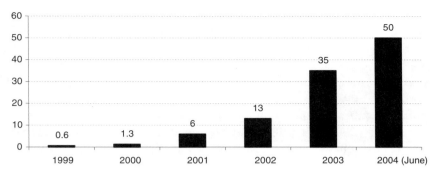

Figure 8.1
The growth of Little Smart subscribers in China, 1999–2003 (in millions).
Source: H. Liu (2004: 1).

The success of Little Smart is, in some ways, a uniquely Chinese phenomenon that would have been impossible without a combination of three factors:

1. *The enormous size of the low-income market*, including migrant workers and petty entrepreneurs in the cities, the growing population of pensioners who need a "life line," and laid-off state-owned-enterprise workers, many of whom are seeking new jobs and have an increased demand for information.
2. *The telecom regulatory regime* which stripped fixed-line operators of their mobile-phone license, just as mobile telephony was becoming the main growth industry.
3. *The central role of UTStarcom*, a private firm founded and managed by former Chinese students in the United States. The company now has headquarters on both sides of the Pacific Ocean. UTStarcom was the first to deploy Little Smart and is still playing a leading role in the business.

Little Smart plays a vital role in meeting the informational needs of the urban underclass, many of whom continued to be pager subscribers in the late 1990s, even as better-off urbanites began to adopt cellular phones.[8] Unfortunately for them, many local pager companies are abandoning the business and their less wealthy users due to the spread of cell phones, but more importantly because of a lack of institutional support for pager operators.[9] This, however, did not mean the disappearance of the low-end market. Rather, the steep decline of pager subscription is among the factors that have contributed to the phenomenal success of Little Smart.

Little Smart brings wireless technology to the less wealthy, many of whom have considerable information needs coupled with limited purchasing power and limited physical mobility. Prepaid cell-phone cards and SMS offered partial solutions to this problem, but still left out a large proportion of the population. While an ordinary cellular phone could easily cost RMB 2,000 (US$ 121.8), China's per capita annual outlay on transportation and communication was only RMB 626 (US$ 75.6) in the urban areas during 2002 (H. Liu 2004: 18). Meanwhile, survey results showed that close to one-quarter of potential new mobile subscribers hoped to spend less than RMB 1,000 on new handsets (BDA China 2002). What was needed for the majority of these potential low-end users was really inexpensive voice telephony with some degree of mobility. As noted by UTStarcom:

Despite its geographical size, China tends to have an extremely localized culture in the sense that the majority of work and social activities for citizens revolve around one's immediate environment. According to a recent survey from the *China Post*, 80 percent of the population spends 80 percent of their time within the city limits, suggesting that the lion's share of demand for mobility solutions will generally remain local. This made the PAS solution all the more attractive for Chinese citizens. (UTStarcom 2003: 3)

The result is that Little Smart has become an appropriate solution for not only high-density urban areas but also small towns in the countryside. It is often adopted in rural areas as a substitute for fixed-line services, with the added advantage of localized mobility (UTStarcom 2003: 3). It is therefore an effective means of increasing teledensity among low- and middle-income populations.

Little Smart is also deployed by fixed-line operators as a way to extend landline services.[10] Because PAS was initially designed to cover small areas, such as a building or residential complex (*Global Entrepreneur* 2004), it has several technical advantages for cost reduction. First, it uses switches for fixed-line network and "requires no modification to the central switching office, nor does it require investing in mobile switching hardware" (Frost and Sullivan 2003: 7). Secondly, it can be scaled to fit areas of different user density, which varies greatly within and around China's city space (Frost and Sullivan 2003: 7). Thirdly, it is relatively easy to set up – taking three to four months to deploy in a city of 10–12 million potential users (Frost and Sullivan 2003: 5).

In this new role as quasi-mobile service provider, fixed-line operators gain new business opportunities. They can attract more subscribers – 50 million of them by June 2004 – while only needing to make minor changes to the existing network. Lots of labor can be saved laying the last mile of copper. This is exactly what China Telecom needed after it lost its mobile communication division. As Mr Ying Wu, CEO of UTStarcom China says, "The key was that when China Telecom lost their mobile business, they lost their growth point . . . We saw a golden opportunity: Sure, the top 20 percent income earners will be cellular subscribers. But that leaves the middle 50 percent – 650 million people – who need wireless service but for whom affordability is the issue" (Kuo 2003). In this business model, fixed lines meet the needs of users with low income and low mobility, cell phones meet the needs of those with high income and high mobility, and Little Smart caters for people with low income and moderate mobility (Frost and Sullivan 2003: 2).

The most serious challenge for Little Smart has less to do with the technology, the market, or the lack of institutional support than with the strong resistance of mobile-phone operators and the vicissitudes of telecom policymaking in China. China Mobile and China Unicom both campaigned hard against Little Smart due to fear of cut-throat price competition at the low-end of the market.[11] Moreover, central decision-makers in Beijing did not show clear support in the beginning because they were not sure if China should adopt this "outdated" technology that had already proved unsuccessful in Japan.[12]

Under these circumstances, the spread of Little Smart followed a model that later came to be known as *nongcun baowei chengshi* ("countryside surrounding cities"), drawing on Mao's famous military strategy during the communist revolution. Although Little Smart began in small cities and towns, rather than the countryside, this metaphor captures the path of the service, which was not available in large urban centers until much later in its development. This was not a deliberate choice by UTStarcom or China Telecom. In part, it was because the former Ministry of Post and Telecommunications imported PAS technology in 1996 in the hope that it would help enhance teledensity in China's mountainous areas (Liu 2004: 3–4). Resistance to Little Smart was considerably lower in these areas, and national decision-makers were also more willing to tolerate it as a temporary experiment in small cities, where teledensity was low anyway.

According to Duncan Clark of BDA China, the process of regulating Little Smart has gone "from a policy of 'grow quietly, but grow' to one of almost no regulation at all" (BDA China 2002). The government's current approach is to "'neither support nor hinder.'"[13] This policy of non-intervention effectively counters the efforts of China Mobile and China Unicom to stop the spread of Little Smart, leading to phenomenal growth as reflected in the increase of UTStarcom's gross revenue from about US$ 200 million in 1999 to almost US$ 2,000 million in 2003 (H. Liu 2004).

The success of the Little Smart product has attracted several domestic companies into the market, effectively cutting UTStarcom's market share to about 70 percent and creating an entire industry in services and handsets (H. Liu 2004). UTStarcom has taken this new competition well because its corporate strategy has been one of significant diversity from the beginning, with its broadband networking systems being one of the more recent fast-growing revenue sources.[14] By 2003, 25 different manufacturers were producing Little Smart handsets (Kuo 2003). There are now more than a hundred Little Smart models, some equipped with advanced features, such as digital cameras, flashlights, Internet browsers, and color LCD display (M. Liu 2003), as well as UTStarcom's own dual-mode handset which allows customers to switch between Little Smart and GSM/CDMA.[15]

At the center of the Little Smart boom is UTStarcom, a global operation with a localized focus. UTStarcom has built its strength at both global and local levels by utilizing global networks to develop high-tech products, explore new markets, and raise significant external investment from stock markets and venture capitalists, while simultaneously developing familiarity and understanding of local conditions in its places of operation.

With its increasing reliability and multi-functional handsets, Little Smart is beginning to attract high-end users, but it is still largely perceived as a poor person's mobile phone due to its humble origins which were characterized by poor reception (H. Liu 2004: 18). In the early days, users had to walk in circles or wave their handset in an open space to maintain the signal, and the device did not work in moving vehicles, including bikes. In some places, people still refer to Little Smart as "*Wei Wei* Call" because its use was associated with having to repeat "*Wei? Wei?*" (Hello? Hello?) due to the poor signal (Kuo 2003). It is understandable that from a business viewpoint some would like to change this stereotype by making Little Smart more appealing to high-end users. However, from the perspective of larger social benefits, it is important for the Chinese market to have telecom solutions that bring low-cost mobility to the masses. This was, after all, the primary motivation for the importation of PAS in 1996 (H. Liu 2004: 1).

WLL for Low-Income Groups in India
Wireless local loop (WLL) is a general category of limited mobility services like the personal handyphone system (PHS) in Japan and its Little Smart variant in China. In the Indian context, the development of WLL followed

technological and institutional trajectories that led to a significant but slow growth pattern which reflects India's specific market dynamics, policy framework, and priorities at this early stage of wireless deployment. The WLL standard most widely used in India is called corDECT, a modified version of the European digital enhanced cordless telephony (DECT) standard. The corDECT standard was created by the Indian company Midas Ltd, a spin-off of the Indian Institute of Technology (IIT) in Madras (O'Neill 2003: 88). With corDECT's lower per-line cost of US$ 275–345 or Rs 12,000–15,000 (as opposed to the average cost of US$ 345–575 or Rs 15,000–25,000 per line for fixed-line telephone), some telecom operators have started to provide voice and Internet connectivity using WLL in certain areas (Jain and Sridhar 2003: 275). Local access loop may also be provided using CDMA at an even lower cost of less than US$ 230 (Rs 10,000; Jain and Sridhar 2003: 275), although some analysts are suspicious about such a claim for CDMA (see O'Neill 2003).[16] According to McDowell and Lee (2003: 376), the introduction of WLL was an attempt by "Mahanagar Telephone Nigam Limited (MTNL) to introduce wireless services, an initiative which the Telecom Regulatory Authority of India (TRAI) had first rejected." However, in January 2001, the Department of Telecommunication released its guidelines for the licensing of WLL services. This new set of policies both facilitated and constrained the application of WLL in the country (Jain and Sridhar 2003: 276).

The price advantage of this so-called "poor man's mobile telephone" has led to impressive growth of WLL in India (O'Neill 2003). The total number of subscribers grew from about 100,000 in September 2001 (O'Neill 2003) to more than 2 million in 2003 (Jain and Sridhar 2003) and then to 7.55 million by March 2004 (Department of Telecommunication 2003–4). Although the distribution is more concentrated in cities, a survey in April–May 2002 in parts of the Kumaon hill region of northern India showed "a rapid increase in deployment, although patchy" of WLL (O'Neill 2003: 92).

While India's WLL exists on a smaller scale than China's Little Smart, the Indian authorities and telecom operators arguably attach more importance to its universal service implications than their Chinese counterparts. The development of WLL was identified as the first of the two main solutions to enhance teledensity in low-income areas, the other being a universal service obligation fund (Jain and Sridhar 2003: 275). According to O'Neill (2003: 89), there was a "fierce debate in and outside parliament" about whether basic service providers should be allowed to build and sell WLL, during which the question of why bureaucracy and vested

commercial interests had been allowed to inhibit the provision of telephony to the rural population was raised. Therefore, when the Department
of Telecommunication endorsed WLL in 2001, it also indicated that the
main policy objective was to expand service availability to all parts of the
country at affordable rates and that allowing basic service licensees to
provide WLL service was just a means to achieve that end (McDowell and
Lee 2003: 371).

Economic barriers, however, remain significant in the deployment of the
WLL system in India. For one thing, the Indian population, especially in
rural areas, has very low income, and parts of the country do not have a
stable electricity supply (O'Neill 2003: 85). The WLL service is not particularly cheap either, when compared to income levels. For example, according to O'Neill (2003: 85), a senior professor in India earns Rs 5,000 (US$
115) per month. A corDECT-based WLL handset produced domestically
cost between Rs 5,000 (US$ 115) and Rs 12,000 (US$ 275) in November
2001, while those imported from South Korea cost Rs 4,000–9,000 (US$
92–207). The rent for a CDMA-based WLL phone in New Delhi was Rs
17,000 (US$ 390) per handset, with a security deposit of Rs 10,000 (US$
230), and a monthly rental of Rs 600 (US$ 14). The relatively high price
of the so-called "poor man's mobile telephone" led to "very public battles"
in mid-2002 (O'Neill 2003: 92–93).

The WLL market is full of competition. Cellular-license holders
attempted to stop basic service providers from building WLL and they still
hope to limit the WLL capacity in order to maintain their market share.
Moreover, the competition remains fierce between the "domestic" standard of corDECT and the "foreign" standard of CDMA, which has been
elevated from a pure matter of technology and market economics to a
politically charged debate, with the fixed-line operators supporting
corDECT, on the one side, and the cellular operators being more receptive
to CDMA, on the other. A legal challenge, for example, was launched by
cellular operators against the TRAI's decision to allow fixed-line operators
to provide a WLL service, which ended up in the Telecom Dispute Settle-
ment Appellate Tribunal (O'Neill 2003: 89). The January 2001 Department
of Telecommunication WLL Guidelines, as a result, reflected compromises
between the two sides, taking certain measures to be "fair to the cellular
license holders," including permitting cellular operators to provide a fixed-
line service based on the GSM network infrastructure (McDowell and Lee
2003: 376). The guidelines also specified that "spectrum for the WLL
services was to be made available on a first-come first-served basis. In
exchange, providers had to roll-out networks simultaneously in urban,

semi-urban, and rural areas" (McDowell and Lee 2003: 376). This stipulation again indicated the full intent of the Indian government to use WLL to meet its universal service obligations. But it also means potentially more disputes if the rollout is slower and less successful in semi-urban and especially rural areas.

Given the intensity of the ongoing debate and the high stakes involved in the WLL market, the Indian government also decided to allow technical neutrality, rather than requiring any specific standards for WLL licenses, thus creating uncertainty in technical standards (Jain 2001: 674). Moreover, the corDECT-based WLL system is mostly for subscribers to gain limited mobile-phone connectivity in one place, usually defined as a residence or business (McDowell and Lee 2003: 376), rather than in a citywide scope as in the Chinese case. All these non-price-related conditions can act to slow down investment decisions in India's WLL market.[17]

The Modified Grameen Model in Uganda

The mobile payphone system in Uganda is essentially a replication of the Grameen model with some modifications to suit the local environment.[18] Currently, there are at least two separate formal programs in the system. The first, MTN villagePhone, is offered through a collaboration between the Grameen Technology Center, MTN Uganda (the second national operator licensed in 1998), and a number of local micro-finance institutions. As occurs in the Bangladeshi case, the micro-finance institutions lend funds to new village-phone operators to buy a starter kit (costing about US$ 230) which includes a mobile phone, SIM card, prepaid airtime card, business cards in the local language, and an advertising sign for the enterprise with phone rates. MTN Uganda also gives operators the prepaid minutes at a discounted rate. The program began in 2003 and had over 1,300 operators in its first year, providing services in more than 18 districts. About a hundred new businesses are added every month and usage levels exceed initial projections by 25 percent.

The second is Celtel Uganda's "community phone" program, also offered in collaboration with a micro-finance company, the Foundation for International Community Assistance (FINCA). Local operators lease a mobile-phone kit from Celtel and retail airtime to callers at US$ 0.28 for local calls and US$ 1.28 for international calls. They pay a monthly leasing fee to FINCA and airtime charges to Celtel. This program is also reportedly extremely successful with revenues growing at nearly twice (10 percent monthly on average) the rate of growth in mobile subscriptions.

Organizational and journalistic reports on these systems in Uganda indicate that the results have been positive: operators enjoy increased income, greater ability to cater for family needs, and expansion of their business networks (most are engaged in additional enterprises, such as dressmaking, farming, or retailing). Power problems are solved by the use of car batteries to recharge the phones. The community also enjoys access to telephony. The uses of village phones by callers in Uganda include business transactions, communicating with family, and checking the prices of agricultural goods. In some communities, people gather at a micro-operator's store to listen and call into radio phone-in programs.

Mobile Payphone Franchises in South Africa
Mobile payphones are provided in South Africa by Vodacom as a strategy to meet its universal service license obligation.[19] One account (BiD Challenge 2004) has it that the idea of introducing mobile payphones arose when Vodacom deduced from its customer records that subscribers with high usage levels were probably selling time to friends who did not own mobile phones. It was a simple matter to identify these individuals and offer them the opportunity to provide this service on a more formal basis. Vodacom runs a "phone shop" franchise whereby individuals buy the necessary equipment from Vodacom for about US$ 3,450, with an additional investment by Vodacom of about US$ 3,950.[20]

Vodacom's contribution goes into the provision of a (Vodacom-branded) modified shipping container in which at least five cellular phones can be set up. Franchises provide their service in pre-approved locations determined by Vodacom in conjunction with the Independent Communications Authority of South Africa (based on population density, proximity to a high-profile location, distance from other phone shops, and accessibility of power source), and at government-mandated prices which are less than one-third of the cost of regular mobile-phone calls (this initially required a subsidy from Vodacom, but a recent price increase has removed the need for the subsidy). Franchisees sell prepaid airtime that they obtain at a discount from Vodacom. These prepaid minutes are credited directly to the phones rather than implemented through purchase of a phone card. Vodacom Community Services began this program in 1994 and has since enrolled about 1,800 franchisees with over 4,400 phone shops and 23,000 mobile-phone lines throughout the country.

Some operators use the modified containers; others use their own alternative accommodation. Franchisees must meet a number of criteria, including nationality (South African), age (at least 21 years), business skill and interest, access to start-up funds, and knowledge of a feasible site. The program provides business training but is otherwise not involved in the day-to-day running of the enterprise. The number of people applying to run franchises exceeds what the program is currently able to support, indicating that the high set-up costs are no barrier to business entrepreneurs. In this sense, the program has been extremely successful, but it is likely that the business opportunity is being enjoyed more by those who have access to seed money, rather than by poorer people who have no source of funds to start a business.

The telephone sets are designed for the context of the phone shops (they resemble traditional landline phones), and Vodacom has entered into arrangements with a national supplier of phones to develop phones that are attuned to the changing needs of the phone shop (for example, the programming of call time, display of time remaining, fax and data capability). The containers housing the phone shops have also been modified over time to provide greater security and/or comfort for users, employees, and owners, as well as to withstand environmental wear and tear (however, this adds to the cost of the unit).

The performance of the community phones has been impressive. Shop operators in good locations can sell over a hundred hours of airtime per month and make up to US\$ 1,190 net revenue. Other benefits include employment generation,[21] expansion of access to telephony for disadvantaged communities, and greater efficiency for professional workers who use the service for their business activities. The phone shops have become social centers as well as sites for other enterprises to take advantage of the flow of human traffic.

Not surprisingly, MTN (a major competitor in mobile telephony) has subsequently begun to provide a competing service in MTN-branded containers. Unofficial versions of the phone shops also exist, some of which mislead customers with the false promise of low per-unit pricing, while they actually end up paying more than twice the charges of the official service. For customers, however, this is sometimes a trade-off between traveling further to find an official phone shop and using a closer, but ultimately more expensive, unofficial one.

Grassroots Mobile Payphone Initiatives in Ghana

There are currently at least three models of the mobile payphone system in Ghana, all operating on the Spacefon Areeba network, hence their popular designation as "Space-to-Space."[22] The first, and precursor of the trend in Ghana, was offered by Spacefon Areeba in response to interconnection problems with the Ghana Telecom network.[23] To overcome this problem, Spacefon provided telecenters with GSM desktop telephone sets fitted with a Spacefon SIM card to enable telecenter users to bypass Ghana Telecom when they wanted to make calls to Spacefon subscribers. The second model of service provision emerged when these GSM desktop phones found their way into the hands of entrepreneurial individuals who began to offer this service outside the telecenter system – in kiosks, convenience stores, hair salons, and other roadside setups. Spacefon was reportedly not averse to this development, especially since it all functions on a prepaid basis. Thus, a new industry in telephone service provision was created. Most of these entrepreneurs began operating during the last six months of 2004 and there are supposedly already about 25,000 such operators around the country.

There is no direct relationship between Space-to-Space operators and Spacefon Areeba. These operators simply use their own funds to purchase the telephone handset from shops that sell telephone equipment. However, according to one operator interviewed in Accra, the phone sets are currently designed in such a way that only Spacefon SIM cards (and therefore only Spacefon prepaid cards) can be used with them.[24] The initial outlay for the enterprise is approximately 7 million cedis (about US$ 800), which is the cost of the telephone set, including 1,000 free units. Once the free units are depleted, operators top up with 3,000-unit prepaid cards, which cost 900,000 cedis (about US$ 100). There is also some minor expense incurred for a table and/or shed, some chairs, a shade, and stationery.

Being the dominant network, calls to Spacefon Areeba subscribers from these roadside service providers cost less than calls to other networks. Space-to-Space calls cost between 1,500 and 2,000 cedis (about 17–22 US cents) per unit, while Space-to-other-networks calls cost about 2,500 cedis (about 28 US cents). Depending on which network is being accessed, one unit may provide 1–4 minutes of talk-time.

As regards the social and economic impacts of this system, it seems intuitive that it not only provides expanded access to telephony, but also addresses serious telecommunication policy failures (especially in ensuring interconnectivity), and provides a source of income for several individuals. There may be questions, however, about the sustainability of these enterprises as mobile-phone subscriptions rise. In addition, the proliferation of such outlets may create excessive competition and spread the system thinly, making it difficult for any single entrepreneur to develop a viable enterprise.

A third variation of the mobile payphone system began in early 2005 when Spacefon Areeba outdoored two versions of a service it calls "i-Tel 'Pop.'" The first is a manned service station where users can make wireless phone calls from payphones. The second is a mobile service provided by bicycle riders. The main difference between this and the Space-to-Space service is that it is institutionally organized. However, entrepreneurs have the option to run the bicycle service as an employee of Spacefon or as a personal enterprise. Rather than modify its anti-competitive stance and make access to and from its networks easier for rival mobile-phone subscribers, the incumbent telecommunications provider, Ghana Telecom, has also announced the future launch of its own mobile payphone system, ONE4ALL (Mobile Africa 2005b).

Despite their similarity to Grameen's villagePhone program (VPP), mobile payphone systems in African countries have some unique characteristics. For example, in recognition of local conditions, appropriate adaptations have been implemented for the Ugandan program. Unlike the Grameen program, women are not the primary target for participation as payphone operators.[25] In addition, in the case of MTN villagePhone, alongside Grameen, several local micro-finance institutions (currently, FINCA Uganda, the Foundation for Credit and Community Assistance Uganda [FOCCAS], Support Organization for Micro-Enterprise Development [SOMED], Uganda Women's Financial Trust, and Uganda Microfinance Union) are involved in the provision of credit to operators, a critical requirement for the program to be able to operate on a viable scale in Uganda. MTN has an open invitation for interested micro-financiers to become partners. Consequently, whereas there is a standard two-year term for borrowers in Bangladesh, MTN villagePhone operators may be offered loan terms ranging from six months to two years, depending on which micro-finance institution they are affiliated with. Furthermore, the Grameen VPP system was not based on the use of prepaid airtime, which meant that there were fairly high labor requirements for accounting; this is not the case in Uganda where the prepaid system further facilitates the smooth running of the system.

There are also important differences between the mobile payphone systems in the three countries reviewed here, which could have implications for their long-term impact and viability. The services offered in Uganda and South Africa are targeted at rural locations and therefore indicate a proactive scheme to extend telecommunications to these underserved areas. In fact, a Norwegian company – UnoPhone Uganda – recently entered the market with the aim of targeting and marketing the village-phone enterprise to rural residents. In order to ensure sustainability, Vodacom requires community-phone operators in South Africa to be located at reasonable distances from each other. However, the South African system is probably less accessible to poor entrepreneurs because they have to source funds on their own, unlike the Ugandan system which is supported by micro-finance institutions. Indeed, Skuse and Cousins (2005: 8) observe that "many poor households are excluded from the possibility of starting small ICT-based enterprises such as cellular phone services by virtue of the start-up costs and small profits."

Conversely, in Ghana, there is little involvement of any formal organization in the system, and it appears that it is highly concentrated in busy urban centers.[26] Space-to-Space operators do not enjoy the prepaid airtime discounts that operators in Uganda and South Africa receive. Mobile payphone services introduced by the major telephone-service providers are not connected to any attempts to improve teledensity in under-served areas. Rather, in the case of Ghana Telecom, it is a reaction to perceived competitive pressures from the Space-to-Space innovation, and in the case of Spacefon Areeba, it is an attempt to capitalize on an obviously successful model that the company did not appreciate when it first implemented its telecenter strategy. Arguably, the institutional support enjoyed by mobile payphone operators in South Africa and Uganda could lead to more systematic and focused benefits to both operators (in terms of long-term sustainability of their enterprise) and users (in terms of provision of access in the more deprived areas). What is nevertheless significant about the street-side vendor model in Ghana is that it has for the most part emerged at the grassroots level and shows how end-users can appropriate mobile telephony. Of the three systems, the Ugandan seems to offer the most opportunity as a business proposition for poor people because of the low start-up cost and inherent link to micro-credit facilities, in addition to facilitating the expansion of telecommunication services to rural areas.

Family Life and Mobile Phones in Chile

In 2002, a research team from the Institute of Sociology of the Catholic University of Santiago conducted a study on the diffusion and use of information and communication technologies in Chile (Garcia et al. 2002). At that time, about a third of the Chilean population were users of mobile phones (5 million people), and 20 percent were connected to the Internet. The researchers found that the working environment was the most conducive to the diffusion of the mobile phone. For professionals, it was considered an essential tool that enabled the spread of the networked business form of organization. Mobile phones allowed one to perform work in the time needed to travel from one location in the city to another, given the inefficient transport system of a large metropolis such as Santiago. It was also widely used in the frequent waiting times in people's daily lives, such as the waiting rooms of various agencies and services.

Yet, this is mainly the case for workers with some level of autonomy in their tasks and time. For those people executing routine work, the mobile phone is mainly a useful means of communication in their social and family life. In this regard, the presence of children in the family is the decisive factor in the intensity of use of the mobile phone. Thus, for families without children, mobile-phone use for family connections is very limited. For those with children younger than 13, the mobile phone becomes significant to ensure permanent communication with the children for safety reasons. For those with children between 13 and 20, mobile-phone use intensifies, since children have fairly independent lives and parents use the mobile connection to maintain what the researchers label "an illusion of control," thus formally maintaining family order under the new cultural conditions of youth emancipation. These observations are not very different from the results of studies in developed countries (see chapters 3 and 4). The reason is that the majority of people observed in this study are from middle-class backgrounds, which, in the Chilean situation, have cultural patterns not too dissimilar to Europe or the United States.

In contrast, other studies of low-income families in Chile show the relevance of the phone as a tool of connectivity for the whole family, instead of being a means of communication for intra-family relationships. Thus, ethnographic research conducted by Ureta (2004) studied the relationship between mobile telephony and the spatial mobility of low-income families. The author presents a qualitative study of 20 families in a recently created neighborhood in the vicinity of Santiago de Chile, more specifically in the borough of Renca, called villa Tuacapel Jimenez II. The focus is on urban poverty, a modern phenomenon which is a mixture of "on one hand, the old and well-known face of the absence of opportunities and innumerable difficulties to satisfy an important number of basic needs, but, on the other . . . They are integrated, symbolically and as consumers, to society, although they are still segregated in social and spatial terms" (Raczynski and Serrano 1999; quoted in Ureta 2004: 3).

The families under discussion have, in fact, very limited mobility, commonly reduced to the strictly necessary. The main reasons for this can be found, first of all, in financial restrictions: even public transport is expensive for the majority of their budgets and the distances to be covered are considerable since these families have been displaced to new neighborhoods located in peripheral regions of the metropolitan area. On the other hand, the perception of risk also leads to a low degree of mobility. In addition, while the reasons related to "the simple disinterest in being outside, in any other place or in the street" (Ureta 2004: 8) appear frequently, we can consider that there is a built discourse to justify the lack of mobility.

In this context, the mobile phone appears to be a non-mobile device for the great majority of the families under study, who almost always use it as a fixed telephone line. Indeed, the former is a substitute for the latter because there is no landline facility in the houses. Thus, the possession of a telephone in the home means a slight reduction in the need to move. First of all, there is no longer a need to use the public phones available in the neighborhood, and, secondly, the mobile phone enables a greater degree of communication and coordination as it allows not only outgoing calls, as public phones do, but also incoming calls. It should also be noted that only in a minority of houses is there more than one mobile. The handset is considered a family device which, therefore, is used collectively. Mobiles stay at home unless the whole family goes out and, mainly for security reasons, the handset must stay as close as possible to the mother, the central node of the family network.

Nevertheless, the possession of a mobile phone does not significantly change the mobility of the members of the family, a result already discussed in chapter 3 (see p. 89). This means, in the specific case under study, that "family members, and obviously their mobiles, stay at home, as much as they used to" (Ureta 2004: 13). Ureta quotes Diego, a 39-year-old man: "[the mobile phone] is here in the house, it didn't move, [it] is as if [it] was here, fixed . . . the telephone does not move from here, we never move it away anyway, because we do not go out" (2004: 13). Despite this, the adoption of the device is accompanied by a discourse of actual mobility by their owners, mainly influenced by publicity, which does not correspond with the actual use of the technology. Therefore, the ideology of the appropriation of the device is different from the process of appropriation itself, as also happened in the early stage of adoption of mobile phones in Europe (see Castells et al. 2004: 78–82).

In view of the information so far presented, and given the available statistics,[27] the families studied by Ureta (2004) would have prepaid mobile phones. What is more, prepaid subscribers in the country tend to use their mobile phones less in comparison with postpaid subscribers.[28] Despite the strict control of expenditure that is achieved by the prepaid payment system, it has been observed that spending on mobile phones takes a significant slice of consumers' disposable income. According to one report, the average household in Chile now spends more on telecommunications than on water and electricity combined (as noted by MacDermot 2005). Retailers, particularly those selling goods such as clothing and footwear, have for a few years muttered about the "mobile-phone effect" depressing sales (MacDermot 2005). Within the geographic boundaries of Latin America, the next section illustrates the significance of Wi-Fi technology for increasing access to telecommunications in developing countries.

Wi-Fi Internet for Development: The Latin American Experience

We have already seen in this chapter that as long as there are not – enough – guaranteed profits, and following the same economic logic that discourages other basic service suppliers from investing, private telecommunication wired operators are not interested in rural or remote areas. The characteristics of these areas put them at a disadvantage that is not easy to overcome. However, wireless technologies seem to bring new opportunities for development.

More precisely, when talking about wireless technologies in their capacity for data traffic, we need to refer to the Internet. Access, as previously stated, is the main point of discussion in connection with developing countries, in contrast with "access on the move" which is the main interest in high-income economies. In this sense, we will pay particular attention to wireless local area network (WLAN) technologies, which, by allowing drastic reductions in network deployment costs, particularly for last-mile connectivity in low-density areas, are fundamentally changing the cost structure of Internet operation. In our discussion, we will follow the work of Galperin (2005) and Galperin and Girard (2005) which consider the opportunities that the aforementioned technologies can bring to sustainable Internet diffusion in the rural areas of the developing world, and particularly of Latin America, thanks to the creation of adequate conditions for a bottom-up deployment model which does not follow a preconceived plan created by any external agent, i.e., a telecommunications operator or government agency. Therefore, flexible infrastructures arising from the community will be better linked to local needs and attributes. Galperin (2005: 50) clearly identifies the elements that make this possible:

WLAN technologies create an alternative . . . better suited to the challenges of extending Internet connectivity to rural areas in the developing world. Because of the cost advantages associated with wireless, the use of unlicensed spectrum, and the lack of significant economies of scale in network deployment and management, infrastructure investments in Wi-Fi networks are within the reach of a variety of local actors – from private entrepreneurs to municipal governments to agricultural cooperatives – that better understand local conditions.

Local actors are often organized as community networks that can be created specifically to bring ICT services to their members or, alternatively, can extend their activities to cover this objective. This organizational model is, in a number of cases, the only way in which there is the possibility of mobilizing specific local resources unavailable to private or public operators. Local resources typically consist of skills and labor which are provided in kind by community residents (at a very small opportunity cost in many cases given current under-employment levels), but could also involve other assets such as information about demand patterns (which services to offer, where, for how much, and so on) and rights of way for cables, antennae, or other network and terminal equipment (Galperin and Bar 2005; Galperin and Girard 2005).

Looking in depth specifically at the technology, Galperin (2005) describes different factors that explain its success and its limitations. Wi-Fi, first of all, can deliver high bandwidth without the wiring costs. Secondly, there is widespread industry support for the Wi-Fi standard which aided a swift drop in price, as long as there is compatibility between the devices involved. A third key to its success lies in the generalized, but not total, lack of a regulatory framework: Wi-Fi networks have blossomed on unlicensed bands; namely, thin slices of radio spectrum reserved for low-power applications in which radio devices can operate on a license-exempt basis. This has allowed a wide variety of actors to build WLANs without any of the delays and expenses traditionally associated with obtaining a radio license from telecommunications authorities. On the other hand, the major drawback of Wi-Fi is the short signal range. Even though point-to-point connections have been made over several kilometers, Wi-Fi networks typically extend for a few hundred meters at most, making the technology generally unsuitable for long-haul transmissions. Nonetheless, related technologies are emerging to address this problem, notably 802.16x (also known as WiMax).

There are a number of specific projects in which the wireless Internet has been introduced in rural or remote areas of developing countries. In the specific case of Latin America, few experiences are analyzed in the literature. However, pilot projects related to the installation of Wi-Fi-supported telecenters tend to appear in newspapers; for example, the specific case of El Chaco in Ecuador, a remote village located in the rainforest which, from January 2005, has been connected to the Internet thanks to a cooperative project in which local communities seem to have been involved (ICA 2005).

In this section, we will briefly discuss the characteristics of a project developed in Peru (Jhunjhunwal and Orne 2003). In the Chancal–Huaral valley, located 80 kilometers north of Lima, Peru, a pilot project brought the Internet to a previously existing community network which seems to be creating spillovers into the whole valley. The project involves a total of 14 interconnected telecenters, 12 of them – those located outside the main city – through Wi-Fi and VSAT (very small aperture terminal) technologies. They use only free open-source software and affordable computer equipment (Belo 2004). Following the analysis of Galperin (2005), the main point to be highlighted here is the full involvement of an existing community organization, the Board of Users of the Huaral river, which is not only the owner responsible for the maintenance of the telecenters but also manages an agricultural information system based on data collected by its members and specifically designed for their needs. The Board is a cooperative arrangement of the 17 irrigation commissions spread throughout the valley, which are in turn composed of the farmers themselves (about 6,000 in total). Faulty communication among these geographically dispersed entities often creates difficulties for efficient water management. Some telecenters are located within educational institutions, making services available to teachers and students in the community.

Based on the performance of this project, as well as on other elements of the telecommunications landscape in Latin America, Galperin and Girard (2005) identify three main regulatory barriers to the development of wireless community networks in the area. First, there is the question of the availability of radio spectrum, particularly the designation of bands for the operation of wireless equipment on a license-exempt basis, which is not guaranteed in all countries; secondly, there are restrictions on the operation of VoIP services on a public basis, as opposed to a private network; and thirdly, interconnection policies discourage network extension in low-income rural areas (Galperin and Girard 2005).

Finally, and from a regulatory perspective, Galperin (2005) considers that there is much to be done to facilitate the deployment of WLANs in Latin America, whether by private entrepreneurs, public entities, or local organizations. In this sense, one important issue to be considered is the network development model associated with universal service funds. These funds were established in Latin America during the period of telecom liberalization in the 1990s for the extension of basic voice telephony services, but more recently have been expanded to include advanced ICT services such as broadband Internet. In Chile, for example, the Telecommunications Development Fund was modified in 2001 to allow funding of a program that has so far financed almost 300 rural community telecenters through competitive tenders. In Colombia, a similar program (called Compartel) has to date financed the installation of over 900 community telecenters across the country. In Ecuador, where it is estimated that 30,000 rural communities lack access to basic ICT services, resources are being allocated for the installation of community telecenters in 5,000 of them.

For example, the aims of the Colombian Compartel Program of Social Telecommunications are to create multipurpose telecenters for community development and, among other goals, to provide rural community telephony. The technologies used are, as in a number of other cases, a combination of available tools. Thus, there is a predominant use of satellite solutions (80 percent of the project), while cellular and WLL represent, respectively, the remaining 18 percent and 2 percent. More specifically, wireless solutions are run under the 2.5 GHz and 5.5 GHz bands, as long as these have been declared free for use by the Ministry of Communication (Compartel 2004). Indeed, satellite communications, specifically VSAT technology, have contributed to the expansion of connectivity in rural areas. An important part of the bidding to grow telecommunication services in rural areas of Latin America (Brazil, Chile, Colombia, and Peru) has been won by companies that use VSAT technology (Proenza 2005).

A caveat, however, must be introduced here: while increased use of universal service funds to finance telecenters in under-served rural areas is encouraging, these programs are generally linked to traditional top-down development. The result is that, as the evidence from the programs in Colombia and Chile reveals, most contracts are awarded to large operators or service aggregators, with little participation from local entrepreneurs and community organizations (Galperin 2005).

Summary

Connectivity via telecommunications is an essential prerequisite for development in our globalized world. The connectivity gap is one of the most formidable obstacles to developing countries and poor regions linking up with the dynamic global economy and with the global communication networks that offer access to information, education, and services. Wireless communication technologies offer the possibility of leapfrogging the stages of development and avoid the cost and time involved in setting up fixed-line infrastructure in currently disconnected, or poorly connected, territories. However, our observations document the excessive optimism that surrounds this new magic bullet of development. There are still major issues to be solved in the provision of wireless communication and the use of mobile phones and the wireless Internet: issues of investment, of deployment, of user education, and of affordability for most of the population. It is still necessary to improve the fixed telecommunications infrastructure and to find ways of effecting complementarity between different communication technologies. The role of regulators is still indispensable to ensure quality, as well as universal access to the network, be it fixed or wireless, for the population. This is particularly important for the diffusion of broadband and its promise of new services dependent on transmission capacity.

Under current conditions of a scarcity of resources and the shortsighted competitive strategies of many operators, we have observed the surge of grassroots initiatives and innovative schemes to access wireless communication through unexpected avenues. We have documented the need and desire of people in developing countries, as well as of companies and institutions, to affirm their right to connectivity, using the flexible technology that wireless represents. The evidence we have been able to unearth shows the pervasive diffusion of wireless communication in all spheres of social life and economic activity. We also appreciate the ability of people and communities to adapt the technologies to their own possibilities and to shape them around their specific communication goals. Wireless communication is no panacea for development. But developmental projects, from all corners of the planet, are embracing the potential of new technology and are using it for their own purposes according to what they are able to achieve.

Conclusion: The Mobile Network Society

In the early years of the twenty-first century, the uses of wireless communication have been transformed by its users as communication networks have diffused globally. From a professional communication device catering for an upscale market, mobile devices have become mass-consumer products, woven into the communicative practices of hundreds of millions of people everywhere in the world. From an advanced technology reserved to advanced countries, it has become the technology of choice for developing countries in order to reduce their connectivity gap. From a mobile substitute for voice communication, it has evolved into a multimodal, multimedia, portable system of communication, which is gradually absorbing most of the functions of the fixed-line phone to the point where there are more wireless phones in the world than fixed-line phones and the gap between the two technologies continues to increase.

Young people have been the drivers of the diffusion of wireless communication technology in developed countries, and they have invented, created, and adapted new communicative uses. Those companies in the industry that have been sensitive to the young users' initiatives have transformed their spontaneous innovations into new technologies, new products, and new business models, in a virtuous circle of interaction between active consumers and receptive business. When business has not been able to respond to the moving culture of mobile users, it has been overtaken by competition or, as in the United States, has lagged behind countries of similar economic level in expanding the market.

Youth culture has found in the mobile phone an appropriate tool to express its demands for "safe autonomy," ubiquitous connectivity, and self-constructed networks of shared social practice. From this core of young users, and from the professional world that first adopted the mobile phone, mobile communication has permeated the entire range of activities and experiences in society. Because the first users are the shapers of the technology itself, the youth culture and the professional culture have framed the forms and content of wireless communication.

In developing countries, as we saw in chapter 8, communities and individuals have adapted to the constraints of poor infrastructure and low income by using a wide array of strategies that have opened up new technological avenues, new business models, and new regulations, from Wi-Fi communities to the village phone and mobile systems for recharging batteries in areas without electricity. Wireless communication has become the heart of the communicative practice of many people. Thus, a number of social practices, values, and organizational patterns have emerged from this interaction between mobile communication and society.

The identification of these emerging patterns has been the object of our book. The evidence we have collected and analyzed can be summarized by pointing to a number of social trends that, together, signal the formation of pervasive networks of mobile communication as a key component of the broader social structure that characterizes our world, the network society. We will not repeat here the empirical observations and analytical commentaries that have been presented throughout this book. Rather, we will focus on what seem to be the main social processes resulting from the observations conducted in a variety of cultural and institutional contexts. These trends interact with each other, so that listing them sequentially is tantamount to an excessively schematic presentation of a highly complex development. Yet, for the sake of simplicity, we will present each one of these distinct trends separately before elaborating on their relationships.

In general terms, we verify again in this study the observation that technology does not determine society: it *is* society, and can only be understood in social terms as a social practice. This means that the uses of wireless communication are fundamentally shaped and modified by people and organizations, on the basis of their interests, values, habits, and projects. However, at the same time, the specific characteristics of the technology – in this case microelectronics-based, digital, wireless communication – enable, enhance, and innovate in the realm and content of communication by extending the domain of what is feasible. Because communication is the fundamental process of human activity, the modification of communication processes by the interaction between social structure, social practice, and a new range of communication technologies, constitutes indeed a profound social transformation.

As stated above, several of the trends that seem to be most significant in transforming communicative practices have been observed primarily among young users of wireless communication. However, we believe that this is not only a function of the correspondence between youth culture and the logic of wireless communication. It is also an expression of the easier acceptance and greater capacity of the younger generation to adopt, adapt, use, and innovate new communication technologies. Thus, because they use these technologies more frequently, better, and faster, they reveal potential uses for the technology more rapidly. Youth culture is at the cutting edge of cultural and technological innovation, without prejudging the merits of this innovation.

Empowered by new communication technologies, youth culture sets behavioral trends that influence people of all ages. Furthermore, because the young generation in all societies is expected to be the future, it is highly likely that they will carry with them into their mature years the habits and practices that we have observed, or at least a modified version of their current patterns of behavior. Thus, we may well consider the youth of today as the harbingers of the mobile network society, although we are aware that the age-group context will indeed modify their communicative behavior in the future. In sum, by observing current mobile communication practices among young people, we may take a glimpse into the future, however distorted by the age bias, without venturing into the unreliable terrain of forecasting the coming society.

With all these caveats in mind, we will now synthetize the most salient, empirically observed trends in the practice of communication which is enabled and enhanced by digital wireless communication technology. We began this book by raising a number of key questions on the social implications of mobile communication. After reviewing the evidence, here are some of the answers.

Safe Autonomy

The subjects of communication processes considerably enhance their autonomy by using wireless communication systems. By this, we mean autonomy vis à vis spatial location, time constraints, and, to a large extent, social and cultural norms. This autonomy is both individual and collective. It may refer to a person or to an organization or to a social group or to a social network or to a social movement. The key issue is that the subject of communication enhances its control over the communication process.

To be sure, this autonomy should not be understood in absolute terms. Access to an effective technological infrastructure, affordability of cost, literacy in the uses of the system, degree of freedom of communication vis à vis regulatory authorities, legal environments, and the like, are impediments to unfettered communication which continue to operate within the new technological environment. Yet, even these limits are often challenged by the new communication technologies, forcing a redefinition of the public space of communication in institutional and cultural terms.

Mobile communication is seen to facilitate the combination of autonomy and safety by making the individual free to relate to the world at large, while still relying on his or her infrastructure of personal support. This has been observed particularly in the case of the family, as children and young people can be by themselves or with their peers without losing permanent contact with home. In fact, mobile communication seems to be contributing to the smoothing of tensions in the hyperactive family by allowing everybody (including children) to have his or her own daily schedule, while coordinating the sharing of family time. This pattern of "safe autonomy" also characterizes other sets of interpersonal relations as well as instrumental practices (for example, the professional worker who is always on the move, yet in touch with his or her base office).

Relentless Connectivity

The key feature in the practice of mobile communication is connectivity rather than mobility. This is because, increasingly, mobile communication takes place from stable locations, such as the home, work, or school. But it is also used from everywhere else, and accessibility operates at any time. So, while in the early stages of wireless communication it was a substitute for the fixed-line phone when people were on the move, mobile communication now represents the individualized, distributed capacity to access the local/global communication network from any place at any time. This is how it is perceived by users, and this is how it is used. With the diffusion of wireless access to the Internet, and to computer networks and information systems everywhere, mobile communication is better defined by its capacity for ubiquitous and permanent connectivity rather than by its potential mobility.

Networks of Choice

Mobile communication has considerably improved the chances, opportunities, and reach of interpersonal sociability and shared practices. People – particularly, but not only, young people – build their own networks of relationships, usually on the basis of their face-to-face experiences, interests, and projects, and then keep them constantly open by using wireless communication, more often than the fixed-line Internet. Thus, peer groups become reinforced in this hybrid space of interaction of physical, online, and wireless communication.

But the technology also allows for a rapidly changing network, adding individuals to or deleting individuals from the network, according to the evolving projects and moods of each individual in the network. So that networks expand, overlap, and are modified following a decentralized multiple entry/exit structure of communication. An extremely malleable pattern of communication follows, highly sensitive to the evolution of orientations among the participants in the communication process. Thus, at the same time, we observe stepped-up communication, the increased rootedness of electronic communication in face-to-face experience, and the complete dependency of the composition of the communication networks on the desires of the communicating subjects. Social choice, including communication choice, continues to be framed by institutions and social structure. But within these obvious limits, wireless communication considerably enhances the choice of interlocutors, and the intensity and density of interaction.

Instant Communities of Practice

One of the most important communicative practices we have observed is the emergence of unplanned, largely spontaneous communities of practice in instant time, by transforming an initiative to do something together into a message that is responded to from multiple sources by convergent wills in order to share the practice. This is, of course, most evident in flash political mobilizations, some of which we have analyzed in this book. But it is not limited to sociopolitical uses. It is manifested as well in professional projects, in cultural experiences, in countercultural expressions, in party going, in family reunions, in the celebrations of sports fans, in religious gatherings, and so on. In other words, the general trend observed in our societies for ad hoc groupings to take precedence over formal structures of interaction and participation, be it family traditions, civic associations, or political parties, finds its technological platform in this capacity to call for action or for meeting or for sharing – in instant time.

It is important to emphasize that these communities can only be formed if the message aimed at constituting them resonates in a network of affinity. In other words, communities of practice, in the mobile society as elsewhere, express the latent existence of common interests and/or values. But on the basis of this latent structure, communities of practice can be formed instantly by a message that strikes a chord along a network of receptive subjects. We currently lack the evidence to evaluate the resilience of these instantly formed communities over time. However, it is an important question, particularly for the understanding of new forms of social mobilization, which should be taken up by researchers.

The Blurring of the Social Context of Individual Practice

Wireless communication does not transcend space and time, as is often stated in terms of an apparently commonsense observation. It blurs, rather than transcends, spatial contexts and time frames. And it induces a different kind of space – the space of flows – made of the networked places where the communication happens, and a new kind of time – timeless time – formed from the compression of time and the desequencing of practices through multitasking. This is what emerges from a number of studies on the social dimensions of wireless communication. These studies show that there is a new spatial context and a new time dimension in which the communication takes place. It is the space and time of the communicating individuals, which is a material form, as material as any other space and time, but it is chosen by the communicating subjects.

Furthermore, since communication is at least bilateral and potentially multiple (networks of wireless communication), the time/spatial context is formed by the frame chosen by the initiator of the communication, the frame of the solicited communicator, and the set of relationships objectively existing between the two or more time/spatial contexts. Besides, not only are time and space blurred (not eliminated, but blurred) but organizational contexts and social practices are often mixed. This is the case with communication that takes place in airports or stations with the family, office, and friends. Or the multiple use of mobile devices from the car, from trains, or from waiting rooms. Or dating on the move. Or the multimedia use of the mobile device (image taking and sending, audio retrieving and playing, data transmission, interpersonal communication) mixed in chosen time/space contexts.

So the system of mobile communication enables the blurring, mixing, and recomposing of a variety of social practices in a variety of time/space contexts. But the blurring process is not undetermined. It is centered on the communicating individual. So it is an individually centered production of the material and social process of communication. In so doing, networks of individual interaction tend to free themselves from organizations,

institutions, norms, and material constraints, on the basis of personal convenience and suitability to individual projects. As a result, there is an extraordinary strengthening of the culture of individualism (meaning the
primacy of individual projects and interests over the norms of society or
reference groups) in material terms. Therefore, individualism rather than
mobility is the defining social trend of the mobile society because it not
only allows users to communicate on the move, but also to communicate
from immobility, as is shown in a number of important studies on the benefits of mobile communication in enhancing the communicative capacity
of disabled persons.

One consequence of this development is that traditional norms of cour-
tesy have to be redefined in the new context. Since people build their own
private space by simply ignoring others around them, a new m-etiquette
(and its implicit norms of cultural domination) is struggling to be adopted,
specifying when it is proper to isolate oneself from the social environment
and when it is not, when it is acceptable to expose one's personal life in
the middle of an audience of strangers and when it is not, or when pupils
can talk to or e-mail their buddies in the classroom and when they cannot.

In sum, the blurring of time, space, and activities into a new frame of
chosen time, space, and multipurpose communication dematerializes
social structure and reconstructs it around individually centered networks
of interaction. This is not the fading away of time, but the emergence of
chosen time, and of compressed time, to fit the multitasking of communication. This is not the end of distance, but the definition of interaction
in a space of communication flows structured around spatial nodes of
opportunity. And this is not the confusion of all social practices, but the
constitution of a set of practices around the interests, values, and priorities of each individual. It is the blurring of the pre-existing social structure
of communication, but it is also the relentless definition of new channels
and forms of communication. More important than communication on
the move is the rise of moving communication patterns.

Access to the Wireless Network as a Source of Personal Value and as a Social Right

We know that the value of a network increases exponentially with its size
and the intensity of interaction between its nodes. From the observation
of social behavior in wireless communication networks, we also verify
this general rule of network logic. Users become dependent on mobile
communication very quickly. They tend to be always on, and find ways to

reduce the cost of communication. When government regulations, technological standards, and business pricing systems favor the diffusion of wireless communication, it becomes explosive. People in general, but particularly young people, find a major source of personal value in wireless communication. And they go to extraordinary lengths to make an effort to access the network. Thus, prepaid cards and call-time rentals have led the diffusion of use in developing countries and among low-income segments of the population in advanced countries. In China, the Little Smart phone systems have made major inroads among working people, far from the trendy professionals of Shanghai, as our fieldwork among migrant workers in the Pearl River Delta shows. In Japan, i-mode became a major success by tailoring not only habits and needs, but pricing systems to its young user population. And in Europe, the relative affordability and flexibility of mobile-phone payment systems explain to a large extent the fast diffusion of wireless communication. In contrast, the US example of misguided competitive strategies, lack of communicating standards, and misunderstanding of the pricing needs of young users and the poorer segments of the population have hampered the diffusion of the technology, with potential serious consequences down the line, in terms of the learning curve and service availability, for both companies and users in general.

In sum, wireless communication technology seems to be the most rapidly adopted technology, and the one that most users have quickly found indispensable to their lives, particularly among young people. As soon as regulatory, technological, and affordability obstacles are overcome, there is an explosion in usage. This places a serious burden on regulators because, in the absence of an affirmative policy favoring diffusion, those countries, areas, and people left behind will clearly suffer from lack of connection to a fundamental network. It is also clear that when wireless communication and the Internet come together, as in the experience of i-mode in Japan, and in new developments in Korea, the effect of increasing communication is amplified. We can even say that the blockage in Internet diffusion that Japan and other Asian countries were experiencing is being solved through wireless Internet connections. However, given the technical and business difficulties experienced by WAP and mobile Internet access in Europe and the United States, it becomes increasingly clear, by looking at patterns of social use, that the true convergence of wireless communication and the Internet will be the critical question in the next phase of the Information Age.

A few years ago, unwarranted expectations frustrated the technological and business promise of this convergence. It is clear that public policy should now inform and guide the choices to be made in this regard, as they condition the entire development of communication capacity for people and societies at large. Under these conditions, it becomes essential that equality of access to the network is assured as a condition for the full participation of all citizens in the network society. What the right to schooling and access to public libraries was a century ago, is today the right to affordable, reliable access to the mobile communication network on which our shared experience is already based.

Users Are the Producers of Content and Services

As stated above, the observed experience of wireless communication shows that people adapt the technology to their needs and interests. They invent new uses, and even a new language (see below), circumvent regulations, quickly find better, cheaper available pricing schemes for themselves, and build networks of communication for purposes and uses that were never in the cards of technologists and business strategists. This fully replicates the experience of the Internet, but it is even more significant because the first users of the Internet were highly sophisticated, while the bulk of innovative users of wireless communication technology are kids and young people with no special technical skills, although they are, indeed, already a part of the network society, being fully acquainted with the new technological paradigm. People find uses and, when they are able to, invent new services and create new content (for example, mass image swapping, texting, and so on), and when they do not find the services and content they want, they vote with their thumbs by not using what is offered. Similarly, it seems that any opportunity to access public services, and deal with these services in a new way through wireless communication, is met with great interest from users. But public bureaucracies are just scratching the surface of their delivery possibilities, and are certainly concerned about altering their routine.

Altogether, it seems that the users of wireless communication are indeed the producers of the content, but their ability is limited, and this will stall the development of the technology on the borders of public bureaucracy and business-as-usual. The implication is that both business and the public sector should find it in their interest to follow the innovations of users, not just by surveying them, but by interpreting the signals of innovative behavior in their uses, and then responding to the latent demand with a full array of services. In other words, supply should follow demand, which is clearly not the case nowadays, unless we think that demand is what pollsters and marketers think it is.

Consumerism, Fashion, Instrumentality, and Meaning

Too often the use of mobile devices has been perceived and interpreted as a function of consumerism, oriented by fashion. Evidence seems to indicate that, while there is some truth in this observation, this is a very narrow interpretation of users' behavior, once wireless communication has diffused broadly in a given social context. Mobile communication is used for all kinds of purposes, many of them highly instrumental, in professional work, in the organization of the everyday life of the family, in sustaining networks of sociability, in commercial transactions, in gathering and forwarding information, in sharing music, in producing and diffusing images, in sociopolitical mobilizations, and the like.

Also, use of the mobile phone is no more dependent on consumption or oriented toward status-seeking than, let us say, use of the automobile. However, as in the case of the automobile, design and customization are important. And in the case of young people, marks of personal identity (colors, ring-tones, shape, ornaments), as well as fashion trends in the design of the handset, certainly comprise one of the dimensions present in the use of the technological device. But it is not the dominant one. What is relevant is that the device itself, and its technological attributes, have meaning for the users. This is part of the process of individual expression, of the construction of identity by appropriating a new technological environment and still feeling one's self. Thus, it is in the relationship between instrumentality and meaning that we find the significance of social uses of mobile communication. It is not just consumption, but the multipurpose practice of communication. It is not just fashion, but identity.

The Transformation of Language

Texting via SMS is changing language through its widespread use in wireless communication. In an evolutionary view, we are seeing a new case in which the adoption of new technology affects language itself, including vocabulary and grammar rules in people's practice. And this practice ultimately affects the common language, and language itself.

In some cases, new forms of written expression are the signs of subcultures and the expression of innovation from users. In fact, creative uses of language become a form of personal and group expression. But, in most cases, it is the result of the simple adaptation of language to the format and limits of the technology, including strategies to reduce the cost of transmission. We are already at the point where new texting-oriented vocabularies can be listed for different languages, on the basis of observed practice.

Furthermore, the multimedia capacity of wireless communication technology (MMS) displays a multimodal form of communication, with text, image, and audio being produced and transmitted from multiple locations. Observation shows that the combination of these different modes of communication, particularly by young users, is creating new forms of meaning, characterized by the mixture of methods of assigning meaning, for example by using texting mainly for personal commentary or for emphasis, while sound and images are thought to be self-explanatory. The merging of text and audiovisual media is now diffused to every context of communication through the distributed communicative capacity of wireless technology. Since language is closely related to the formation of culture (systemic production and communication of meaning), we are clearly in a process of cultural transformation associated with the spread of wireless communication, although the limited amount of academic research on the subject precludes us from knowing the contours and directions of this transformation.

Communication Autonomy, Information Networks, and Sociopolitical Change

Wireless communication extends the realm of autonomy in relation to the mass media that characterizes the Internet. Horizontal networks of communication and multiple sources of information on a global scale create the possibility of a largely autonomous diffusion and exchange of information, ideas, and initiatives. Wireless communications amplify this autonomy by the ability to create networks of information by connecting known individuals in instant or chosen time from any location to any location. This, effectively, has created a high-volume communication medium that is, at the same time, personalized: in most cases, messages are received from known correspondents. This is to be distinguished from the mass distribution of messages from one sender to many receivers using the capability of the mobile device. The chain, networking logic from identifiable sources increases the credibility of the message, and becomes an effective way of propagating information that is deemed reliable. The case studies presented here of instances of sociopolitical mobilization based on the autonomous diffusion of information, bypassing official sources and the mass media, illustrate the potential of wireless communication to transform the political landscape.

However, two points are important in evaluating this phenomenon. On the one hand, rumors of all kinds, often unfounded or manipulated, can propagate easily once they find an entry point into the chain of communication. The social psychology of rumor, which has an old tradition in academic research, will be useful in understanding the workings of well-known mechanisms in a new technological context. On the other hand, the sociopolitical consequences, while strengthening the autonomy of people vis à vis established institutions, do not necessarily point in the direction of the democratization of society, as the power of grassroots mobilization is subjected to the perils of demagogy. Thus, our statement, based on observation, is value neutral: wireless communication considerably increases the information and communication power of people at large, making them more independent of formal sources of information.

One direct consequence (which some governments, from the Philippines to Spain, have already noticed) is that governments and media corporations must be more careful in their attempts to manipulate information or frame public opinion, since the backlash can be rapidly amplified in the form of a storm of counter-information, to which is added public outrage. With the power of relatively free communication in the hands of the majority of the population, politics, because it is predominantly media politics, will never be the same.

The Social Problems of a Wireless World

Every sociotechnical context raises a set of specific social problems. A number of studies have identified several major issues in the emerging communication society. The first one is the wireless communication divide. If the new system is geared toward perpetual contact, and this comprises the large majority of the population, lack of access to the system becomes a very serious social handicap, often combined with other sources of inequality. In addition to income and education, age seems to be a serious factor in amplifying the divide, as older people are less familiar with the technological systems and their related services. Besides, physical disabilities (for example, sight or finger dexterity) appear to be real obstacles to the user friendliness of the communication device. Indeed, some companies have introduced mobile devices with larger characters and buttons, specifically designed for the older age group. However, currently, most communication devices seem to have in mind the trendy young person or the educated, middle-aged, physically able professional. Territorial location, particularly in developing countries, but also in less-populated regions of advanced countries, has also been seen to be an obstacle to access, thus requiring a determined public policy to ensure universal access to a medium of communication that has quickly become indispensable to daily life.

New vulnerabilities have also been found with the fast diffusion of wireless communication. Viruses can now spread at much higher speed and in a much broader range, disrupting the lifelines of society. Multitasking and the practice of communication in all contexts and situations are creating new dangers, such as the increase in automobile accidents that can be directly attributed to using the mobile phone while driving. Legislation and policing do not seem to be enough of a deterrent, so that this observed social problem will probably require the design and diffusion of new wireless communication technologies specific to each context or, as in some current visionary designs, integrated into wearable outfits, with programmed capability to automatically switch on and off depending on context. Widespread anxiety about the health consequences of living surrounded by weak electromagnetic fields (regardless of inconclusive results in the continuing scientific debate on the matter) reactivates the secular, latent conflict between humans and their machines.

The slow emergence of new rules of contextual behavior (or m-etiquette) is also creating tensions and interpersonal conflicts in the uses of communication. A number of analysts have also referred to the renewed danger to privacy of a wireless system that is always on, from a continually identifiable location, which will allow constant monitoring of people's whereabouts, except when they switch off their connection, thus forcing them to choose between privacy and communication.

We have not really found in the academic sources we have consulted specific discussion of the security issues involved in wireless communication (for example, the easy communication between terrorists or criminals, the use of the mobile phone to activate explosives from a distance), but anecdotal evidence from journalistic reports clearly points to growing concern about these issues, which reminds us that the power of technology does not eliminate human ills. Rather, it amplifies them, unless they are treated at their social roots.

Mobile Communication and the Network Society

Mobile communication extends and reinforces the technological platform of the network society, a society whose social structure and social practices are organized around microelectronics-based networks of information and communication. Mobile communication devices power the network enterprise and allow the office on-the-run, the mobile worker, the decentralization of production and management in the business world, and direct connection between public services and their users. Similarly, the family in transition that characterizes our society has found in mobile communication a useful tool to manage the autonomy of its individuals, including increasingly independent children, while maintaining its protective environment. At the same time, the complex organization of the activities of everyday life requires a flexibility of arrangements between family members that is greatly facilitated by the possibilities of constant communication.

The networking of civil society is now accentuated by wireless communication technologies. Media politics is extended into new forms of communication. Multimodal cultural expressions are now integrated into a ubiquitous form, deepening the presence of a culture dominated by a pervasive system of electronic communication made of text, images, and sounds in a relentless interactive exchange. In other words, the mobile communication society deepens and diffuses the network society, which came into existence in the past two decades, first on the basis of networks of electronic exchange, next with the development of networks of computers, then with the Internet, powered and extended by the World Wide Web. Wireless communication technologies diffuse the networking logic of social organization and social practice everywhere, to all contexts – on the condition of being on the mobile Net.

This is neither a vision nor a commonsense technological statement. It is what emerges from the observations that we have been able to gather at this point from a variety of cultural and institutional settings. At the same time, we have observed the diversity of the uses of wireless communication, depending on levels of development, culture, institutions, and business strategies. Yet, while accounting for this diversity, we want to emphasize that wireless communication technology does have powerful social effects by generalizing and furthering the networking logic that defines human experience in our time.

Notes

Chapter 1   The Diffusion of Wireless Communication in the World

1. An important footnote to this analysis is that a wide variety of official and unofficial sources have been used to derive a picture of wireless communication around the world. Some of the trends discussed here are emerging trends, and as such there are few official or academic sources of data. Another point to note is that research and analysis of wireless communication have been more prevalent in some regions and countries than in others (perhaps in line with the nature and level of adoption). Therefore, we have very rich data on some countries (for example, Europe and the Asian Pacific) and rather limited data on others (for example, North America). To supplement the limited official and scholarly work, we have relied on secondary sources, such as newspaper reports and press releases. These reported surveys have generally been conducted by recognized professional research institutions and therefore can be considered fairly reliable. Notwithstanding these limitations, the data do give us some idea of what is known in the public sphere about the diffusion of wireless communication in the societies examined.

2. Unless otherwise stated, all statistical references to numbers of subscribers, teledensity, and growth rates are based on ITU data for 2004 (see www.itu.int).

3. See Banerjee and Ros (2004) and Ling (2004) for concise discussions of global patterns in fixed and mobile telephony.

4. See online statistical annex (appendix 3) for more details.

5. Analog phones are considered the first generation of mobile-phone technology (1G); digital cellular phones, the second generation (2G); and digital devices with high-speed broadband capacity, the third generation (3G).

6. See online statistical annex (appendix 4) for ITU data on the global distribution of high-speed mobile networks.

7. Mobile communication growth trends in the countries and regions covered in this chapter are available in the online statistical annex (appendixes 9–62).

8. There were, however, some shifts, such that Finland, Norway, and Sweden no longer have the highest penetration rates. This may be partly due to the prepaid phenomenon that can distort penetration figures if inactive accounts are counted as active. Finland, in particular, has only about 2 percent prepaid subscriptions, whereas other countries in the region have 50–80 percent prepaid (Kelly et al. 2002).

9. The contribution of prepaid systems to this process cannot be estimated as such data are not available for Russia.

10. See also the online statistical annex (appendix 20) for detailed data relating to Latin American countries.

11. Our own elaboration, based on the ITU (www.itu.int). Only Venezuela had a negative growth tax (0.14 percent).

12. Based on ITU data (www.itu.int).

13. Countries considered are Bolivia, Brazil, Chile, Ecuador, Honduras, Mexico, Nicaragua, Panama, Paraguay, Peru, and Venezuela.

14. Following the economist Leo Sztutman, as reported in Lobato (2004).

15. The period considered runs from December 1994 to November 2004.

16. Data refer to 2003; see online statistical annex (appendix 26).

17. Figures from the Ministry of Public Management, Home Affairs, Posts, and Telecommunications (MPHPT) of Japan. Available at http://www.soumu.go.jp/joho_tsusin/eng/Statistics/telephone.html.

18. Fieldwork, summer 2002 and December 2003 to January 2004.

19. Of the 11 million mobile-phone users, about 6.4 million subscribe to Smart Communications Inc. and 4.6 million to Globe Telecom.

20. According to the Telecom Regulatory Authority of India (TRAI; *Business Line* 2004).

21. According to the Cellular Operators Association of India (McDowell and Lee 2003: 375).

22. Computation based on Department of Telecommunication (2003–4: 111).

23. The reasons for decline are not necessarily related to the rise in the use of mobile phones. For example, the net number of fixed lines in South Africa fell between 2000 and 2002 due to line disconnections as a result of nonpayment of bills (Gillwald 2003; Makhaya and Roberts 2003).

24. It is also possible that some mobile-phone subscription numbers are inaccurate. In South Africa, for example, one research company concluded that subscriptions have been routinely overstated by up to 20 percent (*Business Report* 2004). This is a

result, for example, of people switching networks due to seasonal promotions, having multiple accounts, or of tourists buying prepaid phones while on a trip.

25. See Minges (2005) for an excellent discussion of these problems as they relate to the mobile Internet.

26. Available data relate only to North America, Europe, and parts of Asia.

27. Note Minges's (2005) important point that statistics on the number of messages sent may be inflated by the double counting of messages sent and received, the inclusion of text messages sent from PCs to mobile phones, and the inclusion of large amounts of spam. A potentially more useful indicator – how many mobile users are text messaging – is rarely reported.

28. This report did not include Russia.

29. Unfortunately, Minges (2005) does not cover the Philippines. Note also the high usage levels in Singapore, although SMS is more than twice as expensive as in Japan and about five times more expensive than in Korea.

30. This list comprises commercial hotspots verified by Intel as compatible with its Intel Centrino technology.

31. The practice of marking public places with symbols to indicate the location of an open Wi-Fi network.

32. The emergence of municipal Wi-Fi projects has created conflict between city administrations and commercial providers who fear that these free-access systems will unfairly compete with their services. For example, Philadelphia's plans to become the first major city to implement citywide wireless Internet access has come under attack (Dao 2004).

33. For example, Dekimpe et al. (1998) in an early global analysis found that poverty was associated with lower rates of initial diffusion as well as growth rates. Analyzing the MENA region, Varoudakis and Rossotto (2004) found that income (per capita income) elasticity of demand for mobile phones was greater than one.

34. See ICRA Information Services (2001), Kelly (2001), and Littlechild (2004) for discussion and comparison of RPP and CPP pricing systems.

35. Classification of countries by income is based on categories in the ITU World Telecommunication Database (www.itu.int).

36. $n = 39$; bilateral significance at 1 percent; calculations made with SPSS 12.0.

37. Countries considered for the analysis are the following 39: Argentina, Australia, Austria, Belgium, Brazil, Canada, Chile, China, Colombia, Czech Republic, Denmark, Egypt, Finland, France, Germany, Greece, Hong Kong, Hungary, India, Ireland, Italy, Japan, Malaysia, Mexico, the Netherlands, New Zealand, Norway,

Poland, Portugal, Russia, Singapore, South Africa, Spain, Sweden, Switzerland, Turkey, UK, US, and Venezuela.

Chapter 2   The Social Differentiation of Wireless Communication Users: Age, Gender, Ethnicity, and Socioeconomic Status

1. See, for example, Bucholtz (2002) for a discussion of the complexities associated with defining the concept of youth.

2. There is limited gender-disaggregated data on ICT in general (ITU 2003; Minges 2003), making it impossible to make definitive statements about global trends in mobile communication ownership or access by gender. The ITU currently only tracks gender proportions in telecommunication administrations and Internet users for selected (mainly developed) countries (available at http://www.itu.int/ITU-D/ict/statistics/). Individual countries, primarily in Europe, have some gender-disaggregated national-level data. Even where such data are available, however, variables may differ. For example, some studies report the proportion of mobile communication users that are male or female; others report the proportion of each gender that own or use mobile devices. Finally, in most countries, especially developing countries, data on gender proportions are simply not collected. Hafkin (2003, 2005) suggests that data on gender and ICT tend to be available in wealthier countries and countries where gender digital gaps are small, and unavailable in poorer countries as well as those where gender digital gaps are large. A 2005 report released by Orbicom (Sciadas 2005) addresses this deficiency to some extent, although it provides more information on Internet access and use than on mobile communication devices. It is notable that the Orbicom researchers encountered serious problems collecting gender data, "including lack of consistent gender statistics in a large number of countries, lack of common definitions and concepts and a mixture of both public and private sources" (UNESCO 2005).

3. Note, however, that these findings are not statistically significant.

4. See, for example, a survey by Harris Interactive reported in Sidener (2005).

5. The authors do caution that, although there was a broad representation of different demographics, including gender, their sample was skewed toward females with mobile phones (Samuel et al. 2005: 45).

6. Fieldwork included a survey among Norwegian students of ninth grade (compulsory schooling).

7. Online survey with 1,274 valid cases, developed in 2002 among teenagers (14–17 years old) and young adults (18–22 years old) of both sexes.

8. Note that all these studies were conducted by one mobile-phone operator – Cingular Wireless – and appear to be based on the company's subscriber base.

9. Note that some commentators (for example, Ankers 2004) question the claim that men make mostly business-related calls. The extent to which responses to questions on this matter are influenced by social desirability is unclear.

10. Note, however, that among young teens and old people, the gender proportions are almost equal.

11. This research appears to refer to the UK, although the reports do not make this clear.

12. It should be considered that these findings may have been influenced by the definition of terms such as "trendy," "style," and "status."

13. Fieldwork included a survey among Norwegian students of ninth grade (compulsory schooling).

14. See the section in chapter 4 entitled "Strengthening of Individual Identity and the Formation of Fashion."

15. Interviews in China, December 2003.

16. Fieldwork in Mainland China and Hong Kong, summer 2002 and spring 2004.

17. Where "regularly" is defined as used within the past three months (McKemey et al. 2003).

18. It is important to note that this is an online survey without random sampling. It targets subscribers of Monternet, China Mobile's mobile Internet service, which accounts for most but not all of China's mobile Internet market.

19. For instance, Uganda, Rwanda, Namibia, and Botswana (ITU 2004, referring to 2003).

20. Personal observation in Ghana, February 2005.

21. SIM Cards currently cost about US$ 25 in Ghana.

22. Personal observation in Ghana, January 2005.

23. It should be noted that the definition of African Americans and Latinos in the United States as ethnic minorities is becoming increasingly inaccurate as their populations grow. However, in socioeconomic terms they continue to be viewed as minorities.

24. See Hofstede (1991), for example, for more information on this concept of national cultures.

Chapter 3 Communication and Mobility in Everyday Life

1. Interviews in Beijing, July 2005. A taxi driver was told to keep his mobile phone on all day when he was working for the Annual Assembly of the National People's

Congress. Another low-ranking government official had to stay ready for orders from his supervisor even during weekends.

2. Observations in China, Europe, and the United States, summer 2003 to spring 2005.

3. In Zhang's (2001) ethnographic account, she frequently observed that migrant entrepreneurs from Wenzhou used pagers and cell phones while working and living in Beijing.

4. A study by UK's Oftel reached similar findings that: "The main feature of [these] low income groups as a whole was their high use of telecoms to maintain family links and resulting large spend as a proportion of the total household budget. Telecoms was considered as vital as gas and electricity and the vast majority had fixed phones at home and/or mobile phones, and were satisfied with their telephony provision" (Oftel 2003b: 6).

5. For evidence of penetration in the private sphere in some selected Latin American countries, please see chapter 1 (p. 15).

6. See chapter 8 for a detailed discussion of the shared use of mobile telephones, both within the family and outside its boundaries, in developing countries.

7. See chapter 4 for detailed information on the uses to which children of different ages put the mobile telephone.

8. The surveys were conducted in 2000, 2001, and 2002, involving nationally representative samples of 2,555, 2,816, and 2,333 respondents, respectively.

9. For Bulgaria, see Varbanov (2002: 131); for Norway, see Ling (2004); for Finland, see Oksman and Rautiainen (2002) and Kasesniemi (2003); for the UK, see Eldridge and Grinter (2001) and Grinter and Eldridge (2001); for Sweden, see Weilenmann (2003); for Italy, see Fortunati and Manganelli (2002); for Germany, see Höflich and Rössler (2002); for Belgium, see Lobet-Maris and Henin (2002).

10. See Licoppe and Heurtin (2002) in connection with France, and Ling (2004) in connection with Norway.

11. A similar finding is described in Ito and Okabe (2003: 12–14).

12. See Haddon (2000), Oksman and Rautiainen (2002), Crabtree et al. (2003), Lacohée et al. (2003), Geser (2004), Ling (2004) for different European countries; Cingular Wireless (2003) for the United States; and Lemish and Cohen (2005b) for Israel.

13. See, for example, http://www.nodedb.com/europe/ (September 2004).

14. As reported at www.ayuntamientojun.org/policia.htm.

15. For Bormujos, see Aecomo (2004); for Fuenmayor, see www.fuenmayor.org/webb/sms.html; for Zaragoza, see Zaragoza (2004).

16. Personal observation in Barcelona, Spain, spring 2005.

17. See http://www.laflecha.net/canales/ciencia/200406161/ (news published June 16, 2004).

18. See   http://www.diariodenavarra.es/edicionimpresa/noticiaB.asp?not=A12ART 112209Aandvin=andseccion=economiaanddia=20040302andvf= (news published March 2, 2004).

19. "Car2Car Communication Consortium" is an initiative by a number of European automotive companies whose aims are, among others, to create and establish an open European industry standard for a car-to-car communication system based on wireless LAN components, to guarantee European-wide inter-vehicle operability, and to push the harmonization of these standards worldwide (information retrieved September 20, 2005 from www.car-to-car.org).

20. Galperin (2005), however, describes the need for specific licenses in Peru.

21. City "hotzones" refer to those networks deployed to cover a spatial scope within the city (e.g., downtown areas); see Vos (2005).

22. See chapter 8 for discussion of the use of Wi-Fi for development.

23. Our source is http://www.nodedb.com/europe/ (accessed September 2005). In one year, the growth of free-access nodes gathered in this webpage was 62.2 percent (in September 2004 the number of reported hotspots was 4,567).

24. Our sources are http://iblnews.es/noticias/05/107107.html (published May 7, 2004) and http://www.consumer.es/web/es/noticias/nuevas_tecnologias/2004/05/ 06/99675.php (published June 7, 2004).

25. This study also found that only one-third of elderly people with disabilities used a mobile phone daily.

26. The company Owasys (www.owasys.com) has developed a mobile device for blind people. Mobile-phone operators in other countries also offer solutions for blind users. However, the extent of adoption is unknown.

27. See http://www.textually.org/textually/archives/002427.htm (accessed on September 28, 2005).

28. Moore and Rutter (2004) is one of the very few collections on mobile entertainment.

29. In this sense, Moore (2003: 65) found in his fieldwork that "The use of mobiles has become a leisure activity to rival others in contemporary social life."

30. In Europe, by 2001, the most successful service was ring-tones (estimated revenues of US$ 1,666 million), followed closely by mobile games (US$ 830 million revenue). The more aggressive European markets are Scandinavia, the UK, and Italy. Market analyst reports are still forecasting multi-billion dollar industries for mobile

games, mobile music, and other forms of mobile entertainment (see Wiener 2003).

31. An interesting branch developing now is the convergence of communication devices under good usability constraints/patterns (Heilman and White 2003).

32. For example, according to ITU (2002a: 18), there were only about 3,000 official Docomo i-mode sites, but the number of unofficial sites was 53,000.

33. See   http://europa.edu.int/comm/publications/booklets/eu_glance/19/en.pdf (accessed September 12, 2005).

34. Some information relating to Spain and Europe in this field can be found at the Spanish Association of Mobile Communications webpage (www.aecom.org). For information of a more general nature, see the Wireless World Forum webpage (www.w2forum.com).

35. Fieldwork in China, 2005.

36. Field observation in Hong Kong, February 2005.

37. Fieldwork in Hong Kong, January 2005.

38. Interview with secondary school teachers in Hong Kong, March 2005.

39. Personal communication with a telecom specialist who visited London in May 2005.

40. See also the section "The Management of Autonomy vis à vis Security" in chapter 4 (p. 146).

41 See the section "Mobility at Work" above (p. 78).

42. For the private sphere and schools, see Swartz (2004); for the work sphere, see the articles in the issue of *Surveillance and Society* 1 (2) (2003) on "Work" (www. surveillance-and-society.org).

43. Interview in Shenzhen, March 2005.

Chapter 4   The Mobile Youth Culture

1. "Text messaging was an accident. No one expected it. When the first text message was sent, in 1993 by Nokia engineering student Riku Pihkonoen, the telecommunication companies thought it was not important. SMS – Short Message Service – was not considered a major part of GSM. Like many technologies, the power of text – indeed the power of the phone – was discovered by users. In the case of text messaging, the users were the young or poor in the West and East" (Agar 2003: 169).

2. A "boom call" is a short signal call not intended to be answered. It has no cost for the sender or the receiver.

3. As Agar (2003) explains, the European mobile-phone system early defined a transnational standard, thanks to the leadership of the Nordic countries. In contrast, US wireless communication was, by 1992, a "crazy-paving of licenses covering the country" (2003: 40) which did not allow easy connection between users of different operators.

4. In December 2003, the European average cost of sending an SMS was 0.12 euros (plus taxes; Crédit Suisse/First Boston, quoted in Grenville 2004).

5. The data referred to the cities of La Paz, El Alto, Cochabamba, and Santa Cruz.

6. Personal observation in Ghana, February 2005.

7. For the 18–23 age group, these include SK Telecom's "TTL," KTF's "Na," and LG Telecom's "Khai." For the 13–18 age group, these include SK Telecom's "Ting," KTF's "Bigi," and LG Telecom's "Khai Holeman."

8. The same idea can also be found in Lobet-Maris and Henin (2002: 111) for Belgium; Ling (2002) for Norway; and S-D. Kim (2002) for South Korea.

9. Taking this further, several organizations are already developing high-tech wireless tracking devices targeted at parents (see Schwartz 2002).

10. We grouped the original categories into two: expressive (romance and filtering, jokes and general chat) which account for 59 percent of the 794 SMS studied; and instrumental (social arrangements, work/study, travel/journey, sport, other), which accounts for the remaining 41 percent.

11. Personal observation in Spain, 2004.

12. Observation of Smith et al. (2003) for UK teenagers (16–18 years-old); also personal observation (2003) for young Chinese migrant workers.

13. Approximately US$ 0.0038 for each SMS.

14. Fieldwork 1999–2000 in Norway among students of ninth grade (compulsory schooling).

15. The predominant language used by Generation Txt is a shorthand form of "Taglish," the urban *lingua franca* that combines Tagalog, English, and Spanish. In order to type faster, "Where are you?" becomes "WRU"; and "See you tonight" becomes "CU 2NYT." This aspect will be developed in chapter 6.

16. Following Moore (2003), there are seven factors conditioning and shaping the entertainment uses of mobile communication: (a) access and affordability; (b) age and context appropriateness; (c) clarity of payment; (d) compactness and coolness; (e) complexity; (f) convergence; and (g) fun and usefulness.

17. Quoted sources are Office for National Statistics (2000) and Charlton and Bates (2000).

18. See the conference program at: http://www.21cms.com/DMM/200404/ CN108%20Marketing%20to%20Teenagers.pdf (accessed July 1, 2004).

19. See "KFC m-coupon boosts chicken wrap sales," *21 Communications* (available at http://www.21cms.com/case_studies/kfc.pdf; accessed on February 17, 2006).

Chapter 5 The Space of Flows, Timeless Time, and Mobile Networks

1. By increased reproductive choice, we refer to the possibility for women to decide when and how to have children, using different options, such as birth control to delay pregnancy, in vitro fertilization to have a child later in life (and without a partner if they wish), or just by adopting. Altogether, it changes the fixed pattern between the biological cycle and the lifecycle of women. For this and other aspects of "timeless time," see Castells (2000b: ch. 7).

Chapter 6 The Language of Wireless Communication

1. Examples of SMS language in different European countries and in the United States can be found in the online statistical annex (appendix 65).

2. See the online statistical annex (appendix 65E).

3. Examples quoted are from http://www.textually.org/picturephoning/archives/ 002604.htm (posted on December 16, 2004; accessed February 2005).

4. See the online statistical annex (appendix 65).

Chapter 7 The Mobile Civil Society: Social Movements, Political Power, and Communication Networks

1. Following the original People Power movement that overthrew Ferdinand and Imelda Marcos in 1986.

2. It is unclear, though, to what extent the organizers of Poor People Power, the "political operatives" of Estrada, were relying on mobile phones at the time.

3. National Statistical Coordination Board, the Philippines. See figure at http:// www.nscb.gov.ph/sna/2002/4q-2002/2002per4.asp (accessed on June 12, 2004).

4. This website (www.erap.com) was full of pro-Estrada material when we accessed it on June 3, 2004. However, in July and August 2005, when we tried several times to access it again, we found that the site is no longer in operation and the domain name is up for sale.

5. In 2000 and 2001, even if there had been mobile-phone surveillance systems, they would have been still too primitive to be used during large-scale political movements such as People Power II.

6. For this, some analysts would even characterize him as "unrealistic, foolhardy" (Rhee 2003: 95).

7. Min-Kyung Bae, head of the Cyber Culture Research Association in Seoul, quoted in Demick (2003).

8. The estimates of Nosamo members vary from 70,000 (*Korea Times* 2002) to 80,000 (Demick 2003).

9. It is, however, difficult to find other actual short mobilization messages in the primary and secondary sources, which is rather different from the news accounts of the People Power II movement in the Philippines.

10. The casualties included the agent of the Grupo Especial de Operaciones (GEO-Policia Nacional) who died in Leganes, Madrid.

11. Text reproduced in Cadenaser.com (2004), Cué (2004), Rodríguez (2004), de Ugarte (2004), and, among others, Sampedro Blanco et al. (2005). This last source also brings together up to ten SMS messages sent in Madrid and Barcelona the evening of March 13.

12. According to sources from mobile-phone companies Vodafone and Amena, and informants from the telecommunication sector as reported in the media (for example, Campo Vidal 2004; Delclos 2004; Francescutty et al. 2005).

13. Personal communications with members of the China IT Group.

14. See also the ordinance at http://www.isc.org.cn/20020417/ca38931.htm (accessed on June 29, 2004).

Chapter 8   Wireless Communication and Global Development: New Issues, New Strategies

1. Brewer (2005: 3) goes one step further and argues that "rural connectivity is significantly more likely to be viable" when the focus is put on non-mobile endpoints.

2. This is not to say that mobility is not an important benefit. For some types of subscribers, particularly those whose jobs require them to be on the move (for example, taxi drivers and migrant workers), mobility may be as important as connectivity.

3. Personal observation in Ghana, February 2005.

4. The Grameen model will not be discussed here since there is an extensive literature on the topic (see, for example, Bayes et al. 1999; Richardson et al. 2000; Moni and Ansar 2004). These assessments of the village payphone system in Bangladesh generally conclude that users have enjoyed important benefits, such as increased consumer surplus, as a result of reduced communication costs, and

improved access to business information, while service providers have gained additional income (up to 40 percent of household income), as well as social and economic empowerment, especially in gender terms. The most important contribution of the system, however, appears to be its role in facilitating the flow of remittances from overseas or city-based relatives.

5. China Mobile and China Unicom each had around 21 million new subscribers in 2003 (see *Global Entrepreneur* 2004).

6. Information obtained from interview with middle-ranking manager at UTStarcom in Shanghai, July 2005.

7. Interview with two Little Smart subscribers in Shanghai, January 2004.

8. Fieldwork June–August 2002 and March 2005 in Guangdong Province; January 2004 in Shanghai; and July 2005 in Beijing.

9. Interviews with senior China Unicom manager and former regional pager operator in Guangdong Province, July 2002.

10. Not by wireless operators as an inexpensive mobile phone, though, as was practiced in Japan.

11. Interview with middle-ranking manager at UTStarcom, Shanghai, July 2005.

12. Interview with middle-ranking manager at UTStarcom, Shanghai, July 2005.

13. MII's new minister, Wang Xuedong , quoted in Kuo (2003).

14. Interview with middle-ranking manager at UTStarcom, Shanghai, July 2005.

15. For further information, see www.utstarcom.com.

16. In China, CDMA-based WLL technology is already in use, for example, in the southern city of Shenzhen.

17. Another example illustrating how a poorly defined policy structure may thwart telephone growth was the 2002 Finance Ministry warning that anyone with a phone line (both fixed line and mobile phone) would have to pay more tax. This announcement caused much debate and quickly led to more than 2 million people closing their telephone subscriptions (O'Neill 2003: 86).

18. Details of the system in Uganda are derived from the following sources: Grameen Foundation USA (n.d.), Ulfelder (2002), Dot-ORG (2004), *Frontlines* (2004), and MTN (2005).

19. Vodacom also attempted an alternative approach to extending access to underserved communities by providing "transportables" (mobile-phone handsets) to faculty and administrators at educational institutions in disadvantaged areas. These phones were intended to be made accessible to any student who wished to use them, simply by approaching the person placed in charge of the handset. However, this

system was discontinued following the observation that students were not being given the required access.

20. Details of the community payphone system in South Africa are derived from Vodacom South Africa (n.d.), Hamilton (2003), Reck and Wood (2003), and BiD Challenge (2004).

21. Most operators employ additional staff to run the shops.

22. Details of the mobile payphone systems in Ghana are derived mainly from personal observation and interviews during a field visit in January 2005. Additional sources are Ajao (2005b), Cudjoe (2005), Day (2005), and Mobile Africa (2005a).

23. These problems have been attributed to deliberate attempts by Ghana Telecom to inhibit the performance of Spacefon, which had become the premier mobile-phone provider in the country.

24. Although, since the phones are not manufactured by Spacefon, technically it should be possible to replace the SIM card with that of any other cell-phone company, thus enabling access to in-network calls for subscribers on all networks.

25. However, the target population for most of the participating micro-finance institutions is female, which is likely to lead to a high percentage of female operators.

26. In Accra, one can find Space-to-Space operators located within inches of each other.

27. In 2003, up to 80 percent of mobile subscriptions in the country were prepaid subscriptions. See chapter 1 and online statistical annex (appendixes 22, 23, 36, and 37).

28. The monthly average minutes of use in Chile is higher for postpaid subscriptions (more than 250 minutes per month) than for prepaid (around 50 minutes per month). See online statistical annex (appendix 26).

References

3G Americas (n.d.) Short messaging services (SMS) for success. http://www. 3gamericas.org (accessed May 22, 2004).

A. T. Kearney (2001) Mobinet index no. 3. http://www.atkearney.com (accessed May 21, 2004).

——— (2002) Mobinet index no. 5. http://www.atkearney.com (accessed March 20, 2004).

——— (2004) Mobinet index 2004. http://www.atkearney.com (accessed August 17, 2004).

Aarnio, A., Enkenberg, A., Heikkilä, J., and Hirvola, S. (2002) Adoption and use of mobile services: empirical evidence from a Finnish survey. *Proceedings of the 35th Hawaii International Conference on System Sciences.* csdl.computer.org/comp/ proceedings/hicss/2002/1435/03/14350087.pdf (accessed July 13, 2005).

Abascal, J. and Civit, A. (2000) Mobile communication for people with disabilities and older people: new opportunities for autonomous life. Paper presented at the 6th ERCIM Workshop on User Interfaces for All, Florence, Italy, October 25–26.

Adelman, J. (2004) U say u want a revolution. *TIME Asia* 164 (2): July 12. http:// www.time.com/time/asia/magazine/article/0,13673,501040712–660984,00.html (accessed August 7, 2005).

Aecomo (2004) Innova Telecom ofrece sus servicios de software M-SMS al ayuntamiento de Bormujos. Press release 17/09. www.aecomo.org/content.asp? ContentTypeID=2andContentID=2073andCatId=162andCatTypeID=2 (accessed September 15, 2005).

African Connection Center for Strategic Planning (2002) Senegal rural ICT market opportunity report. http://www.infodev.org/projects/telecommunications/ 351africa/SENEGAL%20Rural%20Market%20Assessment%20-%20FINAL.pdf (accessed March 18, 2005).

Agar, J. (2003) *Constant Touch: A Global History of the Mobile Phone.* UK: Icon Books.

Agbu, J-F. (2004) From "koro" to GSM "killer calls" scare in Nigeria: a psychological view. *CODESRIA Bulletin* 3/4: 16–19.

Ajao, D. (2005a) Mobile communications in Africa: 1G, 2G (GSM and cdmaOne), 3G and 4G. http://www.mobileafrica.net/technologies.php (accessed June 8, 2005).

—— (2005b) The "space to space" phenomenon in Ghana. http://www.mobileafrica.net/news-africa.php?id=484 (accessed July 26, 2005).

Alden, J. (2002) Competition policy in telecommunications: the case of the United States of America. ITU Workshop on Competition Policy in Telecommunications, Geneva, November 20–22, document: CPT/05.

Alford, C. F. (1999) *Think No Evil: Korean Values in the Age of Globalization*. London: Cornell University Press.

Alfredson, K. and Vigilar, R. (2001) The rise and fall of Joseph Estrada. CNN.com World, May 2. http://edition.cnn.com/2001/WORLD/asiapcf/southeast/04/22/estrada.profile/ (accessed June 12, 2004).

Allardyce, J. (2002) Youthful outlook: tapping into the teen market. *Rural Communications*, September/October: 54–56.

AME Info (2004) IDC presents predictions for Gulf States telecoms markets, January 17. www.ameinfo.com.

ANATEL (2004) Brasil supera a marca de 100 milhoes de telefones fixos e movies. Press release of the National Agency of Telecommunications, ANATEL, December 15. http://www.citel.oas.org/newsletter/2004/diciembre/release_15_12_2004(2)pdf (accessed March 11, 2005).

Anckar, B. and D'Incau, D. (2002) Value creation in mobile commerce: findings from a consumer survey. *Journal of Information Technology Theory and Application*, 4 (1): 43.

Andersson, P. and Heinonen, K. (2002) Acceptance of mobile services: insights from the Swedish market for mobile telephony. SSE/EFI Working Paper Series in Business Administration, no. 2002: 16.

Andrade-Jimenez, H. S. (2001) Technology changing political dynamics. *IT Matters*, January 29. http://itmatters.com.ph/news/news_01292001a.html (accessed June 3, 2004).

Andrejevic, M. (2005) The world of watching one another: lateral surveillance, risk and governance. *Surveillance and Society*, 2 (4): 479–497. www.surveillance-and-society.org.

Anfuso, D. (2002) Cell phone usage continues to rise, April 22. http://www.imediaconnection.com/news/581.asp (accessed June 4, 2004).

—— (2003) Study shows buying power of youth, September 8. http://www.imediaconnection.com/news/1958.asp (accessed June 4, 2004).

Ankeny, J. (2004) Guys and dials. *Wireless Review*, 21 (7): 14.

Apoyo (2004) Perfil de la juventud 2004. APOYO Opinión y Mercado Bolivia SA.

Archibold, R. C. (2003) Protest groups planning for Republican Convention. *New York Times*, August 10: 29.

—— (2004) Days of protests, vigils and street theater (thongs, too). *New York Times*, August 26: B7.

Arias Pando, D. (2004) Comunicaciones móviles: un agente consolidado al servicio de las sociedades Latinoamericanas. Communication given at the II Forum, AHCET Mobiles, October, Rio do Janeiro, Brazil. www.ahcetmovil.com/agenda/ponencias/2004/IIAMOVIL/d5/DanielArias.ppt (accessed December 15, 2004).

Arillo, C. T. (2003) *Power Grab*. Manila: Charles Morgan. http://www.erap.ph/highlights/grab.htm.

Arnett, J. J. (1995) Adolescents' uses of media for self-socialization. *Journal of Youth and Adolescence*, 24 (5): 519–533.

Arnold, W. (2000) Manila's talk of town is mobile. *New York Times*, July 5. http://partners.nytimes.com/library/tech/00/07/biztech/articles/05talk.html (accessed June 5, 2004).

Arvidsson, A. (2004) On the "pre-history of the panoptic sort": mobility in market research. *Surveillance and Society*, 1 (4): 456–474. www.surveillance-and-society.org.

Ashurst, M. (2001) Africa's ringing revolution. *Newsweek*, August 27: 14–18.

Associated Press (2003a) Teen uses cell phone to foil abduction, August 1. http://www.foxnews.com (accessed September 13, 2004).

—— (2003b) Terror attacks believed linked to *al-Qaida*, 20 November.

—— (2004) Calif. man accused of stalking via GPS, September 4. http://story.news.yahoo.com/news?tmpl=story&cid=528&ncid=528&e=4&u=/ap/20040905/ap_on_re_us/gps_stalking (accessed September 7, 2004).

Atkinson, C. (2004) The biz: "Vibe," "Blender" can now dial in content. *Advertising Age*, 75 (April 5): 14, 29.

Bagalawis, J. E. (2001) How IT helped topple a president. *Computer World*, January 30. http://wireless.itworld.com/4273/CW_1–31–01_it/pfindex.html (accessed June 3, 2004).

Bajpai, R. (2002) Culture (bartering for mobile phone service). *Fortune*, 145 (8): 56.

Baker, P. M. A. and Bellordre, A. (2003) Factors affecting adoption of wireless technologies: key policy issues, barriers, opportunities for people with disabilities. *ITD Journal*, 9 (2). http://www.rit.edu/~easi/itd/itdv09n2/baker.htm (accessed July 27, 2005).

Ball, K. and Webster, F. (eds) (2003) *The Intensification of Surveillance: Crime, Terrorism and Warfare in the Information Age*. London: Pluto Press.

Banerjee, A. and Ros, A. J. (2004) Patterns in global fixed and mobile telecommunications development: a cluster analysis. *Telecommunications Policy*, 28: 107–132.

Banks, C. J. (2001) The third generation of wireless communications: the intersection of policy, technology, and popular culture. *Law and Policy in International Business*, 32 (3): 585.

Bar, F. and Galperin, H. (2005) Geeks, cowboys, and bureaucrats: deploying broadband, the wireless way. Paper prepared for The Network Society and the Knowledge Economy: Portugal in the Global Context, March 4–5, Lisbon. http://www.presidenciarepublica.pt/network/home.html (accessed September 15, 2005).

Barendregt, B. (2005) The ghost in the phone and other tales of Indonesian modernity. *Proceedings of the International Conference on Mobile Communication and Asian Modernities*, pp. 47–70. Hong Kong, June 7–8.

Barnes, S. and Huff, S. (2003) Rising sun: i-mode and the wireless Internet. *Communications of the ACM*, 46 (11): 79–84.

Barthold, J. (2004) Go fish (with your cell). *Telecommunications Americas*, 38 (13): 26, 28.

Batista, E. (2003) She's gotta have it: cell phone. *Wired News*, May 3. http://www.wired.com/news/culture/0,1284,58861,00.html (accessed July 2, 2004).

Bautsch, H., Granger, J., Karnjate, T., Khan, F., Leveston, Z., Niehus, G. et al. (2001) An investigation of mobile phone use: a socio-technical approach. http://www.cae.wisc.edu/~granger/IE449/IE449_0108.pdf (accessed March 20, 2004).

Bayes, A., von Braun, J., and Akhter, R. (1999) Village pay phones and poverty reduction: insights from a Grameen Bank initiative in Bangladesh. Center for Development Research Discussion Papers on Development Policy, University of Bonn. www.telecommons.com/villagephone/Bayes99.pdf (accessed December 21, 2004).

BBC (2000a) Cardinal Sin tells Estrada to quit, October 11. http://news.bbc.co.uk/1/hi/world/asia-pacific/967115.stm (accessed June 14, 2004).

——— (2000b) Manila on alert after bombings, December 31. http://news.bbc.co.uk/1/hi/world/asia-pacific/1093607.stm (accessed August 3, 2005).

——— (2001) Uganda's 'beeping' nuisance, January 23. http://news.bbc.co.uk/1/hi/world/africa/1132926.stm (accessed July 17, 2005).

——— (2003a) Estrada's fall from hero to villain, July 30. http://news.bbc.co.uk/2/hi/asia-pacific/1063976.stm (accessed August 3, 2005).

——— (2003b) Phone users become picture savvy, February 12. http://news.bbc.co.uk/1/hi/business/2752113.stm (accessed April 6, 2004).

——— (2004a) Mobile usage shows gender split, June 2. http://news.bbc.co.uk/go/pr/fr/-/1/hi/technology/3766643.stm (accessed July 12, 2005).

—— (2004b) Mostar bridge opens with splash, July 23. http://news.bbc.co.uk/go/pr/fr/-/2/hi/europe/3919047.stm (accessed July 23, 2005).

—— (2004c) Political sparks are flying in Italy after the prime minister's office sent text, June 11. http://news.bbc.co.uk/1/hi/world/europe/3798017.stm.

—— (2004d) Tribes take to wireless web, March 3. http://news.bbc.co.uk/go/pr/fr/-/2/hi/technology/3489932.stm (accessed July 25, 2005).

BDA China (2002) Mobile subscribers in China 2002. http://www.bdachina.com/content/research/reports/mobile_subscribers/en (accessed November 7, 2003).

Beaubrun, R. and Pierre, S. (2001) Technological developments and socio-economic issues of wireless mobile communications. *Telematics and Informatics*, 18: 143–158.

Beck, U. (1992) *Risk Society: Towards a New Modernity*. London: Sage.

Becker, M. and Port, B. (2004) At GOP Convention: a technological battle. *New York Daily News*, 3 September. http://pqasb.pqarchiver.com/nydailynews/687890221.html?did=687890221&fmt=abs&fmts=ft&date=sep+3%2c+2004&author=maki+becker+and+bob+port+daily+news+staff+writers&desc=protesters+click+with+new+media+to+mobilize (accessed September 8, 2004).

Bennett, C. and Regan, P. M. (2004) Editorial: surveillance and mobilities. *Surveillance and Society*, 1 (4): 449–455. www.surveillance-and-society.org).

Berniker, M. (2004) China tempers 3G expectations. *EWeek*, April 28. http://www.eweek.com/print_archive/0,1761,a=125605,00.asp (accessed May 5, 2004).

BiD Challenge (2004) Local entrepreneurs run mobile phone shops. http://www.bidchallenge.org/examples/phone_shops (accessed July 25, 2004).

Biddlecombe, E. (2003) Wi-Fi grows, but profits don't. *Wired News*, December 16. http://www.wired.com/news/wireless/0,1382,61618,00.html?tw=wn_story_related (accessed September 5, 2004).

Bociurkiw, M. (2001) Revolution by cell phone. *Forbes*, September 10: 28.

Boyle, M. (2002) Wi-Fi U.S.A. (electronic version), November 25. *Fortune*, 146 (11): 205.

Breure, A. and van Meel, J. (2003) Airport offices: facilitating nomadic workers. *Facilities*, 21 (7/8): 175–179.

Brewer, E. A. (2005) Technology insights for rural connectivity. Research report prepared for the International Workshop on Wireless Communication and Development: A Global Perspective, Marina del Rey, California, October 7–8, Annenberg Research Network on International Communication. http://arnic.info/workshop05/Brewer_RuralConnectivity_Sep05.pdf.

Bridges.org. (2003) Evaluation of the SATELLIFE PDA Project, 2002: testing the use of handheld computers for healthcare in Ghana, Uganda, and Kenya. http://www.

bridges.org/satellife/evaluation_pda_project_28_February_2003.pdf (accessed June 18, 2005).

Brier, N. R. (2004) Coming of age. *American Demographics*, November 1. www. demographics.com.

Brown, B., Green, N., and Harper, R. (eds) (2002) *Wireless World: Social and Interactional Aspects of the Mobile Age*. London: Springer.

Brynin, M., Raban, Y., and Soffer, T. (2004) The new ICTs: age, gender and the family. In e-Living: Life in a Digital Europe, an EU Fifth Framework Project. http://www. euroscom.de/e-living (accessed July 14, 2005).

Bucholtz, M. (2002) Youth and cultural practice. *Annual Review of Anthropology*, 31: 525–552.

Buckingham, D. (2004) The media literacy of children and young people: a review of the literature. London: Ofcom. http://www.ofcom.org.uk/advice/media_literacy/medlitpub/medlitpubrss/ml_children (accessed July 19, 2005).

———, Davies, H., Jones, K., and Kelley, P. (1999) *Children's Television in Britain*. London: BFI.

Business in Africa (2004) Telecoms operators strike gold in Africa, September. http://www.businessinafrica.net/features/telecoms/349759.htm (accessed February 16, 2005).

Business Line (2004) Mobiles overtake landlines: subscriber base crosses 44 million. *Hindu Business Line*, New Delhi, November 9. http://www. thehindubusinessline.com/2004/11/09/stories/2004110901810700.htm (accessed March 4, 2005).

Business Report (2004) Cellular industry grossly overstated, December 3. http://www.businessreport.co.za/index.php?fSectionId=561&fArticleId=2332643 (accessed February 25, 2004).

Cadenaser.com (2004) El artifice de la protesta en la sede del PP empezó enviando un mensaje a diez amigos. Entrevista recogida en el programa "Hoy por hoy." www.cadenaser.com/articulo.html?xref=20040316csrcsrnac_4&type=Tes (accessed September 20, 2005).

Campo Vidal, M. (2004) *11M–14M: la revuelta de los móviles*. Canal Sur–TVC-Lua Multimedia; DVD Documentary 48' (transmitted on TV3–TVC, May 19, 2004).

Caporael, L. R. and Xie, B. (2003) Breaking time and space: mobile technologies and reconstituted identities. In J. E. Katz (ed.), *The Social Context of Personal Communication Technology: Machines that Become Us*, pp. 219–231. New Brunswick, NJ: Transaction.

Capra, F. (2002) *The Hidden Connections: Integrating the Biological, Cognitive, and Social Dimensions of Life into a Science of Sustainability*. New York: Random House.

Carpenter, S. (2004) Pirate radio to moor at Republican Convention. *Los Angeles Times*, August 27: E1.

Carrasco, M (2001) Los ranchillos: diversidad cultural en el mundo globalizado. Presentation at the IV Congreso Chileno de Antropología, November 19–23, University of Chile.

Cartier, C., Castells, M., and Qiu, J. L. (2005) The information have-less: inequality, mobility, and translocal networks in Chinese cities. *Studies in Comparative International Development*, 40 (2): 9–34.

Castells, M. (2000a) Materials for an exploratory theory of the network society. *British Journal of Sociology*, 51 (1): 5–24.

—— (2000b) *The Rise of the Network Society*, 2nd edn. Oxford: Blackwell.

—— (ed.) (2004) *The Network Society: A Cross-cultural Perspective*. London: Edward Elgar.

——, Fernández-Ardèvol, M., Qiu, J. L., and Sey, A. (2004) The mobile communication society: a cross-cultural analysis of available evidence on the social uses of wireless communication technology. Research report prepared for the International Workshop on Wireless Communication Policies and Prospects: A Global Perspective, held at the Annenberg School for Communication, University of Southern California, Los Angeles, October 8–9, pp. 78–82. http://arnic.info/workshop04/MCS.pdf.

——, Tubella, I., Sancho, T., Diaz de Isla, M. I., and Wellman, B. (2003) *La societat xarxa a Catalunya*. Barcelona: Rosa dels Vents and Open University of Catalonia.

Cellular-news (2002) Coke pulls off SMS campaign success in China, October 31. http://www.cellular-news.com/story/7806.shtml (accessed February 8, 2006).

CellularOnline (2004) Europe SMS love affair set to continue through 2004, May 2. http://www.cellular.co.za/news_2002/031502-europe_sms_love_affair_set_to_co.htm (accessed August 7, 2005).

Chang, S-J. (2003.) The Internet economy of Korea. In B. Kogut (ed.), *The Global Internet Economy*, pp. 262–289. Cambridge, MA: MIT Press.

Chango, M. (2005) Africa's information society and the culture of secrecy. *Contemporary Review*, 286 (669): 79–81.

Charlton, A. and Bates, C. (2000) Decline in teenage smoking with rise in mobile phone ownership: hypothesis. *British Medical Journal*, 321: 1155.

Charski, M. (2004) Ad push to get Latinos wired to their cells. *Marketing y Medios*, October 1: 16–17. http://marketingymedios.com/marketingymedios/creative/article_display.jsp?vnu_content_id=1000654034 (accessed February 21, 2005).

Cheil Communications (2003) *Exploring P-Generation*. Seoul, Korea.

Chen, K. and Miles, J. C. (eds) (1999) *ITS Handbook, 2000: Recommendations form the World Road Association (PIARC)*. Artech House ITS Library.

———— and Ramstad, E. (2004) To steer tech market China touts own standards. *Asian Wall Street Journal*, April 23.

Chezzi, D. (2004) Buy and cell. *Macleans*, 117 (43).

China Daily (2003) Cut-price "Little Smart" a big hit in China, August 17. http://www.chinadaily.com.cn/en/doc/2003-08/17/content_255624.htm (accessed February 23, 2006).

China Ministry of Information Industry (2001) *Annual Statistical Report on the Development of Telecommunications in China.*

———— (2005) *Monthly Report*, January.

China Statistical Yearbook [*Zhongguo tongji nianjian*] (1990–2003) Beijing: China Statistics Publications [Zhongguo tongji chubanshe].

Choma, M. and Robinson, D. (2000) Usage and attitudes toward wireless communications in Canada 2000: result highlights. http://www.cwta.ca (accessed March 21, 2005).

Cingular Wireless (2003) Guys still gab more on wireless. *Cingular News* press release, June 23. http://www.cingular.com/about/latest_news/03_06_23 (accessed August 16, 2004).

Clark, D. (2003) From the web to wireless. Paper presented at the Conference on China and the Internet: Technology, Economy, and Society in Transition, Los Angeles: May 30–31.

CNN (2004) GOP Convention protest covers miles of New York, August 30. http://www.cnn.com/2004/ALLPOLITICS/08/29/gop.main/index.html (accessed September 11, 2004).

Coburn, J. (2004) State Senate OKs restricting teenage cell phone use. *Daily Californian*, June 1. http://www.dailycal.org/article.asp?=15385 (accessed June 30, 2004).

Cohen, K. and Wakeford, N. (2003) The making of mobility, the making of the self. www.soc.surrey.ac.uk/incite/AESOP%20Phase3.htm (accessed January 2004).

Colbert, M. (2001) A diary study of rendezvousing: implications for position-aware computing and communications for the general public. *Proceedings of the 2001 International ACM SIGGROUP Conference on Supporting Group Work*, pp. 15–23. Boulder, Colorado.

Colina, C. (2000) Nuevas formas de control social: ¿panopticismo electrónico o seducción post(moderna)? *Anuario ININCO: Investigaciones de la Comunicación*, 1 (11): 25–44. www.revele.com.ve/pdf/anuario_ininco/vol1–n11/pag25.pdf (accessed July 17, 2005).

Collins, J. (2000) Talking teenagers: major wireless players are finally paying attention to the youth market. *Tele.com*, November 13: 38.

Commission for Africa (2005) Information technology, communications and infrastructure seminar. London, January 26. http://www.commissionforafrica.org/english/consultation/bob_geldofs_seminars/discussions/26January2005ITSeminar SummaryofDiscussion.pdf (accessed July 26, 2005).

Communication Today (2001) CMG promises European experience for Canada's SMS market, November 9. http://www.findarticles.com/p/articles/mi_m0BMD/is_211_7/ai_79908402 (accessed February 21, 2005).

Compartel (2004) Compartel programme of social telecommunications: case study. Ministry of Communications, Republic of Colombia/International Telecommunications Union. http://www.itu.int/ITU-D/fg7/case_library/case_study_2/Americas/Colombia_English_.pdf (accessed February 20, 2006).

Continental Research (2001) Mobile phone, July. www.continentalresearch.com.

Cook, J. (2004) Rolling wheat fields are also Wi-Fi country. *Seattle Post Intelligencer*, September 7. http://seattlepi.nwsource.com/business/189699_vivato07.html (accessed September 10, 2004).

Council of Economic Advisors (2000) The economic impact of third-generation wireless technology. http://www.ntia.doc.gov/ntiahome/threeg/ceareportoc t2000.pdf (accessed March 2004).

Country Monitor (2004) Keeping it local. *Country Monitor*, 12 (3): 5 (January 26).

Coutts, P., Alport, K., Coutts, R., and Morrell, D. (2003) Beyond the wireless internet hype: re-engaging the user. www.smartinternet.com.au/SITWEB/publication/files/80_$$$_11151/P04_038_paper.pdf (accessed July 23, 2005).

Coutts, P. J. (2002) Banking on the move: characterising user bottlenecks for m-commerce uptake. Paper presented at the Communications Research Forum, Canberra, October 2–3.

Coyle, D. (2005) Overview. In *Africa: The Impact of Mobile Phones*. Vodafone Policy Paper Series 2: 3–9.

Crabtree, J., Nathan, M., and Roberts, S. (2003) Mobile UK: mobile phones and everyday life (electronic edition). London: Work Foundation. http://portal.acm.org/citation.cf m?doid=500292.

Cronin, J. (2005) Africa counts the cost of making a call, March 6. http://news.bbc.co.uk/go/pr/fr/-/2/hi/business/4277477.stm (accessed March 9, 2005).

Cudjoe, F. (2005) Spacefon: shrewd business vs. ignorant consumer society. http://www.ghanaweb.com/GhanaHomePage/News/Archive/printnews.php?ID=74 474 (accessed February 16, 2005).

Cué, C. E. (2004) *¡Pásalo! Los cuatro días de marzo que cambiaron el país*. Barcelona: Ediciones Península.

CWTA (1998) Who uses wireless phones? Canadians speak out. Press release, May 20. http://www.cwta.ca/CWTASite/english/whatsnew_download/may20.doc (accessed March 21, 2005).

——— (2000) More Canadians are communicating the wireless way. Press release, May 30. http://www.cwta.ca/CWTASite/english/whatsnew_download/may30_00. doc (accessed March 21, 2005).

——— (2001) Canadian small businesses growing more attached to wireless devices. Press release, December 6. http://www.cwta.ca/CWTASite/english/whatsnew_download/dec06_01.doc (accessed March 21, 2005).

——— (2005a) Industry overview. http://www.cwta.ca/CWTASite/english/industry.html (accessed March 10, 2005).

——— (2005b) Canadians now sending 3.4 million text messages per day. Press release, March 22. http://www.cwta.ca/CWTASite/english/whatsnew_download/mar22_05.html (accessed June 2, 2005).

Dano, M. (2004) Wireless entertainment shortchanges women, youth. *RCR Wireless News*, 21 (18): 4 (May 3).

Dao, James (2004) Philadelphia hopes to lead the charge to wireless future. *New York Times*, February 17.

David, K. (2003) Cooking, cleaning and charging the cell phone. Unpublished manuscript, School of Communication, Information and Library Studies, Rutgers University.

——— (2005) Mobiles in India: tool of tradition or change? Paper presented at the 55th Annual International Communication Association Conference, Preconference Workshop, New York, May 26.

Day, P. (2005) Talk is profitable in Ghana, 23 April. http://news.bbc.co.uk/go/pr/fr//2/hi/programmes/from_our_own_correspondent/4473073.stm (accessed April 25, 2005).

Dayal, G. (2004) Yury and his magicbike. *Village Voice*, August 29. www.villagevoice.com/issues/0435/dayal.php (accessed August 31, 2004).

DeJong, J. (2001) Mobile web: "on the move and in touch" is an exciting prospect. But will it ever work? (electronic version). *PC Magazine*, July: 140.

Dekimpe, M. G., Parker, P. M., and Sarvary, M. (1998) Staged estimation of international diffusion models: an application to global cellular telephone adoption. *Technological Forecasting and Social Change*, 57: 105–132.

Delclos, T. (2004) Pasalo. In *El Pais*, Catalan edition, March 16. www.elpais.es/articulo/elpepiautcat/20040316elpcat_16/Tes/P%E1salo (accessed September 20, 2005).

Demars, T. R. (2000) *Modeling Behavior from Images of Reality in Television Narratives: Myth-information and Socialization*. Lewiston, NY: Edwin Mellen.

Demick, B. (2003) Netizens crusade buoys new South Korean leader. *Los Angeles Times*, February 10: A3.

Department of Communications, Information Technology and the Arts (2004) *The Current State of Play*. Australian Government, Department of Communications, Information Technology and the Arts.

Department of Telecommunication (2003–2004) *Annual Report*. Ministry of Communications and Information Technology, Government of India, New Delhi.

Di Justo, P. (2004) Protests powered by cell phone. *New York Times*, September 9. http://tech2.nytimes.com/mem/technology/techreview.html?res=9404E2DE1730F9 3AA3575AC0A9629C8B63 (accessed September 2004).

Diario de Navarra (2004) Alemania cierra un acuerdo para recuperar el peaje por satélite para los camions [Germany closes an agreement to recover the toll for the trucks by satellite], February 3, 2004. http://www.diariodenavarra.es/ edicionimpresa/noticiaB.asp?not=A12ART112209A&vin=&seccion=economia&dia= 20040302&vf=(accessed September 27, 2005).

Digital Times (2003) SK Telecom dominates mobile phone market, December 19 (in Korean). http://news.naver.com/news/ (accessed on June 21, 2004).

Direct Intelligence (2001) Ethnographic study examines consumers' wireless use. *Direct Intelligence*, March.

Dizon, K. (2003) Text messaging makes cell phones even hotter among kids. *Seattle Pi*, September 30. http://seattlepi.nwsource.com/lifestyle/141809_te xting30.html (accessed July 2, 2004).

Donner, J. (2004) Microentrepreneurs and mobiles: an exploration of the uses of mobile phones by small business owners in Rwanda. *Information Technologies and International Development*, 2 (1): 1–22.

——— (2005a) The rules of beeping: exchanging messages using missed calls on mobile phones in sub-Saharan Africa. Paper submitted to the 55th Annual Conference of the International Communication Association, New York, May 26–30.

——— (2005b) The use of mobile phones by microentrepreneurs in Kigali, Rwanda: changes to social and business networks. Research report prepared for the International Workshop on Wireless Communication and Development: A Global Perspective, Marina del Rey, California, October 7–8, Annenberg Research Network on International Communication. http://arnic.info/workshop05/Donner%20_ MobileKigali_Sep05.pdf.

Dossani, R. (ed.) (2002) *Telecommunications Reform in India*. Westport, CT: Quorum Books.

Dot-ORG (2004) Village cellular phones in Uganda for rural income generation and more. *DOTCOMments*, 7. http://www.dot-com-alliance.org/newsletter (accessed June 10, 2005).

Duffy, R. and Zhao, Y. (2002) Short-circuited: communication and working class struggle in China. Paper presented at the China's Media Today and Tomorrow Symposium, University of Westminster, May 14.

Dunlap, K. (2002) Calling all teens: wireless carriers target youths. *Mercury News*, June 15. http://www.siliconvalley.com/mld/siliconvalley/3480214.htm (accessed June 30, 2004).

Dutta, S. and Coury, M. E. (2003) ICT challenges for the Arab world. In S. Dutta, B. Lanvin, and F. Paua (eds), *The Global Information Technology Report 2002–2003: Readiness for the Networked World*, pp. 116–123. New York: Oxford University Press.

Dutton, W. H. and Nainoa, F. (2003) "Say goodbye . . . let's roll": the social dynamics of wireless networks on September 11. *Prometheus*, 20 (3): 237–245.

———, Elberse, A., Hong, T., and Matei, S. (2001) Beepless in America: the social impact of the Galaxy IV pager blackout. In S. Lax (eds), *Access Denied: Exclusion in the Information Age*, pp. 9–32. New York: Palgrave.

———, Gillett, S. E., McKnight, L. W., and Peltu, M. (2004) Bridging broadband Internet divides: reconfiguring access to enhance communicative power. *Journal of Information Technology*, 19: 28–38.

Eckholm, E. (2002) Laid off Chinese protest en masse. *The Associated Press*, March 18.

ECLAC (Economic Commission for Latin America and the Caribbean) (2001) Foreign investment in Latin America and the Caribbean (LC/G.2125–P/E), Santiago, Chile. United Nations, Sales no. S.01.II.G.12.

The Economist (2001) A different way of working. October 13 (electronic version). http://www.economist.com/surveys/displayStory.cfm?Story_id=812006 (accessed September 19, 2004).

——— (2002) Business: the tortoise and the hare – mobile telecoms. March 16, 362 (8264): 76.

——— (2004) How big was the bounce? September 9. http://www.economist.com/world/na/displayStory.cfm?story_id=3177113 (accessed September 13, 2004).

——— (2005a) Calling across the digital divide. March 12: 72.

——— (2005b) Calling an end to poverty. July 9, 376 (8434): 53.

——— (2005c) Less is more: mobile phones and development. July 9, 376 (8434): 11.

——— (2005d) The real digital divide. March 12: 11.

——— (2005e) A spiritual connection. March 10.

EFE (2004) Los españoles enviaron 19,000 millones de mensajes de texto por teléfono móvil en 2003, January 9. http://www.laflecha.net/canales/moviles/200409011/ (accessed September 27, 2005).

Eldridge, M. and Grinter, R. (2001) Studying text messaging in teenagers. Position Paper for CHI 2001 Workshop no. 1, Mobile Communications: Understanding User, Adoption and Design.

Elkin, T. (2003) 18 percent would rather give up TVs than wireless phones: new study tracks growing use of short message service. http://www.upoc.com/corp/news/news-adage.html (accessed May 22, 2004).

eMarketer (2003) More mobile owners turning to text messaging, March 4. http://www.upoc.com/corp/news/news-emarketer.html (accessed May 22, 2004).

Enos, J. (n.d.) Three advertising agencies find success in wireless campaigns for some big players. *mBusiness Daily*. www.mbusinessdaily.com/magazine/story/06_japan (accessed February 26, 2004).

Entner, R. (2003) Third quarter 2002 US wireless forecast 2000–2006 (originally published by the Yankee Group, September 25). http://techupdate.zdnet.com/techupdate/stories/main/Third_Quarter_2002_US_Wireless_Forecast_2000_2006.html (accessed May 15, 2004).

Erard, M. (2004) For technology, no small world after all. *New York Times*, May 6: G.5.

Ericson, R. J. and Haggerty, K. (1997) *Policing the Risk Society*. Toronto: University of Toronto Press.

Eurescom (2001) ICT uses in everyday life: checking it out with the people – ICT markets and users in Europe. Confidential EURESCOM P903 Project Report elaborated by E. Mante-Meijer, L. Haddon, et al. EDIN0161–0903.

—— (2004) e-Living: life in a digital Europe. An EU Fifth Framework Project (IST-2000-25409). http://www.eurescom.de/e-living.

Euromonitor (2003) Cellular and wireless communications systems in the USA, July. Market Monitor Research Database (now Global Market Information Database). http://www.gmid.euromonitor.com.

European Communities (2003) Intelligent transport systems: intelligence at the service of transport networks. European Commission, Energy and Transportation DG, Luxembourg. http://europa.eu.int/comm/transport/themes/network/english/its/pdf/its_brochure_2003_en.pdf (accessed July 2004).

Fairclough, G. (2004) Generation why? The 386ers of Korea question old rules. *Wall Street Journal*, April 14: A1.

Fattah, H. (2003) America untethered (electronic version). *American Demographics*, March 1. http://www.upoc.com/corp/news/UpocAmDem.pdf (accessed May 21, 2004).

FCC (1999) FCC acts to promote competition and public safety in enhanced wireless 911 services. Press release, September 15. http://www.fcc.gov/Bureaus/Wireless/News_Releases/1999/nrwl9040.txt

—— (1999–2003) Annual report and analysis of competitive market conditions with respect to commercial mobile services. www.fcc.gov/wcb/stats.

—— (2002a) Seventh annual report and analysis of competitive market conditions with respect to commercial mobile services. http://hraunfoss.fcc.gov/edocs_public/attachmatch/FCC-02-179A1.pdf (accessed February 2, 2005).

—— (2002b) *Trends in Telephone Service, 2002*. Washington, DC: FCC.

—— (2004) Industry Analysis and Technology Division Wireline Competition Bureau, May.

Felto, J. (2001) Crossed wireless. *American Demographics*, 23 (9): 28.

Fernández-Maldonado, M. (2001) Diffusion and use of new information and communication technologies in Lima. *Journal of Urban Technology*, 8 (3): 21–43 (also quoted in NECG 2004).

Fischer, C. (1988) Gender and the residential telephone 1890–1940: technologies of sociability. *Sociological Forum*, 2: 211–234.

Fischer, L. (2002) A cell phone that's hip enough for teens. *USA Today*, April 30. http://www.usatoday.com/tech/columnist/2002/04/30/fischer.htm (accessed June 6, 2004).

Fitchard, K. (2002) The next tycoon of teen. *Wireless Review*, March 1. http://wirelessreview.com/ar/wireless_next_tycoon_teen/ (accessed September 18, 2004).

—— (2005) Crossing over. *Wireless Review*, July 1. http://wirelessreview.com/mag/wireless_crossing/index.html (accessed August 7, 2005).

Forbes (2002) Wireless phone usage: benchmark study of Forbes.com users, November. www.pdf.forbes.com/fdc/forbes.com_wirelesssurvey.pdf (accessed May 20, 2004).

—— (2005) Cingular Wireless survey reveals that men talk more than women on mobile phones. Forbes.com, June 1. http://www.forbes.com/technology/wireless/feeds/wireless/2005/06/01/wirelessm2c_2005_06_01_tww_0000-0365-tww_200506011548551.html (accessed July 12, 2005).

Ford, N. (2003) Middle East telecoms. *The Middle East*, 340: 45–49.

Fortunati, L. (2002a) Italy: stereotypes, true and false. In J. E. Katz and M. Aakhus (eds), *Perpetual Contact: Mobile Communication, Private Talk, Public Performance*, pp. 42–62. Cambridge: Cambridge University Press.

—— (2002b) The mobile phone: towards new categories and social relations. *Information, Communication and Society*, 5 (4): 513–528.

—— (2003) The mobile phone between orality and writing. Paper presented at the COST 269 Conference on The Good, the Bad and the Irrelevant: The User and the Future of ICTs, Helsinki, Finland. September 3–5.

—— (2005a) The mobile phone between local and global. In K. Nyiri (ed.), *A Sense of Place: The Global and the Local in Mobile Communication*, pp. 61–70. Vienna: Passagen Verlag.

—— (2005b) Mobile phone and the presentation of self. In R. Ling and E. Pedersen (eds), *Mobile Communication and the Re-negotiation of the Social Sphere*, pp. 203–218. London: Springer.

—— and Manganelli, A. M. (2002) Young people and the mobile telephone. *Revista de Estudios de Juventud*, 52: 59–78.

—— and —— (2004) The family, communication and new technologies. In S-D. Kim (ed.), *Mobile Communication and Social Change*, pp. 70–87. Choncheon: Hallym University.

Fotel, T. and. Thomsen, T. U (2004) The surveillance of children's mobility. *Surveillance and Society*, 1 (4): 535–554. www.surveillance-and-society.org.

Fowler, G. A. (2005) In Asia, it's nearly impossible to tell a song from an ad. *Wall Street Journal*, May 31: A1, A10.

Fox, K. (2001) Evolution, alienation and gossip: the role of mobile communications in the twenty-first century. www.sirc.org/publik/gossip.shtml (accessed March 2004).

Francescutty, P., Baer, A., Garcia, J. M., and López P. (2005) La noche de los móviles: medios, redes de confianza y movilización juvenile. In V. F. Sampedro Blanco (ed.), *13-M: Multitudes on Line*, pp. 63–83. Madrid: Ed. Catarata.

Fries, M. (2000) *China and Cyberspace: The Development of the Chinese National Information Infrastructure*. Bochum, Germany: Bochum University Press.

Friginal, R. (2003) From 25 percent to 2005: teledensity to hit 33 percent. Malaya, November 13. http://www.malaya.com.ph/nov13/busi1.htm (accessed June 12, 2004).

Frissen, V. (2000) ICT in the rush hour of life. *The Information Society*, 16: 65–75.

Frontlines (2004) Microloans are helping Uganda villagers get mobile phone service, September. http://www.usaid.gov/press/frontlines/fl_sept04/pillars.html (accessed June 10, 2005).

Frost and Sullivan (2003) The "PAS" phenomenon: revolutionizing local wireless telephony. Frost and Sullivan White Papers, February.

Fulford, B. (2003) Korea's weird wired world, *Forbes*, July 21: 92.

Galperin, H. (2005) Wireless networks and rural development: opportunities for Latin America. *Information Technologies and International Development*, 2 (3): 47–56.

—— and Bar, F. (2005) Diversifying network development: microtelcos in Latin America and the Caribbean. Research report prepared for the International Workshop on Wireless Communication and Development: A Global Perspective; Marina del

Rey, California, October 7–8, Annenberg Research Network on International Communication http://arnic.info/workshop05/Galperin-Bar_Microtelcos_Sep05.pdf.

—— and Girard, B. (2005) Microtelcos in Latin America. In H. Galperin and J. Mariscal (eds), *Information Technology and Poverty Alleviation: Perspectives from Latin America and the Caribbean*. Ottawa: IDRC. http://www-rcf.usc.edu/~hernang/05-Galperin-Girard_23nov.pdf (accessed February 20, 2006).

Gandal, Neil (2002) New horizons: telecommunications policy in Israel in the 21st century. Working paper, Department of Public Policy, Tel Aviv University.

Garcia, B. E. (2004) Teens' cellphone talk not cheap. *Miami Herald*, July 13: 1A.

Garcia, C., Gallo, A., Fernández, J., and Larrain, F. (2002) El cellular en la sociedad chilena: diagnósticy y proyecciones. TIC-ISUC, Santiago, Chile.

Garrett, R. (2004) New technological options for people with physical disabilities through the use of telecommunications equipment. http://www.e-ability.com/articles/telecommunications.shtml (accessed July 20, 2005).

Gartner (2003) Gartner says mobile data revenues in Western Europe to grow 31 percent in 2003. Press release, August 18. http://www.gartner.com/press_releases/pr18aug2003b.html (accessed September 18, 2005).

Gaspar, K. (2001) Once again, an outpouring in the streets brings change to the Philippines. *Sojourners Magazine*, March–April: 15.

Genwireless (2001) Statistics. http://www.genwireless.com/stats.html.

Georges, C. C. (2001) Wired communities and wireless communication of the new millennium. *CTBUH Review*, 1 (2): 1–7.

Gergen, M. (2005) Using mobile phones: a survey of college women and men. Paper presented at the 55th Annual International Communication Association Conference, Preconference Session, Mobile Communication: Current Research and Future Directions, New York, May 26.

Geser, H. (2004) Towards a sociological theory of the mobile phone. http://geser.net/home.html.

Gibbs, C. (2004) SMS to aid protesters at GOP Convention. *RCR Wireless News*, August 30.

Gibson, G. (2004) The Republican Convention is expected to draw hundreds of thousands of protesters eager to air their discontent with the Bush administration. *Baltimore Sun*, August 22: 1F.

Gilbert, D., Lee-Kelley, L., and Barton, M. (2003) Technophobia, gender influences and consumer decision-making for technology-related products. *European Journal of Innovation Management*, 6 (4): 253.

Gillard, P., Bow, A., and Wale, K. (1996) *Ladies and Gentlemen, Boys and Girls: Gender and Telecommunication Services*. Melbourne: Telecommunications Needs Service.

Gillwald, A. (2003) Africa e-usage and access index. http://link.wits.ac.za/research/ Gillwald%20110205.pdf (accessed February 25, 2005).

Gleick, J. (1999) *Faster: The Acceleration of Just About Everything*. New York: Pantheon.

Global Entrepreneur (2004) UTStarcom mengzhidui [The dream team of UTStarcom] *Huangqiu qiyejia* [*Global Entreprenuer*], April, vol. 97.

Goggin, G. (2004) Mobile text. *M/C: A Journal of Media and Culture*, 7. http://www.media-culture.org.au/0401/03–goggin.html.

—— and Newell, C. (2000) An end to disabling policies? Toward enlightened universal service. *The Information Society*, 16: 127–133.

Gottlieb, N. and McLelland, M. (2003) The Internet in Japan. In N. Gottlieb and M. McLelland (eds), *Japanese Cybercultures*, pp. 1–16. New York: Routledge.

Gough, N. and Grezo, C. (2005) Introduction. In *Africa: The Impact of Mobile Phones*. Vodafone Policy Paper Series 2: 2–3.

Graham, S. (ed.) (2004) *The Cybercities Reader*. London: Routledge.

Grameen Foundation USA (n.d.) Village phone program. http://www.gfusa.org/ technology_center/village_phone (accessed June 10, 2005).

Grant, D. and Kiesler, S. (2002) Blurring the boundaries: cell phones, mobility and the line between work and personal life. In B. Brown, N. Green, and R. Harper (eds), *Wireless World: Social and Interactional Aspects of the Mobile Age*, pp. 121–131. London: Springer.

Green, E. and Adam, A. (eds) (2001) *Virtual Gender: Technology, Consumption, and Identity*. London: Routledge.

—— and Singleton, C. (n.d.) "I can't live without my mobile!": Mobile technologies, gender and culture in lives of young black and minority ethnic women and men. http://www.ict.open.ac.uk/gender/papers/green.ppt.

Green, N. (2002) On the move: technology, mobility, and the mediation of social time and space. *The Information Society*, 18: 281–292.

—— and Smith, S. (2004.) "A spy in your pocket?" The regulation of mobile data in the UK. *Surveillance and Society*, 1 (4): 573–587.

Greenspan, R. (2003a) Wireless youth most likely to cut cord. http://www.clickz. com/stats/markets/wireles/article.php/2247081 (accessed June 9, 2004).

—— (2003b) UK texting takes off, July 16. http://www.clickz.com/stats/markets/ wireless/article.php/2236031 (accessed August 16, 2004).

—— (2004a) IM usage nearly doubles. http://www.clickz.com/stats/markets/ wireless/article.php/3400661 (accessed September 7, 2004).

—— (2004b) Mobile phones move beyond voice. http://www.clickz.com/stats/ markets/wireless/article.php/3356241 (accessed September 7, 2004).

Grenville, M. (2004) Operators: Orange France cuts SMS pricing, January 27. http://www.160characters.org/news.php?action=view&nid=914 (accessed July 2004).

Grice, C. and Kanellos, M. (2000) Cell phone industry at crossroads: go high or low? CNET News.com, August 31. http://news.com.com/2009–1033–244415.html? legacy=cnet (accessed August 5, 2005).

Grinter, R. E. and Eldridge, M. (2001) *y do tngrs luv 2 txt msg?* In W. Prinz, M. Jarke, Y. Rogers, K. Schmidt, and V. Wulf (eds), *Proceedings of the Seventh European Conference on Computer-supported Cooperative Work*, ECSCW '01, pp. 219–238. Dordrecht, Netherlands: Kluwer Academic.

Gruber, H. (1999) An investment view of mobile telecommunications in the European Union. *Telecommunications Policy*, 23: 521–553.

Guangzhou Daily (2004) First large-scale survey on mobile Internet in China successfully completed (in Chinese), August 26.

Ha, J-Y. (2002) Diffusion of the use of cell phones as internal office phones (in Korean). *Joongang Ilbo*, January 16. http://news.naver.com/news/read.php?mode= LOD&office_id=025&article_id=0000434286 (accessed March 3, 2004).

Hachigian, N. and Wu, L. (2003) *The Information Revolution in Asia*. Santa Monica, CA: RAND Corporation.

Haddon, L. (2000) The social consequences of mobile telephony: framing questions. Paper presented at the seminar Sosiale Konsekvenser av Mobiltelefoni, organized by Telenor, Oslo, June 16.

——— (2002) Youth and mobiles: the British case and further questions. *Revista de Estudios de Juventud*, 52: 115–124.

Hafkin, N. J. (2003) Gender issues in ICT statistics and indicators, with particular emphasis on developing countries. Keynote address at the Joint UNECE/UNCTAD/ UIS/ITU/OECD/EUROSTAT Statistical Workshop: Monitoring the Information Society: Data, Measurement and Methods, Geneva, December 8–9. http://hdrc. undp.org.in/APRI/Rsnl_Rsrc/Gndr_ICT.pdf (accessed July 14, 2005).

——— (2005) Statistics and indicators: gender and ICT. http://www. un-instraw.org/revista/hypermail/alltickers/en/att-0199/Gender__ICT-Statistics __Indicators-Hafkin.ppt (accessed July 14, 2005).

Hahn, R. W. and Dudley, P. M. (2002) The disconnect between law and policy analysis: a case study of drivers and cell phones. AEI Brookings Joint Center for Regulatory Studies Working Paper 02–7.

Hale, D. and Hale, L. H. (2003) China takes off. *Foreign Affairs*, 82 (6): 36–53 (November/December).

Hamilton, J. (2003) Are main lines and mobile phones substitutes or complements? Evidence from Africa. *Telecommunications Policy*, 27: 109–133.

Hamilton, R. (2003) Community phones connect SA townships http://news.bbc.co.uk/go/pr/fr/-/1/hi/technology/3246732.stm (accessed July 26, 2005).

Han, K-S. (2001) Integrated corporate wired and wireless VPN services (in Korean). *Hankook Ilbo*, May 2. http://news.naver.com/news/read.php?mode=LOD&office_id=038&article_id=0000069130 (accessed March 3, 2004).

Handford, R. (2005) Smartening up or dumbing down? *Mobile Communications International*, 121 (May): 28–33.

Hanware, K. (2005) First "made in Saudi Arabia" mobile coming. *Arab News*, (Jeddah) January 6.

Harper, P. and Clark, C. (2002) Networking the deaf nation. www.crf.dcita.gov.au/papers02/harper%20clark.pdf (accessed July 23, 2005).

Harrington, V. and Mayhew, P. (2001) Mobile phone theft. Home Office Research Study, 235. www.homeoffice.gov.uk/rds/pdfs/hors235.pdf (accessed April 2004).

Harter, B. (n.d.) SMS meets the United States: can SMS achieve the same popularity in the United States as it has overseas? http://www.mbusinessdaily.com/magazine/story/new/05_smsus (accessed June 5, 2004).

Hashimoto, K., Hashimoto, Y., Ishii, K., Nakamura, I., Korenaga, R., Tsuji, D., and Mori, Y. (2000) Survey research on uses of cellular phones and other communication media in 1999 (in Japanese). *Research Bulletin of the Institute of Socio-Information and Communication Studies*, 14: 83–192.

Hauser, C. (2004.) Marchers denounce Bush as they pass GOP Convention Hall. *New York Times*, August 20. http://www.nytimes.com/2004/08/29/politics/campaign/29CND-PORT.html?ex=1114488000&en=f3897bb7d72df891&ei=5070 (accessed April 23, 2005).

Heilman, G. E. and White, D. (2003) On general application of the technology acceptance model. http://mcb.unco.edu/pdfs/WPS/TAM-2.doc.

Henry Fund Research (2003) Wireless communications. http://www.biz.uiowa.edu/henry/Wireless_SP03.pdf (accessed August 14, 2004).

——— (2004) Wireless communications. http://www.biz.uiowa.edu/henry/Wireless_SP04.pdf (accessed August 14, 2004).

Hilado, J. K. F. (2003) "Generation Text." *The Guidon Online*, February 27. http://www.theguidon.com/?get=2003020403 (accessed June 3, 2004).

Hilbert, M. and Katz, J. (eds) (2003) *Building an Information Society: A Latin American and Caribbean Perspective*. Santiago de Chile: Economic Commission for Latin America and the Caribbean.

Himanen, P. (2001) *The Hacker Ethic and the Spirit of the Information Age*. London: Secker and Warburg.

Hjorth, L. (2003) Cute@keitai.com. In N. Gottlieb and M. McLelland (eds), *Japanese Cybercultures*, pp. 50–59. New York: Routledge.

Höflich, J. and Rössler, P. (2002) More than *just* a telephone: the mobile phone and use of the short message service (SMS) by German adolescents: results of a pilot study. *Revista de Estudios de Juventud*, 52: 79–100.

Hofstede, G. (1991) *Cultures and Organizations: Software of the Mind – Intercultural Cooperation and its Importance for Survival*. New York: McGraw-Hill.

Holden, T. J. M. and Tsuruki, T. (2003) *Deai-kei*: Japan's new culture of encounter. In N. Gottlieb and M. McLelland (eds), *Japanese Cybercultures*, pp. 34–49. New York: Routledge.

Holmes, D. and Russell, G. (1999) Adolescent CIT use: paradigm shifts for educational and cultural practices? *British Journal of Sociology of Education*, 20 (1): 69–78.

Horrigan, J. B. (2003) *Consumption of Information Goods and Services in the United States*. Pew Internet and American Life Project. http://www.pewinternet.org/pdfs/PIP_Info_Consumption.pdf.

Horst, H. and Miller, D. (2005) From kinship to link-up. *Current Anthropology*, 46 (5): 755–778.

—— and —— (2006) *The Cell Phone: An Anthropology of Communication*. Oxford: Berg.

Howard, L. (2004) Let's get mobile. http://www.disabilitynow.org.uk/living/equipment/letsgetmobile.htm (accessed July 20, 2005).

Huag, T. (2002) A commentary on standardization practices: lessons from the NMT and GSM mobile telephone standards histories. *Telecommunications Policy*, 26: 101–107.

Hui, Y-M. (2004) State Council plans to merge carriers. *South China Morning Post*, July 6.

Huyer, S., Hafkin, N., Ertl, H., and Dryburgh, H. (2005) Women in the information society. In G. Sciadas (ed.), *From the Digital Divide to Digital Opportunities: Measuring Infostates for Development*, pp.135–196. http://www.orbicom.uqam.ca/projects/ddi2005/index_ict_opp.pdf (accessed January 17, 2005).

IBGE (2003) Diretoria de Pesquisas, Coordenação de Trabalho e Rendimento, Pesquisa Nacional por Amostra de Domicílios. Instituto Brasileiro de Geografia e Estatistica.

—— (2004) Pesquisa Nacional por Amostra de Domicílios. Instituto Brasileiro de Geografia e Estatistica.

ICA (2005) Pilot project for Internet connectivity in rural communities in Ecuador successfully inaugurated. Press release, January, 11. http://www.icamericas.net/documents/ElChaco2_english.doc (accessed February 20, 2006).

ICRA Information Services (2001) The Indian telecommunication industry: the calling party pays system. http://www.icraindia.com/biz-arch/200107flashcpp.pdf (accessed November 17, 2004).

IDC (2003) US youth/young adult wireless subscriber forecast, 2002–2007: in search of the perfect storm. http://www.idc.com/getdoc.jsp?containerId=29822 (accessed June 5, 2004).

——— (2005) IDC finds young adult males more likely than females to purchase mobile phones as fashion statements. Press release. http://www.idc.com/getdoc.jsp?containerId=prUS00022105 (accessed January 17, 2006).

IDC Spain (2004) Mobile messaging spending in CEE to reach $2.9 billion in 2008. http://www.idc.com/spain/pdf/Mobile%20Messaging.pdf (accessed August 7, 2005).

Information Society Research Group. (n.d.) Information society: emergent technologies and development communities in the south. http://www.isrg.info/index.html (accessed October 10, 2005).

In-Stat/MDR (2004a) Mobile data users now representing the mainstream, April 19. http://www.instat.com/press.asp?Sku=IN030933MD&ID=871 (accessed June 5, 2004).

——— (2004b) Youth market represents large opportunity for wireless: Boost Mobile may already have edge, February 3. http://www.instat.com (accessed June 5, 2004).

IPSe (2003) *Third Annual Consumer Report: Survey Results from Research on Mobile Phone Usage*. Tokyo:.IPSe Communication.

Ishii, K. (2004) Internet use via mobile phone in Japan. *Telecommunications Policy*, 28 (1): 43–58.

ISP Planet (2001) US mobile wireless direction, March 28. http://www.isp-planet.com/research/2001/us_mobile_market.html (accessed February 6, 2006).

IT Facts (2005) Americans send 2.5bln text messages a month, January 8. http://www.itfacts.biz/index.php?id=P1103 (accessed March 13, 2005).

Ito, M. (2003a) A new set of social rules for a newly wireless society. *Japan Media Review*. http://www.ojr.org/japan/wireless/1043770650.php (accessed February 24, 2004).

——— (2003b) Mobile phones and the appropriation of place. *Receiver: Mobile Environment*. http://www.receiver.vodafone.com/08/articles/pdf/08_07.pdf (accessed April 5, 2004).

——— (2004) Personal, portable, pedestrian: mobile phones in Japanese life. Southern California Digital Culture Group. Los Angeles: Annenberg Center for Communication.

—— and Okabe, D. (2003) Mobile phones, Japanese youth, and the re-placement of social contact. Paper presented at the Conference Front Stage – Back Stage: Mobile Communication and the Renegotiation of the Public Sphere, Grimstad, Norway.

—— and —— (forthcoming) *Technosocial Situations: Emergent Structurings of Mobile Email Use.*

——, ——, and Matsuda, M. (2005) *Personal, Portable, Pedestrian: Mobile Phones in Japanese Life.* Cambridge, MA: MIT Press.

ITU (2001) 3G mobile policy: the case of Japan. http://www.itu.int/3g (accessed March 29, 2004).

—— (2002a) *Internet for a Mobile Generation.* Geneva: ITU.

—— (2002b) *ITU World Telecommunication Indicators Database, 2002.* Geneva: ITU.

—— (2003) *World Telecommunication Development Report 2003: Access Indicators for the Information Society*, ch. 4. http://www.itu.int/ITU-D/ict/publications/wtdr_03/material/Chap4_WTDR2003_E.pdf.

—— (2004a) African telecommunication indicators 2004. www.itu.int.

—— (2004b) *ITU Internet Reports 2004: The Portable Internet.* Geneva: ITU.

—— (2004c) Shaping the future mobile information society: the case of Japan. http://www.itu.int/osg/spu/ni/futuremobile/general/casestudies/japan.html/(quoted in ITU 2004b).

—— (2004d) *World Telecommunications Indicators Database.* Geneva: ITU.

—— (2005) *World Telecommunication Indicators Database*, 9th edn. Chronological Time Series 1960–2004 (CD-ROM). Geneva: ITU.

Jackson, J. (2004) Speech made at the United for Peace and Justice protest during the 2004 Republican National Convention. August 29. http://www.democracynow.org/article.pl?sid=04/08/30/1453250 (accessed April 23, 2005).

Jain, P. and Sridhar, V. (2003) Analysis of competition and market structure of basic telecommunication services in India. *Communications and Strategies*, 52: 271–293.

Jain, R. S. (2001) Spectrum auctions in India: lessons from experience. *Telecommunications Policy*, 25 (10–11): 671–688.

Japan Ministry of Internal Affairs and Communication (2005) *Japan Statistical Yearbook.* Statistics Bureau and Statistical Research and Training Institute.

Japan MPHPT (2004) *Jyouhou Hakyusho*, White Paper.

Jensen, M. (1998) Wireless in Africa. *Telecommunications*, 32 (4): S6, S8.

Jhunjhunwal, N. and Orne, P. (2003) Left to their own devices: twelve developing-world case studies. In Wireless Internet Institute (ed.), *The Wireless Internet Opportunity for Developing Countries.* http://www.infodev.org/files/838_file_The_Wireless_Internet_Opportunity.PDF (accessed February 20, 2006).

Ji, M. (2005) Rescued from the rubble. *South China Morning Post*, August 3: 6.

Johnson, H. (2003) Gender, technology, and the potential for social marginalization: Kuala Lumpur and Singapore (electronic version). *Asian Journal of Women's Studies*, 9 (1): 60 (accessed July 12, 2005 from Proquest database).

Jones, J. M. (2004) Bush gets small Convention bounce, leads Kerry by seven. Gallup Poll, September 6. http://www.gallup.com/poll/content/login.aspx?ci=12922 (accessed September 13, 2004).

Jorgensen, B. (2002) Kids join the wireless revolution. *Electronic Business*, 28 (11): 29.

Juan, M. (2004) *11/M: La trama completa*. Barcelona: Ediciones de la Tempestad.

Jyothi, K. (2003) SMS divorces. Women's Feature Service, New Delhi, June 2. http://www.proquest.umi.com (accessed July 23, 2005).

Kageyama, Y. (2003) Camera-equipped phones spread mischief. *Associated Press*, July 9.

Kaihla, P. (2001) The Philippines' other revolution. Business2.com, May 8. http://www.business2.com/b2/web/articles/0,17863,513551,00.html (accessed June 4, 2004).

Kasesniemi, E-L. (2003) *Mobile Messages: Young People and a New Communication Culture*. Tampere: Tampere University Press.

———— and Rautiainen, P. (2002) Mobile culture of children and teenagers in Finland. In J. E. Katz and M. Aakhus (eds), *Perpetual Contact: Mobile Communication, Private Talk, Public Performance*, pp. 170–192. Cambridge: Cambridge University Press.

————, Ahonen, A., Kymäläinen, T., and Virtanen, T. (2003) Elävän mobiilikuvan ensi tallenteet: käyttäjien kokemuksia videoviestinnästä [Moving pictures: user experiences with video messaging]. Espoo: VTT Tiedotteinta, Research notes 2204.

Katz, J. E. (1996) The social consequences of wireless communications. In Institute for Information Studies (ed.), *The Emerging World of Wireless Communications*, pp. 91–119. Nashville, TN: Institute for Information Studies.

———— (1998) The social side of information networking (electronic version). *Society*, 35 (2): 402.

———— (1999) *Connections: Social and Cultural Studies of the Telephone in American Life*. New Bunswick, NJ: Transaction.

———— (2003a) Do machines become us? In J. E. Katz (ed.), *Machines that Become Us*, pp. 15–25. New Brunswick, NJ: Transaction.

———— (2003b) Bodies, machines, and communication contexts: what is to become of us? In J. E. Katz (ed.), *Machines that Become Us*, pp. 311–319. New Brunswick, NJ: Transaction.

—— (2004) A nation of ghosts? Choreography of mobile communication in public spaces. In K. Nyiri (ed.), *Mobile Democracy: Essays on Society, Self and Politics*, pp. 21–31. Vienna: Passagen Verlag. http://www.scils.rutgers.edu/ci/cmcs/publications/articles/nation%20of%20ghosts.pdf (accessed July 20, 2005).

—— and Aakhus, M. (2002) Introduction: framing the issues. In J. E. Katz and M. Aakhus (eds), *Perpetual Contact: Mobile Communication, Private Talk, Public Performance*, pp. 1–13. Cambridge: Cambridge University Press.

—— and Rice, R. E. (2003) *Social Consequences of Internet Use: Access, Involvement, and Interaction*. Cambridge, MA: MIT Press.

—— and Sugiyama, S. (2005) Mobile phones as fashion statements: the co-creation of mobile communication's public meaning. In R. Ling and P. Pedersen (eds), *Mobile Communications: Re-negotiation of the Social Sphere*, pp. 63–81. Surrey, UK: Springer.

——, Aakhus, M., Kim, H. D., and Turner, M. (2003) Cross-cultural comparison of ICTs. In L. Fortunati, J. E. Katz, and R. Riccini (eds), *Mediating the Human Body*, pp. 75–83. Mahwah, NJ: Lawrence Erlbaum.

Kauffman, R. J. and Techatassanasoontorn, A. A. (2005) Is there a global digital divide for digital wireless phone technologies? Paper presented at the 2004 Research Symposium on the Digital Divide, Carlson School of Management, University of Minnesota, August, pp. 27–28.

Kellerman, A. (1999) Sociospatial aspects of telecommunications: an overview. In H. Sawhney and G. A. Barnett (eds), *Progress in Communication Sciences, 15*: *Advances in Telecommunication Research*, pp. 217–232. Norwood, NJ: Ablex.

Kelly, T. (2001) Economic aspects of fixed-to-mobile interconnection. ITU Workshop on Interconnection, Sanya City, August 17–19, 2001. http://www.itu.int/osg/spu/presentations/2001/18%20Aug,%20FMI.PPT (accessed August 8, 2005).

——, Minges, M., and Gray, V. (2002) World telecommunication development report: reinventing telecoms. Executive summary. http://www.itu.int.

Kettmann, S. and Kettman, S. (2001) Sermon on the mobile. *Wired News*, May 3. http://www.wired.com/news/wireless/0,1382,43492,00.html (accessed February 15, 2006).

Kibati, M. and Krairit, D. D. (1999) The wireless local loop in developing regions. *Communications of the ACM*, 42 (6): 60–66.

Kim, D-Y. (2002) The politics of market liberalization: a comparative study of the South Korean and Philippine telecommunications service industries. *Contemporary Southeast Asia*, 24 (2): 337–370.

Kim, H-J., Byun, S-K., and Park, M-C. (2004) Mobile handset subsidy policy in Korea: historical analysis and evaluation. *Telecommunications Policy*, 28: 23–42.

Kim, J-M. (2001) Caught in a political Net. *Far Eastern Economic Review*, November 1: 49–50.

Kim, K-D. (1993) *Han'guk sahoe pyondongnon* [Korean Social Change]. Seoul: Nanam.

Kim, M-K., Park, M-C., and Jeong, D-H. (2004) The effects of customer satisfaction and switching barrier on customer loyalty in Korean mobile telecommunication services. *Telecommunications Policy*, 28: 145–159.

Kim, S-D. (2002) Korea: personal meanings. In J. E. Katz and M. Aakhus (eds), *Perpetual Contact: Mobile Communication, Private Talk, Public Performance*, pp. 63–79. Cambridge: Cambridge University Press.

———— (n.d.) President of cyberspace. *Netpolitique*. http://www.netpolitique.net/php/articles/kimsd_art.php3 (accessed May 21, 2004).

King, J. L. and West, J. (2002) Ma Bell's orphan: US cellular telephony, 1947–1996. *Telecommunications Policy*, 26 (3–4): 189–203.

KISDI Report (2003) *Analyses of Consumer Expenditures for and Using Patterns of IT Commodities and Services* (in Korean). Kyunggido: Korea Information Strategy Development Institute.

Kodama, M. (2002) Transforming an old economy company into a new economy success: the case of NTT DoCoMo. *Leadership and Organization Development Journal*, 23 (1/2): 26–39.

———— (2003) Strategic community-based theory of firms: case study of NTT DoCoMo. *Journal of High Technology Management Research*, 14 (2): 307–330.

Kogawa, T. (n.d.) Beyond electronic individualism. http://anarchy.k2.tku.ac.jp/non-japanese/electro.html (accessed April 6, 2004)

Konkka, K. (2003) Indian needs: cultural end-user research in Mombai. In C. Lindhom, T. Keinonen, and H. Kiljander (eds), *Mobile Usability: How Nokia Changed the Face of the Mobile Phone*, pp. 97–112. New York: McGraw-Hill.

Korea Herald (2003) Roh's support group decides not to disband, January 20.

Korea Times (2002) Victory for "Nosamo," makers of president, December 20.

———— (2003) Nosamo opposes assistance to Iraq War, March 24.

Kuo, K. (2003) Little Smart "cell" phone very, very smart in China. *Asia Times*. http://www.atimes.com (accessed May 5, 2004).

Kurvinen, E. (2003) Only when Miss Universe snatches me: teasing in MMS messaging. *Proceedings of the 2003 International Conference on Designing Pleasurable Products and Interfaces*, June 23–26, pp. 98–102. Pittsburgh, Pennsylvania.

Kwon, H. S. and Chidambaram, L. (2000) A test of the technology acceptance model: the case of cellular telephone adoption. *Proceedings of the 33rd Hawaii International Conference on System Sciences*, 1 (1): 1023. Washington, DC: IEEE Computer Society.

http://doi.ieeecomputersociety.org/10.1109/HICSS.2000.926607 (accessed February 20, 2005).

Lacohée, H., Wakeford, N., and Pearson, I. (2003) A social history of the mobile telephone with a view of its future. *BT Technology Journal*, 21 (3): 203–211.

Lahey, L. (2003) Getting the text message? Cross-border, inter-carrier text messaging launches in North America without much fanfare. *Computing Canada*, 29 (5): 18.

Larimer, T. (2000) What makes DoCoMo go? *Time Asia* (November 27) 156, no. 21. http://www.time.com/time/asia/magazine/2000/1127/telecom.docomo.html (accessed February 14, 2004).

Lasen, A. (2002a) A comparative study of the mobile phone: use in public places in London, Madrid and Paris. http://www.surrey.ac.uk/dwrc/Publications/CompStudy.pdf (accessed February 2004).

——— (2002b) The social shaping of fixed and mobile networks: a historical comparison. http://www.surrey.ac.uk/dwrc/Publications/HistComp.pdf (accessed March 20, 2004).

Latour, B. and Woolgar, S. (1979) *Laboratory Life: The Social Construction of Scientific Facts*. Beverly Hills, CA: Sage.

Laurier, E. (2002) The region as a socio-technical accomplishment of mobile workers. In B. Brown, N. Green, and R. Harper (eds), *Wireless World: Social and Interactional Aspects of the Mobile Age*, pp. 46–61. London: Springer.

Lebkowsky, J. (2004) More on SMS-blocking during the RNC. *Smartmobs*, 4 September. www.smartmobs.com/archive/2004/09/04/more_on_smsblo.html (accessed September 13, 2004).

Lee, J. (2002a) I think, therefore IM. *New York Times*, September 19. http://www.nytimes.com/learning/teachers/featured_articles/20020919thursday.html (accessed August 17, 2004).

——— (2002b) Tailoring cellphones for teenagers. *New York Times*, May 30. http://www.teenresearch.com/NewsView.cfm?page_id=127 (accessed February 18, 2006).

Lee, Y., Lee, I., Kim, J., and Kim, H. (2002) A cross-cultural study on the value structure of mobile internet usage: comparison between Korea and Japan. *Journal of Electronic Commerce Research*, 3 (4): 227–239.

Lee, Y-Y., Park, Y-T., and Oh, H-S. (2000) The impact of competition on the efficiency of public enterprise: the case of Korea Telecom. *Asia Pacific Journal of Management*, 17: 423–442.

Lei, Z. (2002) Ordinary beginning reveals three problems in the telecom industry. *Telecom Information Daily*, March 6. http://tech.sina.com.cn/it/t/2002–03–06/105429.shtml (accessed June 30, 2004).

Lemish, D. and Cohen, A. A. (2005a) Mobiles in the family: parents, children and the third person effect. http://www.fil.hu/mobil/2005/Lemish-Cohen.pdf (abstract accessed June 14, 2005).

───── (2005b) On the gendered nature of mobile phone culture in Israel. *Sex Roles: A Journal of Research*, 52 (7/8): 511–521.

Len, S. (2004) President's impeachment stirs angry protests in South Korea. *New York Times*, March 13: A2.

Leonardi, P. M. (2003) Problematizing "new media": culturally based perceptions of cell phones, computers, and the Internet among United States Latinos. *Critical Studies in Media Communication*, 20 (2): 160–179.

Lewis, W. (n.d.) Making the mobile channel work for your youth brand. http://www.mobileyouth.org/news/mobileyouth590.html (accessed July 7, 2004).

Licoppe, C. and Heurtin J-P. (2002) France: preserving the image. In J. E. Katz and M. Aakhus (eds), *Perpetual Contact: Mobile Communication, Private Talk, Public Performance*, pp. 94–109. Cambridge: Cambridge University Press.

Lightman, A. and Rojas, W. (2002) *Brave New Unwired World: The Digital Big Bang and the Infinite Internet*. New York: John Wiley.

Lin, A. (2005) Romance and sexual ideologies in SMS manuals circulating among migrant workers in Southern China. *Proceedings of the International Conference on Mobile Communication and Asian Modernities*, Hong Kong, June 7–8, pp. 141–156.

Lindgren, M., Jedbratt, J., and Svensson, E. (2002) *Beyond Mobile: People, Communications and Marketing in a Mobilized World*. London: Palgrave.

Ling, R. (1999a) I am happiest by having the best: the adoption and rejection of mobile telephony. R&D Report 15/99. Kjeller, Norway: Telenor.

───── (1999b) "We release them little by little": maturation and gender identity as seen in the use of mobile telephone. International Symposium on Technology and Society (ISTAS '99), Women and Technology: Historical, Societal and Professional Perspectives, July 29–31. Rutgers University, New Brunswick, New Jersey.

───── (2000) "We will be reached": the use of mobile telephony among Norwegian youth. *Information Technology and People*, 13 (2): 102–120.

───── (2001) "It is 'in.' It doesn't matter if you need it or not, just say that you have it": fashion and the domestication of the mobile telephone among teens in Norway. In L. Fortunati (ed.), *Il corpo umano tra tecnologie, comunicazione e moda*. Milan: Triennale di Milano.

───── (2002) Adolescent girls and young adult men: two sub-cultures of the mobile telephone. *Revista de Estudios de Juventud*, 52: 33–46.

───── (2004) *The Mobile Connection: The Cell Phone's Impact on Society*. San Francisco, CA: Morgan Kaufmann.

———— and Haddon, L. (2001) Mobile telephony, mobility and the coordination of everyday life. Paper presented at the Machines that Become Us Conference, Rutgers University, April 18–19. http://www.telenor.no/fou/program/nomadiske/articles/rich/(2001)Mobile.pdf (accessed March 2004).

———— and Helmersen, P. (2000) "It must be necessary, it has to cover a need": the adoption of mobile telephony among pre-adolescents and adolescents. Paper presented at the Conference on the Social Consequences of Mobile Telephony, Oslo, Norway, June 16.

———— and Yttri, B. (2002) Hyper-coordination via mobile phones in Norway. In J. E. Katz and M. Aakhus (eds), *Perpetual Contact: Mobile Communication, Private Talk, Public Performance*, pp. 139–169. Cambridge: Cambridge University Press.

Lipp, J. (2003) m-Commerce in Korea. *MFC Insight*. April 18.

Littlechild, S. C. (2004) Mobile termination charges: calling party pays vs receiving party pays. Cambridge Working Papers in Economics 0462. http://www.econ.cam.ac.uk/dae/repec/cam/pdf/cwpe0426.pdf (accessed July 21, 2005).

Liu, G. (2005) Expressing identity and sexuality: comparison between the use of newspaper media and ICTs by Hong Kong sex workers. Paper presented to the International Conference on Mobile Communication and Asian Modernities, Hong Kong, June 7–8.

Liu, H. (2004) Research on market dynamics of Little Smart in China. TeleInfo Institute, China Academy of Telecommunications Research of MII, March.

Liu, M. (2003) Little Smart handset market heats up. http://www.telecomasia.net/telecomasia/article/articleDetail.jsp?id=88183 (accessed May 5, 2004).

Livingstone, S., Van Couvering, E., and Thumin, N. (2004) Adult media literacy: a review of research literature. London: Ofcom. http://www.ofcom.org.uk/advice/media_literacy/medlitpub/medlitpubrss/aml (accessed July 19, 2005).

Lobato, E. (2004) Teles fixas perdem 2,200 clientes por dia. *Folha de São Paulo*, Folha dinheiro, May 23: B1 and B4.

Lobet-Maris, C. and Henin, J. (2002) Talking without communicating or communicating without talking: from the GSM to the SMS. *Revista de Estudios de Juventud*, 57: 101–114.

Long Chen (2002) I am a backstage manipulator of SMS culture [wojiushi duanxinwenhua de muhouheishou]. *New Weekly* [xinzhoukan], Guangzhou, July 15.

Lopez, A. (1998) Watch out, landowners: an ex-communist is taking on agrarian reform. *Asia Week*, July 17. http://www.asiaweek.com/asiaweek/98/0717/nat_4_land.html (accessed June 10, 2004).

Lorente, S. (2002) Youth and mobile telephones: more than a fashion. *Revista de Estudios de Juventud*, 57: 9–24.

Lowman, S. (2005) Youth spend $100bn on phones, April 1. http://www.itweb.co.za/sections/telecoms/2005/0504011200.asp?A=MAW&S=Mobile%20and%20Wireless%20Technology&O=FPT (accessed September 27, 2005).

Luna, L. (2002) Party animals, supermodels, and naked knights. *Wireless Review*, August: 20–29. http://wirelessreview.com/marketingsales/wireless_party_animals_supermodels/.

Lundin, J. (2005) Coordination in mobile police work. Paper presented at the Hong Kong Mobility Roundtable, Hong Kong, June 1–3.

Lynch, G. (2000) US cellular: so cheap, so underappreciated. *America's Network*, 104 (11): 18.

Lyon, D. (1994) *The Electronic Eye: The Rise of Surveillance Society*. Cambridge: Polity Press.

—— (2001) *Surveillance Society: Monitoring Everyday Life*. Milton Keynes, Buckingham: Open University Press.

—— (2002) Editorial. Surveillance studies: understanding visibility, mobility and the phenetic fix. *Surveillance and Society*, 1 (1): 1–7. www.surveillance-and-society.org (accessed February 8, 2006).

—— (2004) Surveillance technologies: trends and social implication. In *The Security Economy*, pp. 127–148. Paris: OECD.

Lyytinen, K. and Fomin, V. A. (2002) Achieving high momentum in the evolution of wireless infrastructures: the battle over the 1G solutions. *Telecommunications Policy*, 26 (3–4): 149–190.

Ma, L. and Zhang, Y. (2005) How to improve the management of the SP market [*SP shichang ruhe zouxiang dazhi*]. *Southern Weekend* [*Nanfang zhoumo*], Guangzhou, China, July 28: A6.

McAteer, S. (2005) (X marks the spot): so Y are games targeted at men? *Wireless Review*, 22 (6): 26. http://www.wirelessreview.com.

McCahill, M. (2002) *The Surveillance Web: The Rise of CCTV in an English City*. Collompton: Willan.

McCartney, N. (2004) Cultural ties shape consumer choices. *Variety*, October 4–10.

MacDermot, F. (2005) Mobile communications in Latin America. In *Latin America: Regional Overview, Country Forecast*, pp. 16–24. The Economist Intelligence Unit.

McDonald, M. (2000) Cell phones are hot with teens. *Detroit News*, June 5. http://www.detnews.com/2000/features/0006/05/b01-68451.htm (accessed June 30, 2004).

McDowell, S. D. and Lee, J. (2003) India's experiments in mobile licensing. *Telecommunications Policy*, 27: 371–382.

McGinity, M. (2004) Weaving a wireless safety net: staying connected. *Communications of the ACM*, 47 (9): 15–18.

McGray, D. (2002) Japan's Gross National Cool. *Foreign Policy*, May/June: 44–54.

Mack, A. M. (2003) Verizon seeks new way to woo teens. *Adweek*, October 13: 13.

McKemey, K., Scott, N., Souter, D., Afullo, T., Kibombo, R., and Sakyi-Dawson, O. (2003) Innovative demand models for telecommunications services. Final Technical Report (R8069). http://www.telafrica.org/.

McVeigh, B. J. (2003) Individualization, individuality, interiority, and the Internet: Japanese university students and e-mail. In N. Gottlieb and M. McLelland (eds), *Japanese Cybercultures*, pp. 19–33. New York: Routledge.

Maddox, B. (2005) Not so far from heaven. *New Statesman*, March 28, 11–12.

Maddox, K. (2003) Report predicts surge in WiFi use. *BtoB Online*, December 8. http://www.btobonline.com/cgi-bin/article.pl?id=11989 (accessed September 5, 2004).

Mader, B. (2003) Prepaid wireless phones a hit in youth market. *Business First of Columbus*, March 3. http://columbus.bizjournals.com/columbus/stories/2003/03/03/focus3.html (accessed June 26, 2004).

Mahan, A. (2005) Prepaid mobile and network extension. In A. K. Mahan and W. H. Melody (eds), *Stimulating Investment in Network Development: Roles for Regulators*, pp. 63–76. Denmark: World Dialogue on Regulation for Network Economics.

Makhaya, G. and Roberts, S. (2003) Telecommunications in developing countries: reflections from the South African experience. *Telecommunications Policy*, 27: 41–59.

Mann, S., Nolan, J., and Wellman, B. (2003) Sousveillance: inventing and using wearable computing devices for data collection in surveillance environments. *Surveillance and Society*, 1 (3): 331–355.

Mann, W. C., Helal, S., Davenport, R. D., Justiss, M. D., Tomita, M. R., and Kemp, B. J. (2004) Use of cell phones by elders with impairments: overall appraisal, satisfaction, and suggestions. *Technology and Disability*, 16 (1): 49–57.

Mante, E. (2002) The Netherlands and the USA compared. In J. E. Katz and M. Aakhus (eds), *Perpetual Contact: Mobile Communication, Private Talk, Public Performance*, pp. 110–125. Cambridge: Cambridge University Press.

——— and Piris, D. (2002) SMS use by young people in the Netherlands. *Revista de Estudios de Juventud*, 52: 47–58.

Marek, S. (2004) Raising the bar on ringtones. *Wireless Week*, May 15. http://www.wirelessweek.com (accessed August 14, 2004).

Mariscal, J. (2004) Regulation of information technologies and inequality. Paper presented at the Regional Workshop on ICT Regulation and Equity, Institute for Connectivity in the Americas (ICA), Montevideo, November 11–12. http://www.regulateonline.org/content/view/289/32/ (accessed March 12, 2005).

────── and Rivera, E. (2005) New trends in mobile communications in Latin America. Research report prepared for the International Workshop on Wireless Communication and Development: A Global Perspective, Marina del Rey, California, October 7–8, Annenberg Research Network on International Communication. http://annenberg.usc.edu/international_communication/workshop05/Mariscal_MobileTrendsinLA_Sep05.pdf (accessed October 13, 2005).

Marketing Week (1998) Spotlight: mobiles still a male enclave. 21 (4): 42.

Marsh, V. (2004) Probado con éxito el primer vehículo europeo sin conductor ni volante. *La Flecha*. http://www.laflecha.net/canales/ciencia/200406161/ (accessed September 27, 2005).

Marth, J. J. (2004) Supercharge, December 7. http://www.cityweekend.com.cn/en/beijing/features/2002_21/Zeitgeist_Shouji (accessed July 22, 2005).

Massey, D. (1993) Power-geometry and a progressive sense of place. In L. Ticker (ed.), *Mapping the Futures: Local Cultures, Global Change*, pp. 59–69. New York: Routledge.

Matambalya, F. and Wolf, S. (2001) The role of ICT for the performance of SMEs in East Africa: empirical evidence from Kenya and Tanzania. ZEF Discussion Papers on Development Policy, no. 42. Center for Development Research (ZEF), Bonn. http://131.220.109.9/index.php?id=192 (accessed March 2, 2005).

Medford, C. (2001) The power of mobility: wireless web access revs up productivity and efficiency at early adopters. *Network World* (electronic version). Expanded Academic ASAP database (accessed May 20, 2004).

Media Awareness (2005) Are you web aware? Text messaging. http://www.media-awareness.ca/ (accessed June 28, 2005).

Meyer, D. (2002) Carriers target youth market via prepaid, unlimited local plans. (electronic version). *RCR Wireless News*. http://www.infotrac.com.

Meyrowitz, J. (2004a) Global nomads in the digital veldt. In K. Nyíri (ed.), *Mobile Democracy: Essays on Society, Self and Politics*, pp. 73–90. Vienna: Passagen Verlag. http://21st.century.phil-inst.hu/Passagen_engl3_Meyrowitz.pdf (accessed July 20, 2005).

────── (2004b) The rise of glocality: new senses of place and identity in the global village. Keynote address at the Conference on The Global and the Local in Mobile Communication: Places, Images, People, Connections, Budapest, Hungary, June 10–12. http://www.fil.hu/mobil/2004/meyrowitz_webversion.doc (accessed July 20, 2005).

MGAIN (2003a) Mobile entertainment industry and culture; IST-2001-38846; D3.2.1.

────── (2003b) Mobile entertainment industry and culture; IST-2001-38846; D4.1.1.

MIC Report (2003) *Annual Report of Electronics and Telecommunications in 2003* (in Korean). Seoul, Korea: Ministry of Information and Communication.

Middle East and Africa Wireless Analyst (2004) Operators seek to cut service providers out of value chain (May 12): 2 (9): 3. http://demo.telecoms.com/pdf/May-2004/24/meawa051204.pdf (accessed February 21, 2005).

Miller, V. (2003) Parking violators can get help. *Iowa City Press-Citizen*, October 28. http://www.press-citizen.com/news/102803parking.html (accessed April 13, 2004).

Min, G. and Yan, X. (2005) The gender's influence on wireless Internet access technology acceptance. Paper presented at the Hong Kong Mobility Roundtable, Hong Kong, June 1–3.

Min, W. (2000) Telecommunications regulations: institutional structures and responsibilities, p. 42. OECD DSTI/ICCP/TISP(99)15/FINAL. http://www.oecd.org/dataoecd/39/32/21330624.pdf (accessed March 8, 2005).

Minges, A. (1999) Mobile cellular communications in the Southern African region. *Telecommunications Policy*, 23: 585–593.

Minges, M. (2003) ICT and gender. Paper presented at the 3rd World Telecommunication/ICT Indicators Meeting, Geneva, January 15–17.

―――― (2005) Is the Internet mobile? Measurements from the Asia-Pacific region. *Telecommunications Policy*, 29: 113–125.

――――, Magpantay, E., Firth, L., and Kelly, T. (2002) Pinoy Internet: Philippines case study. http://www.itu.int.

Ministry of Information Industry (2001) *Annual Statistical Report on the Development of Telecommunications in China*. www.mii.gov.cn.

Mitchell, W. (2003) *Me++: The Cyborg Self and the Networked City*. Cambridge, MA: MIT Press.

Mobile Africa (2005a) ONEtouch introduces ONE4ALL, May 19. http://www.mobileafrica.net/news-africa.php?id=272 (accessed July 26, 2005).

―――― (2005b) GSM desktop phones, July 3. http://www.mobileafrica.net/articles-africa.php?id=17 (accessed July 26, 2005).

―――― (2005c) Mobile phone card currency in Kenya, July 5. http://www.mobileafrica.net/news-africa.php?id=492 (accessed July 26, 2005).

Mobile Data Association (2005) May makes way for all time top text total. Press release. http://www.mda-mobiledata.org/MDA/documents/May05SMSPressRelease.pdf (accessed August 7, 2005).

Mobile Entertainment Forum (2003) MEF white paper on future mobile entertainment scenarios. http://www.m-e-f.org/pdf/MEF-WP-on-Future-ME-Scenarios.pdf (accessed February 3, 2006).

Mobile Streams Ltd (2002) Next messaging: an introduction to SMS, EMS and MMS. http://www.pucp.edu.pe/fac/cing/telecom/6_catedra/messaging.pdf (accessed July 2004).

Mobile Village (2003) Poll: those who don't own mobile phones cite cost barriers. http://www.mobilevillage.com/news/2003.05.09/cellphone-poll.htm (accessed June 6, 2004).

Monge, P. and Contractor, N. (2003) *Theories of Communication Networks*. New York: Oxford University Press.

Moni, M. H. and Ansar, U. M. (2004) Cellular phones for women's empowerment in rural Bangladesh. *Asian Journal of Women's Studies*, 10 (1): 70.

Monitoring (2004) ICT in private life: preliminary survey (in Russian). Department of Administration and Information of the City of Moscow. http://monitoring.iis.ru/groups/16/Document.2004-04-19.6897465810 (accessed April 12, 2005).

Moore, K. (2003) ME and end user practices. MGAIN: Mobile Entertainment Industry and Culture, IST-2001-38846, D3.2.1, pp. 62–86. http://www.mgain.org/mgain-wp3-D321-delivered.pdf (accessed February 14, 2006).

—— and Rutter, J. (eds) (2004) *Proceedings of Mobile Entertainment: User-centred Perspectives*. Manchester: ESRC Centre for Research on Innovation and Competition.

Moschella, D. (1999) For wireless, US culture drives us down our own path. *Computer World*, February 1: 33.

Motsay, E. (2003) Black jack rings up revenues: notably, carriers must target the youth and young adult markets in order to see positive revenue results from mobile gaming (electronic version). *RCR Wireless* (January 6), 22 (1): 10.

MTN (2005) MTN villagePhone. http://www.mtn.co.ug/payphone/villagephone.htm

Mueller, M. and Tan, A. Z. (1997) *China in the Information Age: Telecommunications and the Dilemmas of Reform*. Westport, CT: Praeger.

Mureithi, M. (2003) Self-destructive competition in cellular: regulatory options to harness the benefits of liberalization. *Telecommunications Policy* 27: 11–19.

Murtagh, G. M. (2002) Seeing the "rules": preliminary observations of action, interaction and mobile phone use. In B. Brown, N. Green, and R. Harper (eds), *Wireless World: Social and Interactional Aspects of the Mobile Age*, pp. 81–91. New York: Springer.

Mutula, S. (2002) The cellular phone economy in the SADC region: implications for libraries. *Online Information Review*, 26 (2): 79–91.

Nagamine, Y. (2001) Isolation fears lead to phone addiction. *Daily Yomiuri*, July 13: 6.

National Telecommunications and Information Administration (NTIA) (2002) A nation online: how Americans are expanding their use of the Internet. http://www.ntia.doc.gov (accessed September 7, 2004).

NECG (2004) The diffusion of mobile telephony in Latin America: successes and regulatory challenges. Network Economics Consulting Group (NECG) document prepared for Bellsouth International. Final report. http://www.ustr.gov/assets/ Trade_Sectors/Services/Telecom/Section_1377/2005_Reply_Comments_on_Review_ of_Compliance_with_Telecom_Trade_Agreements/BellSouth/asset_upload_file247_7 125.pdf.

Netsize Group (2005) *The Netsize Guide.* Paris: Netsize Group.

New Media Age (2004) Making money out of mobile: states of play, April 15: S7.

New Media Institute (2004) WiFi clouds and zones: a survey of municipal wireless initiatives. http://www.nmi.uga.edu/research/WiFiCloudsZones-8-10-04.pdf (accessed August 30, 2004).

New Weekly (2002) Special issue on the "Thumb Tribes" of China, July 15.

NewsCanada Online (2003) Whr R U ¿? Text messaging made simple. http://www.newscanada.com.

News.com (2002) Text messaging tests China's freedom, March 5. http://www. monitor.ca/monitor/issues/vol9iss8/netbytes.html (accessed February 18, 2006).

Newsweek (2004) The wireless world, June 7. http://www.msnbc.msn.com/id/ 5092843/site/newsweek/).

Noguchi, Y. (2004) Bringing memory to market: cell phone makers have high hopes of high-definition. *Washington Post*, March 23: E05. http://www.washingtonpost. com (accessed June 10, 2004).

Nokia (2002) Nokia and Cellular One of northeast Arizona sign agreement for GSM/EDGE network delivery. Press release, December 23. http://press.nokia.com/ PR/200212/886670_5.html (accessed July 25, 2005).

NOP Technology (2003) 2003 Wireless LAN benefits study. www.unstrung.com (accessed May 21, 2004).

Nordli, H. and Sørensen, K. H. (2003) Diffusion as inclusion? How adult men and women become users of mobile phones. SIGIS Deliverable Number: DO5. Appendix 2.15, NTNU4, IST-2000-26329SIGIS. http://www.rcss.ed.ac.uk/sigis/public/ displaydoc/full/D05_2.15_NTNU4 (accessed July 12, 2005).

Norris, C. and Armstrong, G. (1999) *The Maximum Surveillance Society: The Rise of CCTV.* Oxford: Berg.

Noticiaswire (n.d.) Teens and the cell phone connection: Verizon wireless offers helpful tips for parents. http://www.noticiaswire.com/consumer_news/verizon/ (accessed June 6, 2004).

NTIA (1999) Falling through the net: defining the digital divide. http://www.ntia. doc.gov/ntiahome/digitaldivide/.

——— (2000) Falling through the net: toward digital inclusion. http://www. ntia.doc.gov/ntiahome/digitaldivide/.

NTT Docomo (2003) NTT Docomo Group subscriber. http://www.nttdocomo.com/files/presscenter/33_subscriber_0402.pdf (accessed March 30, 2004).

O'Connell, B. (1999) Gender differences top new mobile phone trends. *Newsbytes*, August 11.

Öczan, Y. Z. and Koçak, A. (2003) Research note. A need or status symbol? Use of cellular telephony in Turkey. *European Journal of Communication*, 18 (2): 241–254.

OECD (2000) Cellular mobile pricing structures and trends. http://www.oecd.org/dataoecd/54/42/2538118.pdf (accessed February 21, 2005).

——— (2003) Mobile phones: pricing structures and trends. http://www.sourceoecd.org (accessed May 14, 2004).

Oestmann, S. (2003) *Mobile Operators: Their Contribution to Universal Service and Public Access*. Vancouver: Intelecon Research and Consultancy.

Office for National Statistics (2000) *Drug Use, Smoking and Drinking among Teenagers in 1999*. London: ONS.

Oftel (2003a) *Consumers' Use of Fixed and Mobile Telephony*, Q13 May. London: Oftel.

——— (2003b) Telecoms usage amongst low income groups, and identification of any issues specifically related to ethnicity. London: Oftel Qualitative Research, March 2003. http://www.ofcom.org.uk/static/archive/oftel/publications/research/2003/ethnicity0403.pdf (accessed June 16, 2005).

O'Hara, K., Perry, M., Sellen, A., and Brown, B. (2002) Exploring the relationship between the mobile phone and document activity during business travel. In B. Brown, N. Green, and R. Harper (eds), *Wireless World: Social and Interactional Aspects of the Mobile Age*, pp. 180–194. London: Springer.

Okabe, D. and Ito, M. (2003) Camera phones changing the definition of picture-worthy. *Japan Media Review*. www.ojr.org/japan/wireless/1062208524p.php (accessed April, 5, 2004).

Oksman, V. and Rautiainen, P. (2002) I've got my whole life in my hand. *Revista de Estudios de Juventud*, 52: 25–32.

O'Neill, P. D. (2003) The "poor man's mobile telephone": access versus possession to control the information gap in India. *Contemporary South Asia*, 12: 85–102.

Pabico, A. (n.d.) Hypertext revolution. *I Magazine*. http://www.pldt.com/hypertext.htm (accessed June 5, 2004).

Pacheco, I. (2004) *11-M: La respuesta*. Madrid: Asociación Cultural Amigos del Arte Popular.

Paetsch, M. (1993) *Mobile Communications in the US and Europe: Regulation, Technology and Markets*. Norwood, MA: Artech House.

Palen, L. (2002) Mobile telephony in a connected life. *Communications of the ACM*, 45 (3): 78–82.

————, Salzman, M., and Youngs, E. (2000) Going wireless: behavior and practice of new mobile phone users. *Proceedings of the ACM Conference on Computer Supported Cooperative Work* (CSCW 2000), pp. 201–210. Philadelphia, PA. www.cs.colorado.edu/~palen/Papers/cscwPalen.pdf (accessed September 19, 2004).

Pamantalaang Mindanaw (2000) President for impeachment, anyone?, October 16. Mindanao Institute of Journalism. http://www.mindanaw.com/2000/10/16gloria.html (accessed June 10, 2004).

Panos (2004) Completing the revolution: the challenge of rural telephony in Africa. http://www.panos.org.uk/PDF/reports/Panos%20Report%20-%20Completing%20the%20Revolution.pdf (accessed February 13, 2005).

Park, H-Y. and Chang, S-G. (2004) Mobile network evolution toward IMT-2000 in Korea: a techno-economic analysis. *Telecommunications Policy*, 28: 177–196.

Parker, T. (2005) Making a mark in Asia. *Mobile Communications International*, 121, May 4.

Partal, V. and Otamendi, M. (2004) *11-M: El periodismo en crisi*. Barcelona: Edicions Ara Libres.

Passerini, R. (2004) ITU/ICTP Workshop on New Radiocommunication Technologies for ICT in Developing Countries (Africa Region), Trieste, Italy, May 17–21. http://wireless.ictp.trieste.it/ITU_workshop/lectures/passerini/Passerini_09.pdf (accessed November 17, 2004).

Pastore, M. (2001) Wireless aims for widespread appeal. *ClickZ Stats*, February 13, 2003. http://www.clickz.com/stats/markets/wireless/article.php/587701 (accessed August 16, 2004).

Paul Budde Communication Pty Ltd (2004) *Australia: Mobile Communications and Mobile Data*.

Paustian, C. (2002) Youth must be served. *Credit Card Management*, 15 (8): 50, 52–53.

Peha, J. M. (1999) Lessons from Haiti's Internet development. *Communications of the ACM*, 42 (6): 67–72.

Peifer, J. W. (2005) Mobile wireless technologies for rehabilitation and independence. *Journal of Rehabilitation Research and Development*, 42 (2): 7.

Pelofsky, J. (2004) Pace for wireless number switches quickens, September 3. http://www.reuters.com/newsArticle.jhtml?type=technologyNews&storyID=6149398 (accessed September 7, 2004).

Perlin, N. (2003) Operating at the edge of technology. *Technical Communication*, 50 (1): 13–16.

Perry, M., O'Hara, K., Sellen, A., Brown, B., and Harper, R. (2001) Dealing with mobility: understanding access anytime, anywhere. *ACM Transactions on Computer–Human Interaction*, 8 (4): 323–347.

Pertierra, R., Ugarte, E. F., Pingol, A., Hernandez, J., and Decanay, N. L. (2002) *TXT-ING Selves: Cellphones and Philippine Modernity*. Manila, the Philippines: De La Salle University Press.

Petroff, P. (2002) Teens want m-everything – but at what price? *Wireless Week*, April 22. http://www.wirelessweek.com (accessed June 6, 2004).

Pew Internet and American Life Project (2002a) The Internet goes to college: how students are living in the future with today's technology, September 15. http://www.pewinternet.org.

—— (2002b) October 2002 tracking survey by Princeton Survey Research Associates. http://www.pewinternet.org.

—— (2004a) Pew Internet Project Data Memo, May: 28 percent of American adults are wireless ready. http://www.pewinternet.org.

—— (2004b) Pew Internet Project Data Memo, April: 55 percent of adult internet users have broadband at home or work, p. 3. http://www.pewinternet.org.

Philippine Daily Inquirer (2001) Estrada suspends talks with MLF, January 7: 2.

Pigato, M. A. (2001) Information and communication technology, poverty and development in sub-Saharan Africa and South Asia. Africa Region Working Paper Series no. 20. Washington, DC: World Bank.

Plant, S. (n.d.) On the mobile: the effects of mobile phones on social and individual life. www.motorola.com/mot/doc/0/234_MotDoc.pdf (accessed May 20, 2004).

—— (2003a) Wireless and fancy free. *New Statesman*, special supplement: Our Mobile Future, September 15: x. http://www.newstatesman.com/pdf/nsmobilesupp 2003.pdf (accessed January 17, 2006).

—— (2003b) A world of difference. *New Statesman*, special supplement: Our Mobile Future, September 15: ix–x. http://www.newstatesman.com/pdf/ nsmobilesupp2003.pdf (accessed January 17, 2006).

PR Newswire (2004) Prime Minister Koizumi's e-mail magazine now in English, March 10. www.prnewswire.co.uk/cgi/news/release?id=118889 (accessed April 21, 2004).

—— (2005) China reaches 50 million PAS subscribers, April 18. http://www. prnewswire.com/gh/cnoc/comp/159138.html (accessed February 23, 2006).

Proenza, F. (2005) The road to broadband development in developing countries is through competition driven by wireless and VOIP. Research report prepared for the International Workshop on Wireless Communication and Development: A Global Perspective, Marina del Rey, California, October 7–8, Annenberg Research Network on International Communication. http://arnic.info/workshop05/Proenza_Wireless &VoIP_5Oct2005.pdf (accessed February 20, 2006).

Prometeus (2002) Mobile phone for the hearing impaired. *Prometeus Newsletter*, no. 16. http://www.prometeus.org/news/PROMETEUS_Newsletter16.pdf (accessed July 23, 2005).

Puro, J-P. (2002) Finland: a mobile culture. In J. E. Katz and M. Aakhus (eds), *Perpetual Contact: Mobile Communication, Private Talk, Public Performance*, pp. 19–29. Cambridge: Cambridge University Press.

Qiu, J. L. (2004) (Dis)connecting the Pearl River Delta: transformation of a regional telecommunication infrastructure, 1978–2003. Unpublished PhD thesis, University of Southern California.

Raczynski, D. and Serrano, C. (1999) *Nuevos y viejos problemas en la lucha contra la pobreza en Chile* [New and Old Problems in the Fight against Poverty in Chile]. Santiago.

Rae-Dupree, J. (2004) Anthropologist helps Intel see the world through customers' eyes. *Silicon Valley/San Jose Business Journal*, August 16. http://sanjose.bizjournals. com/sanjose/stories/2004/08/16/story5.html (accessed July 23, 2005).

Rafael, V. (2003) The cell phone and the crowd: messianic politics in the contemporary Philippines. *Popular Culture*, 15 (3): 399–425.

Rakow, L. (1992) *Gender on the Line: Women, the Telephone and Community Life*. Urbana: University of Illinois Press.

―――― and Navarro, V. (1993) Remote mothering and the parallel shift: women meet the cellular telephone. *Critical Studies in Mass Communication*, 10: 144–157.

Rambabu, G. (2004) Cellular subscriber base figures overestimated? *The Hindu Business Line*, January 2. http://www.thehindubusinessline.com/2004/01/03/stories/ 2004010302710100.htm (accessed March 4, 2005).

Reardon, M. (2004) Wireless gets workout at RNC. CNET News.com. http://news.com.com/Wireless+tech+gets+workout+at+RNC/2100-1033_3-5330792. html (accessed August 31, 2004).

Reck, J. and Wood, B. (2003) *What Works: Vodacom's Community Services Phone Shops*. Seattle: World Resources Institute.

Reinhardt, A., Bonnet, A., and Crockett, R. O. (2004) Can Nokia get the wow back? *Business Week*, May 31: 18–22.

Reuters (2001a) Japan PM's million-human e-mail. *Wired News*, June 14. www. wired.com/news/politics/0,1283,44528,00.html (accessed August 8, 2005).

―――― (2001b) Manila on alert after blasts. *The Australian*, World section, January 1: 9.

———— (2002) US wireless Internet users reach 10 mln – survey, August 28. http://wirelessreview.com/ar/telecom_us_wireless_internet/ (accessed September 2, 2004).

Revolution (2003) Comment: educating Americans on the benefits of SMS. December 1: 25.

Rhee, I-Y. (2003) The Korean election shows a shift in media power. *Nieman Reports*, 57 (1): 95–6.

Rheingold, H. (2002) *Smart Mobs: The Next Social Revolution*. Cambridge, MA: Perseus.

Rice, R. E. and Katz, J. E. (2003) Comparing internet and mobile phone usage: digital divides of usage, adoption, and dropouts. *Telecommunications Policy*, 27: 597–623.

Richardson, D., Ramirez, R., and Haq, M. (2000) Grameen Telecom's village phone programme in rural Bangladesh: a multi-media case study final report. http://www.devmedia.org/documents/finalreport%2Epdf (accessed December 21, 2004).

Richie, D. (2003) *The Image Factory: Fads and Fashions in Japan*. London: Reaktion Books.

Robbins, K. A. and Turner, M. A. (2002) United States: popular, pragmatic and problematic. In J. E. Katz and M. Aakhus (eds), *Perpetual Contact: Mobile Communication, Private Talk, Public Performance*, pp. 80–93. Cambridge: Cambridge University Press.

Robison, R. and Goodman, D. (1996) *New Rich in Asia: Mobile Phones, McDonalds and Middle-class Revolution*. London: Routledge.

Robles, R. (2005) Presidential tape now Manila's hottest ring tone. *South China Morning Post*, June 17: 11.

Rockhold, J. (2001) Cell phones are, like, cool, May 15 (electronic version). *Wireless Review*, 18: 28–30. http://www.wirelessreview.com (accessed June 4, 2004).

Rodgers, Z. (2004) Fans text to American Idol in record numbers. *ClickZ Internet*, June 4. http://ww.clickz.com/news/article.php/3364031 (accessed June 9, 2004).

Rodini, M., Ward, R. R., and Woroch, G. A. (2003) Going mobile: substitutability between fixed and mobile access. *Telecommunications Policy*, 27: 457–476.

Rodríguez, P. (2004) *11-M mentira de estado: Los tres días que acabaron con Aznar*. Barcelona: Ediciones B.

Rubin, J. (2004) Ruckus RNC 2004 Text Alerts: updates and information on demonstration activities during the Republican National Convention 2004 in New York City. http://www.joshrubin.com/coolhunting/archives/2004/09/rnc_text_alerts.html (accessed April 24, 2005).

Rule, J. B. (2002) From mass society to perpetual contact: models of communication technologies in social context. In J. E. Katz and M. Aakhus (eds), *Perpetual Contact: Mobile Communication, Private Talk, Public Performance*, pp. 242–254. Cambridge: Cambridge University Press.

Ryan, V. (2000) Youth movement: marketing wireless to the teen crowd. *Telephony*, June 26. http://telephonyonline.com/ar/telecom_youth_movement_marketing/ (accessed June 5, 2004).

Salmon, A. (2004) Parties rallying behind Internet in race for votes. *Washington Times*, April 11.

Salterio, L. (2001) Text power in Edsa 2001. *Philippine Daily Inquirer*, January 22: 25.

Sampedro Blanco, V. F. and Martínez Nicolàs, D. (2005) Primer voto: castigo politico y descrédito de los medios. In V. F. Sampedro Blanco (ed.), *13-M: Multitudes on Line*, pp. 24–62. Madrid: Ed. Catarata.

———, Alcalde, J., and Sadaba, I. (2005) El fin de la mentira prudente: colapso y aperture de la esfera pública. In V. F. Sampedro Blanco (ed.), *13-M: Multitudes on Line*, pp. 229–278. Madrid: Ed. Catarata.

Samuel, J., Shah, N., and Hadingham, W. (2005) Mobile communications in South Africa, Tanzania and Egypt: results from community and business surveys. In *Africa: The Impact of Mobile Phones*, pp. 44–52. Vodafone Policy Paper Series 2. http://www.vodafone.com/assets/files/en/AIMP_09032005.pdf.

Sánchez, E., Almazán, C., and Parada, A. (2001) *La telefonia mòbil i els seus efectes en la salut de la informació: informe tècnic*. Barcelona: Agència d'Avaluació de Tecnologia i Recerca Mèdiques.

Sanchez, J. M. (2005) Cronología. In V. F. Sampedro Blanco (ed.), *13-M: Multitudes on Line*, pp. 307–309. Madrid: Ed. Catarata.

Sandvig, C. (2003) Public Internet access for young children in the inner city: evidence to inform access subsidy and content regulation. *The Information Society*, 19: 171–183.

Sarker, S. and Wells, J. D. (2003) Understanding mobile handheld device use and adoption. *Communications of the ACM*, 46 (12): 35–40.

Sawa, K. (2000) Mobile phones silence chatty students, October 22. *Mainichi Daily News*, 12.

ScenarioDNA (2004) ScenarioDNA finds a morphing structure of change for Gen Y. Press release, March 31. http://www.scenariodna.com/newsandevents/033104.html (accessed May 20, 2004).

Schegloff, E. A. (2002) Beginnings in the telephone. In J. E. Katz and M. Aakhus (eds), *Perpetual Contact: Mobile Communication, Private Talk, Public Performance*, pp. 284–300. Cambridge: Cambridge University Press.

Schiano, D., Chen, C., Ginsberg, J. Gretarsdottir, U., Huddleston, M., and Isaacs, E. (2002) Teen use of messaging media. In *Extended Abstracts of CHI 2002*, pp. 594–595. http://home.comcast.net/~diane.schiano/CHI2002.short.ttalk.pdf (accessed June 30, 2004).

Schwartz, E. (2002) For the children. *Infoworld*, August 26: 28. www.infoworld.com.

Sciadas, G. (ed.) (2005) From the digital divide to digital opportunities: measuring infostates for development. http://www.orbicom.uqam.ca/projects/ddi2005/index_ict_opp.pdf (accessed January 17, 2005).

Scott, N., McKemey, K., and Batchelor, S. J. (n.d.) The use of telephones amongst the poor in Africa: gender implications. Paper submitted to *Gender, Technology and Development*. http://www.telafrica.org (accessed March 20, 2005).

Sefton-Green, J. (1998) Being young in the digital age. In J. Sefton-Green (ed.), *Digital Diversions: Youth Culture in the Age of Multimedia*, pp. 1–20. London: UCL Press.

Selian, A. (2004) Mobile phones and youth: a look at the US student market. Preliminary draft. ITU/MIC Workshop on Shaping the Future Mobile Information Society, Seoul, March 4–5. http://www.itu.int/osg/spu/ni/futuremobile/YouthPaper.pdf (accessed April 13, 2004).

Selingo, J. (2004) Hey kid, your backpack is ringing. *New York Times*, March 18: G1.

Sewell, C. (2002) Understand that you will never understand: Cingular Wireless' youth marketing strategy. *Wireless Review*, June 1. http://wirelessreview.com/mag/wireless_understand_understand/index.html (accessed February 21, 2004).

Shachtman, N. (2004) Political protesters hear call with text messaging. *Chicago Tribune*, August 28. http://www.chicagotribune.com/technology/chi-0408280053aug28,1,6383205.story?coll=chi-technology-hed&ctrack=1&cset=true (accessed April 23, 2005).

Shanmugavelan, M. (2004) Mobile Africa must not leave its villages behind. *Contemporary Review*, July, 285: 33–34.

Sharma, C. and Nakamura, Y. (2003) *Wireless Data Services: Technologies, Business Models and Global Markets*. Cambridge: Cambridge University Press.

Sherry, J. and Salvador, T. (2002) Running and grimacing: the struggle for balance in mobile work. In B. Brown, N. Green, and R. Harper (eds), *Wireless World: Social and Interactional Aspects of the Mobile Age*, pp. 108–120. London: Springer.

Sidener, J. (2005) Gadget gap: new gender roles in digital world. *San Diego Union-Tribune*. http://www.signonsandiego.com/news/computing/personaltech/20050110-9999-mz1b10gap.html (accessed March 13, 2005).

Siew, A. (2003) Charge your phone in public. *Computer Times*, January 11. http://it.asia1.com.sg/newsdaily/news007_20030111.html (accessed July 22, 2005).

Silho (2004) Wireless marketing: know your audience. *The Insider*, 1 (11). http://www.theinsidernews.com/issue111 (accessed July 19, 2005).

Silva, J. (2003) Disability groups to push industry on access problems. http://www.hearingloss.org/html/fcc_push.html.

——— (2004) Verizon settles O'Day FCC complaint. Suit alleged failure to make products and services accessible, August 27. http://adawatch.org/VerizonODay.htm.

Simon, E. (2004) Protesters get "txt msgs" to join marches, avoid violence. *Associated Press*, September 1. http://pqasb.pqarchiver.com/ap/689698741.html?did=689698741&FMT=ABS&FMTS=FT&date=Sep+1%2C+2004&author=ELLEN+SIMON&desc=Protesters+get+%27txt+msgs%27+to+join+marches%2C+avoid+violence (accessed September 9, 2004).

Skeldon, P. (2003) Adult-to-mobile: personal services. www.juniperresearch.com (accessed July 2004).

Skog, B. (2002) Mobiles and the Norwegian teen: identity, gender and class. In J. E. Katz and M. Aakhus (eds), *Perpetual Contact: Mobile Communication, Private Talk, Public Performance*, pp. 255–273. Cambridge: Cambridge University Press.

Skuse, A. and Cousins, T. (2005) Managing distance: the social dynamics of rural telecommunications access and use in the Eastern Cape, South Africa. ISRG Working Paper no.1. http://www.isrg.info/ISRGWorkingPaper1.pdf (accessed October 10, 2005).

Slackman, M. (2004) If a protest is planned to a T, is it a protest? *New York Times*, August 22. http://query.nytimes.com/gst/abstract.html?res=F00D12FE3C5A0C718EDDA10894DC404482&incamp=archive:search (accessed April 23, 2005).

Slater, D. and Kwami, J. (2005) Embeddness and escape: Internet and mobile use as poverty reduction strategies in Ghana. ISRG Working Paper no. 4. http://www.isrg.info/ISRGWorkingPaper4.pdf (accessed October 10, 2005).

Slatina, S. (2005) Brand war. *Foreign Policy*, May/June, 148: 92.

Smith, B. (2004a) Measure of data's bottom-line effect. *Wireless Week*, May 1. http://www.wirelessweek.com/article/CA414479?text=measure+of&stt=001%C3%8A (accessed August 16, 2004).

——— (2004b) Personalization: bigger than games? *Wireless Week*, January 15. http://www.wirelessweek.com/article/CA374733 (accessed June 6, 2004).

Smith, H., Rogers, Y., and Brady, M. (2003) Managing one's social network: does age make a difference? In *Proceedings of INTERACT 2003*, pp. 551–558. www.cogs.susx.ac.uk/users/hilarys/papers/interact03.pdf.

Snapshots International (2005) Australia mobile phone services 2005. http://dx.doi.org/10.1337/au330855 (accessed April 2005).

Softpedia (2005) Russia: a new El Dorado for mobile telephony, March 25. http://news.softpedia.com/news/Russia-a-new-El-Dorado-for-mobile-telephony-790.shtml (accessed June 16, 2005).

Sohu-Horizon Survey (Sohu-lingdian diaocha) (2003) October. http://it.sohu.com/2004/02/19/96/article219129623.shtml (accessed June 30, 2004).

Spanish Minister of Internal Affaires (2005) Interior ha resuelto 1,510 expedientes e indemnizado a las víctimas con más de 44 millones un año después del 11-M. Press release, March 2. Madrid, Spain. www.mir.es/oris/notapres/year05/np030202. htm (accessed September 20, 2005).

Spanish Parliament (2004) Comisión de investigación sobre el 11 de Marzo. http://www.congreso.es/pdf/comision_investigacion/04/julio_04.htm (accessed September 2004).

Srivastava, L. (2004a) Shaping the future mobile information society: the case of Japan. ITU/MIC Workshop on Shaping the Future Mobile Information Society, Seoul, March 4–5. Document: SMIS/06. http://www.itu.int.

—— (2004b) Social and human considerations for a more mobile world. Background paper. ITU/MIC Workshop on Shaping the Future Mobile Information Society, Seoul, March 4–5. Document: SMIS/04. http://www.itu.int.

—— and Sinha, S. (2001) TP case study: fixed–mobile interconnection in India. *Telecommunications Policy*, 25: 21–38.

Standard and Poor's (2004) Industry surveys: telecommunications – wireless. http://www.marshall.usc.edu/ctm/Executive%20Education%20files/AMPT/AMPT%20Materials/September%2011-17,%202004/Telecommunications_%20Wireless_9_04.pdf (accessed September 19, 2004).

Steinbock, D. (2003) *Wireless Horizon: Strategy and Competition in the Worldwide Mobile Market Place*. New York: AMACOM.

Stocker, T. (2000) The future at your fingertip, October 3. http://www.tkai.com/press/001004_Independent.htm (accessed April 2, 2004).

Strategis (2005) Telecommunications service in Canada: an industry overview. Section 4. http://strategis.ic.gc.ca/epic/internet/insmt-gst.nsf/en/sf07005e.html (accessed July 8, 2005).

Strøm, G. (2002) The telephone comes to a Filipino village. In J. E. Katz and M. Aakhus (eds), *Perpetual Contact: Mobile Communication, Private Talk, Public Performance*, pp. 274–283. Cambridge: Cambridge University Press.

Sulaiman, T. (2005) Girl's rape filmed by teenagers on mobile. *Times Online*, June 18. http://www.timesonline.co.uk/printFriendly/0,,1-2-1659528-2,00.html.

Sundgot, J. (2003) More mobile minutes in the US. http://www.infosyncworld.com/news/n/3571.html.

Sundqvist, S., Frank, L., and Puumalainen, K. (2005) The effects of country characteristics, cultural similarity and adoption timing on the diffusion of wireless communications. *Journal of Business Research*, 58 (1): 107–110.

Suzuki, A. (2004) Case study 45: Sagamihara, Kanagawa Prefecture (in Japanese). Nikkei BP Government Technology, June 21.

Swartz, N. (2004) Tagging toothpaste and toddlers. *Information Management Journal*, 38 (5): 22.

Syahreza, A. (2004) *Sex on the Phone: Sensasi, Fantasi, Rahasia*. Jakarta: Gagas Media.

Tamayo, G. (2003) Venezuela mini-case study 2003: short message service "convergence" interconnection in Venezuela. http://www.itu.int/ITU-D/treg/Case_Studies/Convergence/Venezuela.pdf (accessed February 15, 2006).

Tan, S. (2001) Wireless text gains appeal (electronic version). *Miami Herald*, June 26. www.upoc.com/corp/MiamiHerald.pdf (accessed June 6, 2004).

Technology Review (2003/4) Wireless for the disabled. 106 (10): 64.

Tedeschi, B. (2004) The ring tone business looks good to record companies – but a do-it-yourself program may cut the profits short. *New York Times*, February 23: C5.

Teenage Research Unlimited (2000) Teens serious in quest for fun. Press release, October 18. http://www.teenresearch.com (accessed June 9, 2004).

———— (2003) Family life line at hand. Press release, February 7. http://www.teenresearch.com (accessed June 9, 2004).

———— (2004) Teens spent $175 billion in 2003. Press release, January 9. http://www.teenresearch.com (accessed June 9, 2004).

Telecom CIDE (2005) Contribuciones sociales y económicas de la telefonía móvil en México según un análisis de las fases de maduración del Mercado. Mexico DF: Telefónica Movistar México.

Telecomworldwire (2004) Gender differences start to show in mobile phone report, June 2. M2 Communications Ltd. http:/www.m2.com. www.findarticles.com/p/articles/mi_m0ECZ/is_2004_June_2.

Terdiman, D. (2004) Text messages for critical mass. *Wired News*, August 12. www.wired.com/news/politics/0,1283,64536,00.html?tw=wn_tophead_3.

TGI Global (2004) TGI global barometer: TGI facts and figures from around the world. http://www.tgisurveys.com/news/Barometer/Barometer_MobilePhones.htm (accessed July 12, 2005).

Thomas, K. (2004) Fixed lines outnumber cellphones in most circles. *Hindu Business Line*, November 26. http://www.thehindubusinessline.com/2004/11/27/stories/2004112701150500.htm (accessed March 4, 2005).

Thompson, J. (2005) Mobile phones and ration cards: refugee resettlement and the Sudanese experience. Paper presented at the 3rd Annual Forced Migration Student Conference, Oxford Brookes University, Oxford, May 13–14. http://www.brookes.ac.uk/schools/planning/dfm/FMSC/P&A/thompson.pdf (accessed July 19, 2005).

TNS (2001a) New study reveals low user awareness of 3G wireless Internet in key European and North American markets, June 18. http://www.tns-global.com (accessed June 4, 2004).

——— (2001b) New survey indicates wireless web penetration highest among young affluent males, February 8. http://www.tns-global.com (accessed May 20, 2004).

——— (2002a) Report of 30-nation study indicates meeting and creating information needs of young and upscale consumers can be instrumental to growth of wireless Internet, May 23. http://www.tns-global.com/corporate/Doc/0/E0LL1FUNNGFK3F05C9NJULKP62/586.htm (accessed June 4, 2004).

——— (2002b) Wireless and Internet technology adoption by consumers around the world. www.tns-global.com (accessed March 20, 2004).

——— (2004) American kids shout, "I want to be a millionaire!" and stress importance of having lots of money, April 19. http://www.tns-global.com/corporate/Doc/0/DC88BMIUDNMKN1QP91MJ301S26/OK-KidsMoney_FINAL-+US.pdf (accessed June 4, 2004).

Tohtsoni, N. J. (2001) Catching up with modern times: federal program provides mobile phones to low-income residents. *Navajo Times*, 40 (32): A1. http://www/proquest.umi.com (accessed July 23, 2005).

Toral, J. (2003) State of wireless technologies in the Philippines. *Proceedings for Closing Gaps in the Digital Divide: Regional Conference on Digital GMS*, pp. 173–177. Bangkok, Thailand, February 26–28.

Torgersen, L., Hertzberg Kaare, B., Heim, J., Brandtzæg, P. B., and Endestad, T. (2005) A digital childhood. http://www.aeforum.org/aeforum.nsf/0/BC4EBC2C44893A708025703100489D63?OpenDocument.

Townsend, A. M. (2000) Life in the real-time city: mobile telephones and urban metabolism. *Journal of Urban Technology*, 7 (2): 85–104.

——— (2002) Mobile communications in the twenty-first century city. In B. Brown, N. Green, and R. Harper (eds), *Wireless World: Social and Interactional Aspects of the Mobile Age*, pp. 62–77. London: Springer.

Trujillo, M. (2003) Abbreviated phraseology: linguists divided on cell phone messages. *Associated Press*, March 7. http://www.upoc.com/corp/news/UpocAP_2.pdf (accessed May 21, 2004).

Tsang, A., Ka, T., Irving, H., Alaggia, R., Chau, S. B. Y., and Benjamin, M. (2003) Negotiating ethnic identity in Canada: the case of the "satellite children." *Youth and Society*, 34 (3): 359–384.

Tsuji, D. and Mikami, S. (2001) A preliminary student survey on the e-mail uses of mobile phones. Paper presented at JSICR, Tokyo, June.

Turchetti, D. (2003) SMS craze unlocks way to youth hearts but tread carefully. *Media*, March 7. http://www.21cms.com/Company/News/News_8 (accessed August 2004).

TWICE (2004) Wireless data use still limited in US. March 22, 19 (7): 14.

de Ugarte, D. (2004) *11M: Redes para ganar una Guerra*. Barcelona: Ed. Icaria.

Ulfelder, J. (2002) Upwardly mobile. *World Link*, 15 (2): 48–51.

UMTS Forum (2004) Benefits of mobile communications for the society. Mobile Market Evolution and Forecast, Report no. 36, June.

UNESCO (2005) Evolution of the digital divide: new report to be released in September, July 13. http://portal.unesco.org/ci/fr/ev.php-URL_ID=19392&URL_DO= DO_TOPIC&URL_SECTION=201.html (accessed July 14, 2005).

United Nations (2002) Towards a knowledge-based economy. Russian Federation: Country Readiness Assessment Report, p. 19. www.unece.org/operact/enterp/ documents/coverpagerussia.pdf (accessed 12 March, 2005).

—— (2003) *World Youth Report: The Global Situation of Young People*. United Nations, Sales no. E.03.IV.7.

Ureta, S. (2004) The immobile mobility: spatial mobility and mobile phone use among low-income families in Santiago, Chile. Paper presented at the 5th Wireless World Conference, University of Surrey, July 15–16.

US Department of Commerce (2002) A nation online: how Americans are expanding their use of the Internet. http://www.ntia.doc.gov/ntiahome/dn/ anationonline2.pdf (accessed May 21, 2004).

——, Bureau of the Census (1989) *Historical Statistics of the United States: From Colonial Times to 1970*. White Plains, NY: Kraus International Publications.

UTStarcom (2003) China telecom case study. www.utstarcom.com.

Uy-Tioco, C. A. S. (2003) The cell phone and Edsa 2: the role of communication technology in ousting a president. Paper presented to the 4th Critical Themes in Media Studies Conference, New School University, New York, October 11.

Valor, J. and Sieber, S. (2004) Uso y actitud de los jóvenes hacia internet y la telefonía móvil. e-business Center PricewaterhouseCoopers and IESE. http://www.iese. edu/es/files/5_9073.pdf (accessed February 14, 2006).

Van Impe, M. (2003) US youth discovers SMS at last. *Nordic Wireless Watch*, January 8. http://www.nordicwatch.com/wireless/story.html?story_id=2631 (accessed June 5, 2004).

Varbanov, V. (2002) Bulgaria: mobile phones as post-communist cultural icons. In J. E. Katz and M. Aakhus (eds), *Perpetual Contact: Mobile Communication, Private Talk, Public Performance*, pp. 126–136. Cambridge: Cambridge University Press.

Varney, D. (2003) British Asians and corporate citizenship. Institute of Directors, London, October 15. http://www.csmworld.org/public/balcc/nationalconf/ presentation/DavidVarney.doc (accessed July 23, 2005).

Varoudakis, A. and Rossotto, C. M. (2004) Regulatory reform and performance in telecommunications: unrealized potential in the MENA countries. *Telecommunications Policy*, 28: 59–78.

Vershinskaya, O. (2002) Mobile communication: use of mobile phones as a social phenomenon – the Russian experience. *Revista de Estudios de Juventud*, 52: 139–149.

Vilano, M. (2003) Wi-Fi is hot but users still warming to it, November 21. http://www.clickz.com/stats/markets/wireless/article.php/3112271.

Visiongain (2004) Bluetooth virus reveals the vulnerability of the mobile phone, August 3. http://newsweaver.co.uk/ewirelessnews/e_article000287464.cfm?x= b3mRlKJ,b2cD3NFD.

Vodacom South Africa (n.d.) Community services. http://www.vodacom.co.za/about/community_franchise.jsp.

Vos, E. (2005) Second anniversary report: Muniwireless.com reports on municipal wireless and broadband projects, July. www.muniwireless.com/reports/docs/July2005report.pdf (accessed September 20, 2005).

VV. AA. (2004) ¡Pásalo! Relatos y análisis sobre el 11-M y los días que le siguieron. Madrid: Traficantes de Sueños. www.traficantes.net (accessed January 2005).

Wachira, N. (2003) Wireless in Kenya takes a village. *Wired News*. http://www.wired.com (accessed February 24, 2005).

Wade, W. (2004) Cell-phone remittances: a P2P import? *American Banker*, 169 (226): 1.

Wakeford, N. and Kotamraju, N. P. (2002) Mobile devices and the cultural worlds of young people in the UK and the US. Executive summary of findings for workshop presentation, Annenberg Center for Communication, University of Southern California, Los Angeles, October 10.

Watanabe, T. (2001) Merits of open-source resolution to resolve a digital divide in information technology. In W. Kim, T. W. Ling, Y-J. Lee, and S-S. Park (eds), *Human Society and the Internet*, pp. 92–99. Berlin: Springer-Verlag.

Watson, I. and Lightfoot, D. J. (2003) Mobile working with connexions. *Facilities*, 21 (13/14): 347–352. www.emeraldinsight.com/0263-2772.htm.

Waverman, L., Meschi, M., and Fuss, M. (2005) The impact of telecoms on economic growth in developing countries. In *Africa: The Impact of Mobile Phones*. Vodafone Policy Paper Series 2: 10–23.

Wei, R. (2004) Expanding the horizon: using the cellphone as a mass medium. Paper submitted to the Technology Division of the ICA for review and consideration for presentation at the 2004 Convention in New Orleans.

Weigun, W. and Shibiao, C. (2004) 1,200 sets of mobile phone charging station ready to provide convenient charging service for passengers, May 18. http://www.chaliyuan.com/english/m8/ (accessed July 22, 2005).

Weilenmann, A. (2003) "I can't talk now, I'm in a fitting room": formulating availability and location in mobile-phone conversations. *Environment and Planning A*, 35 (9): 1589–1605.

—— and Larsson, C. (2002) Local use and sharing of mobile phones. In B. Brown, N. Green, and R. Harper (eds), *Wireless World: Social and Interactional Aspects of the Mobile Age*, pp. 99–115. London: Springer.

Wellman, B. (2002) Little boxes, glocalization, and networked individualism. In M. Tanabe, P. van den Besselaar, and T. Ishida (eds), *Digital Cities II: Computational and Sociological Approaches*, pp. 10–25. Berlin: Springer-Verlag.

—— and Haythornthwaite, C. (2002) *The Internet in Everyday Life*. Oxford: Blackwell.

——, Quan-Haase, A., Boase, J., Chen, W., Hampton, K., Diaz de Isla, E., and Miyata, K. (2003) The social affordances of the Internet for networked individualism. *Journal of Computer Mediated Communication*, 8 (3). http://jcmc.indiana.edu/vol8/issue3/wellman.html (accessed February 7, 2006).

Werbach, K. (2002) Open spectrum: the new wireless paradigm. Spectrum Series Working Paper 6, October. http://werbach.com/docs/new_wireless_paradigm_htm (accessed July 15, 2005).

Wetzstein, C. (2003) Family lifeline at hand. *Washington Times*, February 7. http://www.teenresearch.com (accessed June 9, 2004).

Wheeler, J. O., Aoyama, Y., and Warf, B. (eds) (2000) *Cities in the Telecommunications Age: The Fracturing of Geographies*. New York: Routledge.

Wiener, S. N. (2003) *The State of Mobile Entertainment*. Mobile Entertainment Forum.

Wilhelm, A. G. (2002) Wireless youth: rejuvenating the net. *National Civic Review*, 91 (1): 293–302.

Williams, M. (2005) Mobile networks and foreign direct investment in developing countries. In *Africa: The Impact of Mobile Phones*. Vodafone Policy Paper Series 2: 24–40.

Wilska, T-A. (2003) Mobile phone use as part of young people's consumption styles. *Journal of Consumer Policy*, 26 (4): 441.

Wilson, S. (2003a) Generation Y provides wireless potential, March 20. http://www.imediaconnection.com/content/2570.asp (accessed June 5, 2004).

—— (2003b) Youth play significant role in wireless, August 18. http://www.imediaconnection.com/news/1952.asp (accessed May 2004).

Wired (2003) UnWired: Special *Wired* Report.

Wireless Week (2003) Wi-fi's gender gap, April 1. http://www.wirelessweek.com/article/CA288577?spacedesc=Departments&stt=001 (accessed August 16, 2004).

Wireless World Forum (2002) Mobileyouth2002: sample extract. http://www.w2forum.com (accessed June 5, 2004).

Wolfensberger, D. R. (2002) Congress and the Internet: democracy's uncertain link. In L. D. Simon, J. Corrales, and D. R. Wolfensberger (eds), *Democracy and the Inter-*

net: Allies or Adversaries?, pp. 67–102. Washington, DC: Woodrow Wilson Center Press.

Women with Disabilities Australia (1999) Telecommunications and women with disabilities. http://www.wwda.or.au/telecom.htm.

Wright, B. (2004) Africa's great mobile revolution. *African Business*, May, 298: 15–17.

Xinhuanet (2003) SMS market triggers thumb economy, February 12. http://news.xinhuanet.com/fortune/2003-02/12/content_725629.htm (accessed July 4, 2004).

Yang (2003) Information ministry plans "ubiquitous Korea" by 2007. *Korea Times*, June 10.

Yang, H-D., Yoo, Y-J., Lyytinen, K., and Ahn, J-H. (2003) Diffusion of broadband mobile services in Korea: the role of standards and its impact on diffusion of complex technology system. Unpublished manuscript. http://weatherhead.cwru.edu/pervasive/Paper/UBE%202003%20-%20Yoo.pdf (accessed June 21, 2004).

Yang, S-J. (2003) Korea pursuing global leadership in info-tech industry. *The Korea Herald*, January 20.

Yankee Group (2004) Yankee Group estimates youth market generated $21 billion in carrier revenue in 2003. Press release, February 2. http://www.yankeegroup.com/public/news_releases/news_release_detail.jsp?ID=PressReleases/news_02022004_wms.htm (accessed May 22, 2004).

Yoon, K-W. (2003a) Youth sociality and globalization: an ethnographic study of young Koreans' mobile phone use. Unpublished PhD thesis, University of Birmingham.

———— (2003b) Retraditionalizing the mobile: young people's sociality and mobile phone use in Seoul, South Korea. *European Journal of Cultural Studies*, 6 (3): 327–343.

Yoshii, H., Matsuda, M., Habuchi, C., Dobashi, S., Iwata, K., and Kin, N. (2002) *Keitai Denwa Riyou no Shinka to sono Eikyou*. Tokyo: Mobile Communications Kenkyuukai.

Yu, L., Louden, G., and Sacher, H. (2002) Buddysync: thinking beyond cell phones to create a third-generation wireless application for US teenagers. *Revista de Estudios de Juventud*, 57 (2): 173–188.

Yue, Z. (2003) Mobile phone demonstrates individuality: new expressions of today's hand-phone culture. *Beijing Morning Post*, February 24. http://mobile.tom.com/Archive/1145/2003/2/24-53731.html (accessed April 29, 2004).

Zaragoza (2004) El ayuntamiento de Zaragoza y telefónica móviles ponen en marcha "Ciudad Móvil." Press release, September 30. www.empresa.movistar.es/60/60507103.shtml (accessed September 15, 2005).

Zhang, L. (2001) *Strangers in the City*. Stanford: Stanford University Press.

Zhang, X. and Prybutok, V. R. (2005) How the mobile communication markets differ in China, the US, and Europe. *Communications of the ACM*, 48 (3): 111–114.

Zhao, J. (2002) Why MNCs are so enthusiastic about China? *Economic Daily*, March 28. http://business.sohu.com/29/60/article200436029.shtml (accessed June 30, 2004).

Zhao, Y. (2004) Between a world summit and a Chinese movie: visions of the "information society." *Gazette*, 66 (3–4): 275–280.

——— (forthcoming) Marketization and the social biases of the Chinese "information revolution." In J. Wasko and G. Murdoch (eds), *Cultural Capitalisms: Media in the Age of Marketization*. Cresskill, NJ: Hampton Press.

Zimmerman, P. (2005a) Anatel propõe ampliar para seis meses validade do crédito de pré-pago. [Anatel considers extending validity of pre-paid credit to six months]. *Folha do São Paulo*, September 9. http://www1.folha.uol.com.br/folha/dinheiro/ult91u100175.shtml (accessed February 20, 2006).

——— (2005b) Usuário de celular poderá receber em dinheiro crédito de pré-pago não usado. [Cellular users will be able to receive in money credit for unused pre-paid cards]. *Folha do São Paulo*, September 12. http://www1.folha.uol.com.br/folha/dinheiro/ult91u100218.shtml (accessed February 20, 2006).

Index